Modeling Engine Spray and Combustion Processes

Springer
*Berlin
Heidelberg
New York
Honkong
London
Mailand
Paris
Tokio*

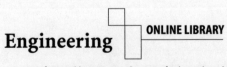

http://www.springer.de/engine/

Gunnar Stiesch

Modeling Engine Spray and Combustion Processes

Springer

Dr.-Ing. Gunnar Stiesch
Universtät Hannover
Instit. f. Technische Verbrennung
Welfengarten 1 A
30167 Hannover
Germany

ISBN 3-540-00682-6 Springer Verlag Berlin Heidelberg New York

Cataloging-in-Publication Data applied for
A catalog record for this book is available from the Library of Congress.
Bibliographic information published by Die Deutsche Bibliothek
Die Deutsche Bibliothek lists this publication in the Deutsche Nationalbibliografie;
detailed bibliographic data is available in the Internet at http://dnb.ddb.de

This work is subject to copyright. All rights are reserved, whether the whole or part of the material is concerned, specifically the rights of translation, reprinting, reuse of illustrations, recitation, broadcasting, reproduction on microfilm or in other ways, and storage in data banks. Duplication of this publication or parts thereof is permitted only under the provisions of the German Copyright Law of September 9, 1965, in its current version, and permission for use must always be obtained from Springer-Verlag. Violations are liable for prosecution under German Copyright Law.

Springer-Verlag is a part of Springer Science + Business Media
Springer-Verlag Berlin Heidelberg 2003
Printed in Germany
http://www.springer.de

The use of general descriptive names, registered names, trademarks, etc. in this publication does not imply, even in the absence of a specific statement, that such names are exempt from the relevant protective laws and regulations and therefore free for general use.

Product liability: The publisher cannot guarantee the accuracy of any information about dosage and application contained in this book. In every individual case the user must check such information by consulting the relevant literature.

Coverdesign: Erich Kirchner, Heidelberg
68/3111   Printed on acid-free paper - 5 4 3 2 1   SPIN 11576457

Preface

The utilization of mathematical models to numerically describe the performance of internal combustion engines is of great significance in the development of new and improved engines. Today, such simulation models can already be viewed as standard tools, and their importance is likely to increase further as available computer power is expected to increase and the predictive quality of the models is constantly enhanced.

This book describes and discusses the most widely used mathematical models for in-cylinder spray and combustion processes, which are the most important subprocesses affecting engine fuel consumption and pollutant emissions. The relevant thermodynamic, fluid dynamic and chemical principles are summarized, and then the application of these principles to the in-cylinder processes is explained. Different modeling approaches for the each subprocesses are compared and discussed with respect to the governing model assumptions and simplifications. Conclusions are drawn as to which model approach is appropriate for a specific type of problem in the development process of an engine. Hence, this book may serve both as a graduate level textbook for combustion engineering students and as a reference for professionals employed in the field of combustion engine modeling.

The research necessary for this book was carried out during my employment as a postdoctoral scientist at the Institute of Technical Combustion (ITV) at the University of Hannover, Germany and at the Engine Research Center (ERC) at the University of Wisconsin-Madison, USA. The text was accepted in partial fulfillment of the requirements for the postdoctoral Habilitation-degree by the Department of Mechanical Engineering at the University of Hannover.

Many individuals have assisted me in various ways while I have worked on this book. First and foremost I would like to thank Prof. Günter P. Merker, the director of the ITV, for supporting my work in every possible respect. Many ideas for this book were obtained from our numerous stimulating discussions about the subject of engine modeling. I am also indebted to Prof. Rolf D. Reitz, the director of the ERC, for inviting me to the University of Wisconsin-Madison and for teaching me many aspects about multidimensional engine codes during my stay at the ERC. Professor Dieter Mewes of the University of Hannover and Prof. Meinrad K. Eberle of the ETH Zurich, Switzerland contributed to this work by their critical reviews and constructive comments.

Special thanks are due to my colleagues and friends both at the University of Hannover and at the University of Wisconsin-Madison for providing an inspiring environment in which to carry out engine research.

I would like to acknowledge Mrs. Christina Brauer, the tracer at the ITV, for her thoughtful assistance with the schematic illustrations and technical drawings contained in this book.

Last but not least, I would like to thank my family for their constant support and understanding while I was working on the manuscript.

Hannover, January 2003 *Gunnar Stiesch*

Contents

Preface V

Contents VII

Nomenclature XI

1 Introduction 1
   1.1 Modeling of Combustion Processes 1
   1.2 Direct Injection Combustion Engines 3

2 Thermodynamic Models 5
   2.1 Thermodynamic Fundamentals 5
   2.2 Single-Zone Cylinder Model 6
      2.2.1 Mass and Energy Balances 6
      2.2.2 Mass Fluxes 8
      2.2.3 Mechanical Work 12
      2.2.4 Wall Heat Transfer 14
      2.2.5 Heat Release by Combustion 16
      2.2.6 Ignition Delay 21
      2.2.7 Internal Energy 23
   2.3 Two-Stroke Scavenging Models 28
   2.4 Empirical Two-Zone Combustion Model 32
   2.5 Typical Applications 35
      2.5.1 Heat Release Analysis 35
      2.5.2 Analysis of Complete Power Systems 35
   References 37

3 Phenomenological Models 41
   3.1 Classification 41
   3.2 Heat Release in Diesel Engines 43
      3.2.1 Zero-Dimensional Burning Rate Function 43
      3.2.2 Free Gas Jet Theory 47
      3.2.3 Packet Models 57
      3.2.4 Time Scale Models 67
   3.3 Gas Composition and Mixing in Diesel Engines 72
      3.3.1 Two-Zone Cylinder Models 72
      3.3.2 N-Zone Cylinder Models 75

Contents

    3.3.3 Packet Models 76
  3.4 Advanced Heat Transfer Models 77
    3.4.1 Heat Transfer Mechanisms 77
    3.4.2 Convective and Radiative Heat Transfer Model 78
  3.5 SI Engine Combustion 88
    3.5.1 Burning Rate Calculation 88
    3.5.2 Gas Composition 91
    3.5.3 Engine Knock 92
  References 96

4 Fundamentals of Multidimensional CFD-Codes 101
  4.1 Conservation Equations 101
  4.2 Numerical Methodology 104
  4.3 Turbulence Models 109
  4.4 Boundary Layers and Convective Heat Transfer 112
  4.5 Application to In-Cylinder Processes 116
  References 117

5 Multidimensional Models of Spray Processes 119
  5.1 General Considerations 119
    5.1.1 Spray Processes in Combustion Engines 119
    5.1.2 Spray Regimes 120
  5.2 The Spray Equation 122
    5.2.1 Equations and Exchange Terms 122
    5.2.2 Numerical Implementation 124
  5.3 Droplet Kinematics 126
    5.3.1 Drop Drag and Deformation 126
    5.3.2 Turbulent Dispersion / Diffusion 128
  5.4 Spray Atomization 130
    5.4.1 Breakup Regimes 131
    5.4.2 Wave-Breakup Model 135
    5.4.3 Blob-Injection Model 137
    5.4.4 Turbulence and Cavitation Based Primary Breakup Model 139
    5.4.5 Sheet-Atomization Model for Hollow-Cone Sprays 146
  5.5 Secondary Droplet Breakup 153
    5.5.1 Drop Breakup Regimes 153
    5.5.2 The Reitz-Diwakar Model 154
    5.5.3 The Taylor-Analogy Breakup Model 155
    5.5.4 The Kelvin-Helmholtz Breakup Model 158
    5.5.5 The Rayleigh-Taylor Breakup Model 159
  5.6 Droplet/Droplet and Spray/Wall Interactions 161
    5.6.1 Droplet Collision and Coalescence 161
    5.6.2 Spray-Wall Impingement 165
  5.7 Fuel Evaporation 171
    5.7.1 Droplet Evaporation 172
    5.7.2 Multi-Component Fuels 174

    5.7.3 Flash Boiling 180
  5.8 Grid Dependencies 181
    5.8.1 Problem Description 181
    5.8.2 Reduction of Grid Dependencies 183
  References 186

6 Multidimensional Combustion Models 193
  6.1 Combustion Fundamentals 193
    6.1.1 Chemical Equilibrium 193
    6.1.2 Reaction Kinetics 196
    6.1.3 Reaction Mechanisms for Hydrocarbon Flames 198
    6.1.4 Combustion Regimes and Flame Types 202
  6.2 Ignition Processes 205
    6.2.1 Ignition Fundamentals 205
    6.2.2 Autoignition Modeling 209
    6.2.3 Spark-Ignition Modeling 213
  6.3 Premixed Combustion 222
    6.3.1 The Flamelet Assumption 222
    6.3.2 Eddy-Breakup Models 222
    6.3.3 Flame Area Evolution Models 224
    6.3.4 The Fractal Model 227
  6.4 Diffusion Combustion 228
    6.4.1 The Characteristic Time Scale Model 228
    6.4.2 Flamelet Models 230
    6.4.3 pdf-Models 238
  6.5 Partially Premixed Combustion in DISI Engines 241
    6.5.1 Flame Structure 241
    6.5.2 A Formulation based on Lagrangian Flame Front Tracking 242
    6.5.3 A Formulation based on the G-Equation 246
  References 249

7 Pollutant Formation 255
  7.1 Exhaust Gas Composition 255
  7.2 Nitrogen Oxides 257
    7.2.1 Reaction Paths 257
    7.2.2 Thermal NO 258
    7.2.3 Prompt NO 260
  7.3 Soot 261
    7.3.1 Phenomenology 261
    7.3.2 Semi-Global Mechanisms 264
    7.3.3 Detailed Chemistry Mechanisms 269
  References 271

8 Conclusions 275

Index 279

Nomenclature

Abbreviations

Bi	Biot number
bmep	break mean effective pressure
CDM	continuum droplet model
CFD	computational fluid dynamics
CI	compression ignition (diesel) engine
CN	cetane number
Da	Damköhler number
DDM	discrete droplet model
DI	direct injection
DISI	direct injection spark ignition
DNS	direct numerical simulation
EGR	exhaust gas recirculation
EOC	end of combustion
EPC, EPO	exhaust port closing, opening
EVC, EVO	exhaust valve closing, opening
HCCI	homogeneous charge compression ignition
imep	indicated mean effective pressure
IPC, IPO	intake port closing, opening
IVC, IVO	intake valve closing, opening
Ka	Karlovitz number
La	Laplace number
LES	large eddy simulation
LHV	lower heating value
M	third body species in chemical reactions
Nu	Nusselt number
ON	octane number
pdf	probability density function
Pr	Prandtl number
RANS	Reynolds-averaged Navier-Stokes equations
Re	Reynolds number
Sc	Schmidt number
Sh	Sherwood number
SI	spark ignition engine
SMD	Sauter mean diameter
SOC	start of combustion

St Stokes number
We Weber number
Z Ohnesorge number

Symbols

af	air-fuel ratio, -
A	area, m^2
b	combustion regress variable, -
B	cylinder bore, m
c	clearance height, m
	flow velocity, m/s
c_D	discharge coefficient, -
c_i	concentration of species i, mol/m^3
c_m	mean piston speed, m/s
c_p	specific heat at constant pressure, kJ/kg K
c_v	specific heat at constant volume, kJ/kg K
C	constant
C_D	drag coefficient
d	diameter, m
D	binary diffusion coefficient, m^2/s
E	total energy, J
E_A	activation energy, kJ/kmol
f	droplet-pdf in spray equation, -
f_R	residual mass fraction, -
f_v	soot volume fraction, -
$\vec{F}$	force per unit mass (acceleration), m/s^2
g	acceleration of gravity, m/s^2
	spec. Gibbs free energy, kJ/kg
G	Gibbs free energy, J
h	convective heat transfer coefficient, W/m^2 K
	spec. enthalpy, kJ/kg
h_{fg}	latent heat of evaporation, kJ/kg
$\tilde{h}_{f,i}^0$	enthalpy of formation, kJ/kmol K
H	enthalpy, J
k	reaction coefficient, mol-cm-s
	thermal conductivity, W/m K
	turbulent kinetic energy, kJ/kg
	wave number ($=2\pi/\lambda$), 1/m
K	equilibrium constant, -
l_I	turbulence integral length scale, m
l_K	Kolmogorov length scale, m
L	latent heat of evaporation, kJ/kg

L_{min}	stoichiometric air-fuel ratio, -
m	mass, kg
MW	molecular weight, g/mol
n	engine speed, rpm
	number of moles, mol
$\dot{n}$	molar flux, mol/m^2 s
N_d	number of droplets, -
p	pressure, kPa
$\dot{q}$	heat flux, W/m^2
Q	heat transfer, J
r	radius, m
r_S	swirl ratio, -
R	gas constant, kJ/kg K
s	spec. entropy, kJ/kg K
s_l	laminar flame speed, m/s
s_t	turbulent flame speed, m/s
S	entropy, J/K
	spray tip penetration, m
t	time, s
T	temperature, K
u	spec. internal energy, kJ/kg
u'	turbulence intensity, m/s
$\vec{u}$	gas velocity vector in a two-phase system, m/s
U	internal energy, J
v	specific volume, m^3/kg
	velocity, m/s
$\vec{v}$	velocity vector in a single-phase system, m/s
	droplet velocity vector in a two-phase system, m/s
V	volume, m^3
V_c	clearance volume, m^3
V_d	displacement volume, m^3
$\dot{w}$	chemical reaction rate, kg/m^3 s
W_t	technical work, J
x_i	x, y, z coordinate, m
y	droplet distortion parameter, -
Y_i	mass fraction of species i, -
Z	mixture fraction, -

Greek Letters

α	thermal diffusivity, m^2/s
	spray angle, rad
χ	scalar dissipation rate, s^{-1}
γ	specific heat ratio c_p/c_v

δ_{ij}	Kronecker delta
δ_l	laminar flame thickness, m
$\delta_{s,w}$	wall soot layer thickness, m
δ_{th}	thermal boundary layer thickness, m
ε	compression ratio, -
	dissipation rate of turbulent kinetic energy k, kJ/kg s
ϕ	fuel/air equivalence ratio, -
η	effectiveness, -
	wave amplitude, m
φ	crank angle, rad
λ	air/fuel equivalence ratio ($= 1/\phi$), -
	wave length, m
Λ	delivery ratio
μ	viscosity, kg/s m
μ_i	chemical potential of species i, kJ/kmol
ν	kinematic viscosity, m^2/s
ν_i	stoichiometric coefficient of species A_i, -
π	pressure ratio p_1/p_0, -
θ	void fraction, -
	half cone angle of hollow cone spray, rad
ρ	density, kg/m^3
σ	normal viscous stress, N/m^2
	surface tension, N/m
Σ	turbulent flame surface density, m^2/m^3
τ_{id}	ignition delay, s
τ_{ij}	viscous stress tensor, N/m^2
ω	angular velocity, rad/s
	wave growth rate, 1/s

Subscripts and Superscripts

a	air
b	burned
bb	blowby
bu	breakup
c	combustion
cyl	cylinder
cv	control volume
d	droplet
$diss$	dissipation
$evap$	evaporation
exh	exhaust
f	fluid; fuel

FF	flame front
g	gas phase
i	species
in	intake
inj	injection
k	ignition kernel
l	liquid phase; laminar
mot	motored
n	normal direction
noz	nozzle
p	particle
R	residuals
ref	reference state
s	soot; spray; surface
stoic	stoichiometric
t	turbulent
th	thermal boundary layer
u	unburned
v	vapor phase
w	wall
$^\circ$	property at standard state
$-$	ensemble-averaged quantity
$\sim$	Favre-averaged quantity; molar quantity

1 Introduction

1.1 Modeling of Combustion Processes

This complex task of improving on combustion engines, that have already reached a very high level of sophistication during their more than 100 year long history, can nowadays be achieved only by a combination of advanced experimental and computational studies. Despite the quantitative uncertainties of numerical simulations that are often greater than those of experiments, the modeling of combustion engine processes has some significant advantages that make its utilization in today's engine development a necessity. In this regard, it is obvious that numerical simulations are especially suited to carry out extensive parametric studies, since they are much more time and cost effective than the alternative construction and investigation of numerous prototypes.

In addition, spray and combustion models offer some further possibilities that might be even more beneficial to the research and development engineer. One important feature is that they allow to output every single variable of a problem at any position in physical space and at any point in time during the process. Such a complete set of information cannot be obtained by experiments for several reasons. First of all it is extremely difficult to apply sophisticated optical measurement techniques to a rapidly oscillating combustion engine without affecting the boundary conditions of spray development and combustion. But even if this task is achieved to a satisfactory degree, there will always remain several areas of interest that are not accessible, e.g. the optically dense region of the fuel spray close to a nozzle orifice, which is very important for the subsequent breakup and atomization of the spray. Moreover, experiments can hardly yield three-dimensionally resolved information. They are usually limited to two dimensions if light-sheets are applied or even to an integrated (zero-dimensional) information for a specific volume.

Another capability of mathematical models is the possibility of artificially separating specific sub-processes from others, that would interact in a real system, or to investigate the effect of unnatural boundary conditions on such processes. These features are also important in order to obtain more detailed information about spray and combustion processes that are still not fully understood today.

For the above reasons, the development and application of mathematical models for combustion engines is invaluable despite their greater uncertainties compared to experimental studies. The plenty of information they provide can help to understand the complex sub-processes occurring in combustion engines and especially the various interdependencies between these processes much better than it could be the case with the sole execution of experiments.

However, it has to be noted that even though the quality of numerical models is steadily improving and their importance will continue to increase, a wide variety and great number of experimental studies will remain necessary as well. Because of their greater accuracy they are needed to both control simulation results at selected positions and timings during the combustion process and also to validate and calibrate new mathematical models and submodels in isolated experiments, e.g. in constant volume vessels where the fuel spray can be investigated in a noncombustible atmosphere.

Depending on the various possible applications, different types of models for engine combustion processes have been developed. Three different model categories are typically distinguished. In an order of increasing complexity and increasing requirements with respect to computer power, these are the thermodynamic (or zero-dimensional) models, the phenomenological (or quasi-dimensional) models, and the multidimensional models utilized in so-called CFD (computational fluid dynamics) codes.

The thermodynamic codes utilize a simplified description of the combustion chamber, that assumes that all contents are ideally mixed at any time. The models are computationally very efficient but cannot provide insight into local processes such as air fuel mixing or pollutant formation. Consequently, this model category is most often applied in applications where short computing times are more crucial than the details of subprocesses. Examples are transient systems analyses and extensive parametric studies, provided that the model is capable of reproducing the effect of the varied engine parameters.

Phenomenological spray and combustion models are more complex than the thermodynamic models, because they typically divide the combustion chamber into numerous different zones that are characterized by different temperatures and compositions. Because of this spatial resolution, important subprocesses can be modeled and the prediction of heat release rates and exhaust emissions as a function of characteristic engine parameters becomes possible. However, the spatial resolution is still relatively coarse in these models, and the turbulent flow field inside the engine cylinder is not solved for. Thus, effects of the engine geometry on combustion can hardly be accounted for, and analyses of subscale processes that would require a higher spatial (and temporal) resolution are limited.

Finally, multidimensional CFD-codes solve the full set of differential equations for species mass, energy, and momentum conservation on a relatively fine numerical mesh and also include submodels to account for the effects of turbulence. As a result, these models are best suited to analyze the various subscale processes of mixture formation and combustion in greater detail. However, they are also much more costly in terms of computer power than the other model categories.

As noted above, especially the more detailed numerical simulation tools are extremely helpful in investigating and analyzing different subprocesses, that proceed simultaneously and interfere with each other. Therefore, they will become even more important in the future, because the trend in engine development leads more and more towards the utilization of direct injection technology. As will be discussed in the following section, this holds true both for diesel and gasoline engines. For these concepts, mixture formation by injection of liquid sprays directly into the combustion chamber and combustion interact in multiple ways, such that improvements of the overall combustion process demand a very detailed understanding of the governing mechanisms. This understanding can be advanced by numerical modeling the spray and combustion processes.

1.2 Direct Injection Combustion Engines

Direct fuel injection technology has become more and more popular both in compression ignition (diesel) and in spark ignition (gasoline) engines in recent years. In diesel applications direct injection (DI) engines are characterized by a superior fuel economy compared to indirect injection pre- and swirl-chamber engines, but also by a disadvantage with respect to vibration and noise control. While the DI technology has been used in the cost-sensitive field of marine and heavy duty diesel engines for a long time, it is only during the past decade that it has become the standard for passenger car diesel engines as well. This trend was – beside other factors – facilitated by the development of modern injection systems that are more flexible and allow significantly higher injection pressures and thus better spray atomization and combustion characteristics than their predecessors.

However, in view of increasing oil prices and a continuously tightened emission legislation the past achievements in diesel engine technology are not sufficient and further improvements will be necessary. In this context significant reductions of particulate matter and nitrogen oxide emissions that have to be realized without penalty in fuel economy are the main focus of current research.

In contrast to diesel engines, spark ignition engines for passenger cars have almost solely relied on external mixture formation mechanisms with carburetors and port fuel injection systems during the past (one prominent exception being the Mercedes Benz 300 SL from 1954). The reasons for this dominance have been the relatively simple control mechanisms ensuring reliable ignition and combustion under all operating conditions. In addition, the stoichiometric combustion of port fuel injection engines offers the possibility of utilizing a three-way catalytic converter as a very effective and affordable means of exhaust aftertreatment.

Only during the most recent past more attention has been paid to direct injection spark ignition (DISI) engines again, with Mitsubishi being the first manufacturer to offer this technology in a series production passenger car in 1997. The main driving force nowadays is the possibility to run DISI engines in the stratified charge mode under part load conditions. Therefore, the characteristic throttling losses of homogeneously operated SI engines are reduced and an improved fuel economy can be achieved. However, it should be noted that in contrast to DI-diesel engines, there are still significant problems in DISI engines that have to be overcome in order to take full advantage of the potential offered by the direct injection technology. In this respect, the control of mixture formation that ensures stable ignition and combustion for all load and speed combinations as well as the development of a reliable and affordable aftertreatment system for lean exhaust gases are most crucial.

So even though diesel and gasoline direct injection engines are at very different development states today, further development and optimizations are mandatory for both engine types. Such engine optimizations are always an extremely complex task however, since there are numerous parallel and interacting subprocesses in the engine taking place instantaneously. This holds true especially for direct injection engines where spray penetration and atomization, droplet evaporation, air-fuel mixing, ignition as well as premixed and diffusion combustion all take place simultaneously within the combustion chamber and influence each other. Hence, in order to optimize the overall efficiency and the exhaust emissions of an engine all these phenomena and their interdependencies have to be taken into account.

2 Thermodynamic Models

The models described in this chapter are called thermodynamic models since they are based on the first law of thermodynamics and mass balances only. The principles of momentum conservation are not considered in this model type and spatial variations of composition and thermodynamic properties are neglected. Thus, the entire combustion chamber of an internal combustion engine is typically treated as a single, homogeneously mixed zone. These assumptions obviously represent a significant abstraction of the problem and prohibit the usage of thermodynamic models in order to study locally resolved subprocesses such as detailed spray processes or reaction chemistry. However, the great advantage of these models is that they are both easy to handle and computationally very efficient. Therefore, they are still widely used in applications where there is only interest in spatially and sometimes even temporally averaged information and where computational time is crucial.

2.1 Thermodynamic Fundamentals

Generally, an open thermodynamic system with transient in- and outflows as shown in Fig. 2.1 can be described by mass and energy balances. The change of mass contained within the control volume is equal to the difference between all entering and exiting mass flows, $\dot{m}_i$ and $\dot{m}_e$ respectively:

$$\frac{dm_{cv}}{dt} = \sum_i \dot{m}_i - \sum_e \dot{m}_e \:. \tag{2.1}$$

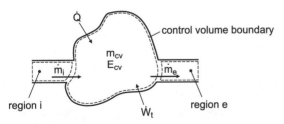

Fig. 2.1. Schematic illustration of an open control volume with transient in- and outflows

The energy balance (first law of thermodynamics) for the control volume becomes

$$\frac{dE_{cv}}{dt} = \dot{Q} + \dot{W}_t + \sum_i (\dot{m}_i h_i) - \sum_e (\dot{m}_e h_e), \qquad (2.2)$$

where $\dot{W}_t$ represents the rate of mechanical work and $\dot{Q}$ is the rate of heat transferred to the system. If changes in kinetic and gravitational potential energies of the mass contained in the system are neglected the change in total energy dE_{cv} is equal to the change in internal energy dU_{cv} and Eq. 2.2 can be written as

$$\frac{dU_{cv}}{dt} = \dot{Q} + \dot{W}_t + \sum_i (\dot{m}_i h_i) - \sum_e (\dot{m}_e h_e). \qquad (2.3)$$

In order to correlate the above change in internal energy of the control volume dU_{cv} to changes in temperature and pressure, that are usually of greater interest, an equation of state is necessary in addition to the above mass and energy balances. For a pure substance i, the mass based specific internal energy $u_i = U_i/m$ is generally a function of temperature and pressure,

$$u_i = u_i(T, p), \qquad (2.4)$$

but it reduces to a function of only temperature for ideal gases. This assumption usually represents a good approximation for combustion gases because of the high temperatures encountered. However, combustion products are not a pure substance but a mixture of various components such as CO_2, H_2O, N_2, O_2, etc.. Hence, the internal energy depends not only on temperature but also on the mixture composition, i.e. the mass fractions Y_i of the each species i within the mixture:

$$u = \sum_i (Y_i u_i(T)) \qquad \text{(ideal gas mixture)}. \qquad (2.5)$$

For a known gas composition, which in combustion engines could be approximated by the equivalence ratio ϕ, Eq. 2.5 can be utilized to solve the energy balance (Eq. 2.3) for the temperature T. Once T is known, the corresponding pressure p of the system can be obtained by the gas law, which for an ideal gas becomes

$$pv = RT \qquad \text{(ideal gas)}. \qquad (2.6)$$

2.2 Single-Zone Cylinder Model

2.2.1 Mass and Energy Balances

The name single-zone model comes from the fact that the contents of the combustion chamber are assumed to be homogeneously mixed at all times in this model. Thus, the thermodynamic state of the working fluid is only a function of time, but it does not depend on the spatial position within the cylinder. With this assumption the entire combustion chamber can be chosen as the control volume for which mass and energy balances are applied according to the general forms given by Eqs. 2.1 and 2.3, respectively.

Figure 2.2 displays a typical combustion chamber of an engine which is surrounded by the liner, piston head, cylinder head and the inlet and exhaust valves that connect the combustion chamber to the manifolds. In addition, the control volume boundary as well as the thermodynamic properties and the mass and energy fluxes relevant to establish the balance equations are included. Following this schematic diagram and introducing the convention, that all mass and energy flows directed into the system have positive values and all flows directed out of the system have negative values, the mass balance for the combustion chamber can be written as

$$\frac{dm_{cyl}}{dt} = \frac{dm_{in}}{dt} + \frac{dm_{exh}}{dt} + \frac{dm_{fuel}}{dt} + \frac{dm_{bb}}{dt},\qquad(2.7)$$

and the energy balance yields

$$\frac{dU_{cyl}}{dt} = \frac{dQ_w}{dt} + \frac{dQ_{chem}}{dt} - p_{cyl}\frac{dV_{cyl}}{dt} + \frac{dm_{in}}{dt}h_{in} + \frac{dm_{exh}}{dt}h_{exh}$$
$$+ \frac{dm_{fuel}}{dt}h_{fuel} + \frac{dm_{bb}}{dt}h_{bb}\qquad(2.8)$$

The various components of Eqs. 2.7 and 2.8 will be discussed in the subsequent sections.

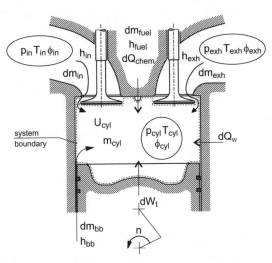

Fig. 2.2. Single-zone cylinder model

2.2.2 Mass Fluxes

Gas Exchange

In thermodynamic combustion models the sole interest in simulating the intake and exhaust processes is to determine the instantaneous mass flow rates through the intake and exhaust valves in order to integrate the trapped mass of gases within the combustion chamber. Specific flow patterns such as swirl or tumble inside the cylinder that can be created by the geometry of the intake manifolds are generally neglected here. The consideration of the effects of these flow patterns on mixture formation, heat release rate and emission formation are subject to more complex phenomenological or CFD-models that will be discussed in later chapters.

The mass flows through the valves can be approximated by a steady flow through an adiabatic nozzle. The mass balance for this system,

$$\frac{dm}{dt} = \dot{m}_0 - \dot{m}_1 = 0, \tag{2.9}$$

and the energy balance,

$$\frac{dE}{dt} = \overset{0}{\cancel{\dot{Q}_w}} + \overset{0}{\cancel{\dot{W}_t}} + \dot{m}_0 \left(h_0 + \frac{c_0^2}{2} + \overset{0}{\cancel{gz_0}} \right) - \dot{m}_1 \left(h_1 + \frac{c_1^2}{2} + \overset{0}{\cancel{gz_1}} \right) = 0, \tag{2.10}$$

can be combined to relate the velocity changes to enthalpy changes,

$$\frac{c_1^2}{2} - \frac{c_0^2}{2} = h_0 - h_1, \tag{2.11}$$

where subscripts 0 and 1 represent the states before and at the smallest cross-section of the nozzle, respectively. The enthalpy difference between these states can be calculated based on the relation for an isentropic process ($ds = 0$) of an ideal gas,

$$h_0 - h_1 = \frac{\gamma}{\gamma - 1} RT_0 \left[1 - \pi^{\frac{\gamma-1}{\gamma}} \right], \tag{2.12}$$

where γ is the ratio of specific heats c_p/c_v and π the pressure ratio p_1/p_0. Further assuming that the cross-sectional area ahead of the valve is large compared to the one in the valve gap, the upstream velocity c_0 reduces to zero. Rearranging Eqs. 2.11 and 2.12 then yields

$$c_1 = \sqrt{\frac{2\gamma}{\gamma - 1} RT_0 \left[1 - \pi^{\frac{\gamma-1}{\gamma}} \right]} \tag{2.13}$$

for the velocity in the valve gap.

The theoretical mass flow rate through a valve can now be estimated as a function of this flow velocity, the cross-sectional area of the valve gap and the gas density:

$$\dot{m}_{th} = A_1 c_1 \rho_1 = A_1 c_1 \rho_0 \pi^{1/\gamma} . \qquad (2.14)$$

However, the actual mass flow rate through an intake or exhaust valve is typically less than this theoretical value for two reasons. The first is that the geometric open area of a cam-driven poppet valve is not constant but depends on valve lift and thus time. Figure 2.3 displays a typical valve lift curve as well as the corresponding flow areas over the camshaft position. The diagram is subdivided into two regions a and b, that represent different constraining cross-sections of the valve configuration. For small and intermediate valve lifts (a), the flow through the valve is limited by the area between the valve head and the seat while for large lifts (b) the area around the valve shaft represents the decisive constraint.

The second reason is that the streamlines of the flow cannot follow the sharp edges on their way and separate from the walls. Figure 2.4 illustrates this effect and indicates that the effective cross-sectional area A_{eff} of the valve is smaller than the theoretical area A_{th}. Since the effective cross-sectional area is hard to obtain fundamentally the flow contraction is usually taken into account by introducing an empirical discharge coefficient c_D, that has to be determined experimentally in steady flow tests for various lifts of a given valve assembly. It is commonly defined as the ratio between the time dependent effective flow area A_{eff} and the maximum geometrical area A_{geo} at maximum valve lift. The latter corresponds to case (b) in Fig. 2.3. A typical course of c_D as a function of valve lift is given in Fig. 2.5. For a fully opened poppet valve the discharge coefficient usually ranges around 0.7.

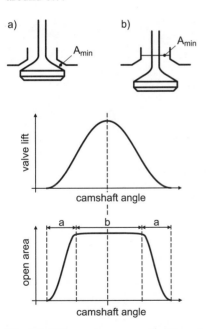

Fig. 2.3. Valve open area as a function of valve lift

With the above definition of the discharge coefficient c_D, the actual mass flow rate through a partly or fully opened poppet valve can now be estimated by the relation

$$\dot{m} = c_D A_{geo} \rho_0 \pi^{1/\gamma} \sqrt{\frac{2\gamma}{\gamma-1} RT_0 \left[1 - \pi^{\frac{\gamma-1}{\gamma}}\right]}. \tag{2.15}$$

Hence, the mass flow rate is a function of the valve geometry, the gas properties and the thermodynamic states upstream and downstream of the valve, that are expressed through the pressure ratio π and the upstream temperature T_0. While the state within the combustion chamber is the solution of the integration of the cylinder mass and energy conservation equations (Eqs. 2.7 and 2.8), the states in the inlet and exhaust manifolds have to be provided as boundary conditions. They can either be obtained experimentally by indicating the pressure and temperature in the manifolds or they can be calculated by modeling the entire gas passage of the engine and its periphery. This can include air filter, compressor, air cooler, throttle and various connecting ducts for the intake side as well as exhaust manifold, turbine, exhaust gas aftertreatment, muffler and connecting ducts in the tailpipe. Since the objective of this text is the modeling of the in-cylinder processes rather than the gas exchange processes a detailed description of the available models is omitted and only a brief summary will be given below.

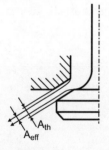

Fig. 2.4. Flow contraction at the valve seat

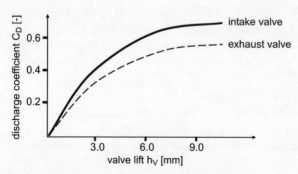

Fig. 2.5. Discharge coefficient vs. valve lift [20]

It has been shown in various studies, e.g. [8, 24], that for turbocharged diesel engines it is usually sufficient to calculate the mass flow rates and pressures based on cycle-averaged data. The various intake and exhaust subsystems are treated as combinations of reservoirs and flow restrictions analogous to the calculation of the mass flow through valves in Eq. 2.15. The turbocharger performance can be described by characteristic maps for both the compressor and the turbine that correlate rotation speed, mass flow rates and pressure ratios. However, for naturally aspirated SI engines the volumetric or scavenging efficiencies of the intake and exhaust systems are strongly affected by transient pressure waves that are traveling through the system. Therefore, it is hardly possible to assume cycle-averaged values for pressure, temperature and mass flow rates within the gas exchange devices. In addition to the mass and energy balances the momentum conservation has to be considered in order to solve for the pressure fluctuations in the intake manifold. This is often done by assuming a one-dimensional flow through the system. A detailed description of these so-called one-dimensional gas dynamics is provided e.g. by Heywood [12] or Ramos [23].

Fuel Injection

For obvious reasons the term dm_{fuel} in Eq. 2.7 applies only to engines with internal mixture formation, i.e. to diesel and direct injection spark ignition engines. For engines with external mixture formation systems, i.e. gasoline or gas engines with port fuel injection or carburetors, this term reduces to zero as a fuel-air mixture enters the cylinder through the intake valve and the respective fuel mass is already included in the inlet mass flow rate dm_{in}.

In internal mixture formation engines the mass flow rate of injected fuel $\dot{m}_{fuel}$ can be estimated by simulation models describing the hydraulics inside the injection system, e.g. [2, 3], but more often it is simply specified by an experimentally obtained injection rate profile over time. A standard single-pulse injection is usually approximated by a trapezoid shape and pre- or post injections can be simplified by triangular shaped profiles. A schematic illustration of such an empirical injection rate profile is shown in Fig. 2.6.

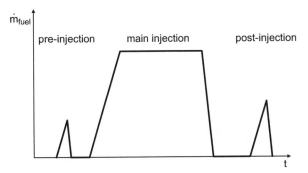

Fig. 2.6. Illustration of an empirical injection rate profile with pre- and post injections

The assumption of a homogeneously mixed combustion chamber in the single-zone cylinder model implies that the fuel that is injected in the liquid phase immediately evaporates once it enters the control volume of the cylinder. This vaporization causes an internal cooling of the cylinder gases because of the latent heat of evaporation of the fuel. However, this process does not have to be modeled explicitly here, since it is already taken account of by the specific enthalpy of the liquid fuel h_{fuel}, that is associated with the injection rate in the energy conservation equation 2.8. It is very small compared to the specific enthalpy of the cylinder gases and therefore causes a drop in the mean specific internal energy and thus in the temperature after mixing. It should further be noted that the fuel enthalpy in Eq. 2.8 does not contain the chemically bound energy in terms of the heat of formation. In this model, the chemical energy released through combustion has to be added to the energy balance separately through the term dQ_{chem} which will be discussed in Sect. 2.2.5.

Blowby

During the high pressure part of the combustion engine process, i.e. during the compression and expansion strokes of a four-stroke engine or between exhaust port closing and opening in a two-stroke engine, a leakage can pass from the combustion chamber through the piston-liner-ring assembly into the crankcase because of the significant pressure difference. This leakage that is commonly termed blowby and referenced as dm_{bb} in Eq. 2.7 causes a drop in the cylinder pressure and hence, a loss in thermal efficiency of the engine. However, it is generally very small compared to the trapped mass inside the cylinder such that it is often neglected in engine simulations. Nevertheless it can be modeled since it turns out to be roughly proportional to the greatest flow resistance in the flow path between the cylinder and the crankcase [21]. When there is good contact between the compression rings and the liner as well as between the rings and the bottom of the grooves, this relevant resistance is determined by the cross-sectional area of the ring gap.

2.2.3 Mechanical Work

The only kind of mechanical work performed on or by the control volume of the combustion chamber is the compression/expansion work that is due to the volume change of the cylinder forced by the rotation $d\varphi$ of the crankshaft:

$$dW_t = -p_{cyl} dV_{cyl}. \tag{2.16}$$

The volume change is solely a function of the cylinder bore B and the change in the vertical piston position dz which is governed by the geometry of the crank mechanism:

$$dV_{cyl} = -\frac{\pi}{4} B^2 dz \ . \tag{2.17}$$

For a standard configuration of the crank mechanism with intersecting axes of the piston and the crankshaft as displayed in Fig. 2.7 the following relations apply:

$$\begin{aligned} c &= a \cdot \sin \varphi \\ z_1 &= a \cdot \cos \varphi \\ z_2 &= \sqrt{l^2 - c^2} = \sqrt{l^2 - a^2 \sin^2 \varphi} \\ z &= z_1 + z_2 = a \cdot \cos \varphi + \sqrt{l^2 - a^2 \sin^2 \varphi} \end{aligned} \tag{2.18}$$

Thus, the derivative of z with respect to the crank angle φ can be expressed as

$$\frac{dz}{d\varphi} = -a \ \sin \varphi - \frac{a^2 \sin(2\varphi)}{2 \ \sqrt{l^2 - a^2 \sin^2 \varphi}} \ , \tag{2.19}$$

where φ is related to time by the rotation speed of the engine:

$$d\varphi = \omega \, dt = 2 \pi n \, dt \ . \tag{2.20}$$

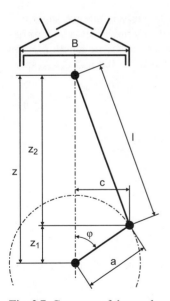

Fig. 2.7. Geometry of the crank mechanism

2.2.4 Wall Heat Transfer

The heat transfer between the cylinder gases and the combustion chamber walls can be due to both convection and solid body radiation which originates from hot soot particles. However, in thermodynamic combustion models only the convective heat transfer mode is considered, since the temporally and spatially variable distribution of soot cannot be solved for with the assumption of an ideally mixed combustion chamber. The effect of radiative heat transfer can only be approximated by an empirical augmentation of the convective heat transfer.

The convective heat transfer rate between the gas and the wall can be described by the Newton's cooling law as

$$\dot{Q}_w = h A \left(T_w - T_{cyl} \right), \tag{2.21}$$

where h is called the convective heat transfer coefficient, A is the surface area and T_w and T_{cyl} are the mean wall and gas temperatures, respectively [15]. The heat transfer coefficient has been derived by many researchers by assuming an analogy with a steady turbulent flow over a solid wall, where the dimensionless temperature gradient at the surface, i.e. the Nusselt number, is described in terms of the Reynolds and Prandtl numbers of the flow, e.g. [1, 17, 30]:

$$\text{Nu} = \frac{hL}{k} = C \cdot \text{Re}^a \cdot \text{Pr}^b. \tag{2.22}$$

In Eq. 2.22 L represents a characteristic length of the problem, and C, a and b are empirical constants that are determined by curve fitting experimental data of wall heat transfer rates.

Probably the most widely used approach of this category is the one suggested by Woschni [30] who assumes an analogy between the cylinder flow and a turbulent flow through a circular tube. Consequently, the characteristic length L is the cylinder bore B and the constants a and b are chosen to be 0.8 and 0.4, respectively. Replacing the Reynolds number by

$$\text{Re} = \frac{\rho c B}{\mu} = \frac{p}{RT} \frac{cB}{\mu}, \tag{2.23}$$

where c is the characteristic flow velocity, Eq. 2.22 now becomes

$$\text{Nu} = \frac{hB}{k} = C \cdot \text{Re}^{0.8} \text{Pr}^{0.4} = C \left(\frac{p}{RT} \frac{cB}{\mu} \right)^{0.8} \text{Pr}^{0.4}. \tag{2.24}$$

Introducing the property relations

$$\text{Pr} = const. \approx 0.74, \qquad \frac{k}{k_0} = \left(\frac{T}{T_0} \right)^x, \qquad \frac{\mu}{\mu_0} = \left(\frac{T}{T_0} \right)^y,$$

and setting the characteristic flow velocity to the mean piston velocity c_m, we can solve for the heat transfer coefficient h:

$$h = C^* \cdot B^{-0.2} \cdot p^{0.8} \cdot c_m^{0.8} \cdot T^{-r}; \qquad r = 0.8 \left(1 + y \right) - x. \tag{2.25}$$

By comparison with experimental data the exponent r of the temperature dependency is found to be $r = 0.53$ and the constant C^* becomes 127.93 W/(m^{2}K). However, the above relation is applicable only for motored engine operation. Since the combustion process during fired operation significantly increases the turbulence level and thus the heat transfer coefficient, Eq. 2.25 has to be corrected by replacing c_m by a velocity term v, that takes into account both the mean piston velocity and the combustion induced turbulence:

$$h = 127.93 \cdot B^{-0.2} \cdot p^{0.8} \cdot v^{0.8} \cdot T^{-0.53}, \qquad (2.26)$$

where

$$v = C_1 c_m + C_2 \frac{V_d T_1}{p_1 V_1}(p - p_{mot}), \qquad (2.27)$$

and index 1 refers to the state at the start of compression, i.e. at IVC, and V_d is the displacement volume. The difference in cylinder pressure between fired and motored engine operation (p-p_{mot}) accounts for the turbulence increase by combustion and the constants C_1 and C_2 are suggested to be:

$$C_1 = \begin{cases} 6.18 + 0.417 R_s : & \text{gas exchange} \\ 2.28 + 0.308 R_s : & \text{compression/expansion} \end{cases}, \qquad (2.28)$$

$$C_2 = \begin{cases} 6.22 \cdot 10^{-3} & \text{m/(s K)}: \quad \text{prechamber engine} \\ 3.24 \cdot 10^{-3} & \text{m/(s K)}: \quad \text{direct injection engine} \end{cases}. \qquad (2.29)$$

Equation 2.28 has been validated for swirl ratios between 0 and 3.

Typical profiles for the cylinder temperature and pressure as well as for the heat transfer coefficient and the resulting heat transfer rates for a turbocharged direct injection diesel engine are plotted in Fig. 2.8 for two different injection timings. It is clearly visible how the heat transfer rate is accelerated at the start of combustion. However, it can also be seen that this effect will be less pronounced if combustion starts well after TDC because the peak cylinder pressure will be much lower in this case.

Several authors have proposed modifications to the Woschni model in order to yield a better agreement with additional experimental data, obtained for different engine types and boundary conditions, e.g. [4, 13]. However, in most cases these are minor variations applying to model constants and the definitions of characteristic flow properties and the general form of Eq. 2.22 that defines the convective heat transfer coefficient is retained.

It was shown by Huber [14] that the effective velocity based on the combustion term in Eq. 2.27 underestimates the heat transfer rate for low loads and for motored operation. Therefore, the velocity is expressed in terms of the current cylinder volume V, the clearance volume V_c and the indicated mean effective pressure imep,

$$v = c_m \left[1 + 2 \left(\frac{V_c}{V} \right)^2 \cdot \text{imep}^{-0.2} \right], \qquad (2.30)$$

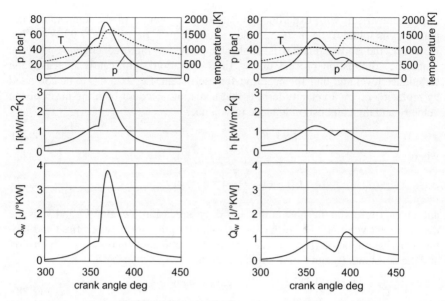

Fig. 2.8. Wall heat transfer in a turbocharged diesel engine. Left: $\varphi_{soc} = 355°\text{ATDC}$, Right: $\varphi_{soc} = 368°\text{ATDC}$ [20]

where the unit of imep is in bar. It is then suggested to input the greater value of either Eq. 2.27 or Eq. 2.30 into Eq. 2.26 to solve for the heat transfer coefficient. Furthermore, a correction of the constant C_2 given in Eq. 2.29 is recommended for wall temperatures greater than 600 K in direct injection diesel engines:

$$C_2^* = 5.0 \cdot 10^{-3} + 2.3 \cdot 10^{-5} \left(T_w - 600\text{K}\right) \quad \text{m/(s K)}. \tag{2.31}$$

2.2.5 Heat Release by Combustion

In thermodynamic cylinder models the rate of heat release by combustion, dQ_{chem}/dt in Eq. 2.8, cannot be derived by a detailed modeling of physical and chemical subprocesses such as fuel evaporation, ignition, flame propagation etc., because these processes are strongly affected by the unresolved spatial temperature and composition distributions. Because of the assumption of an ideally mixed combustion chamber, it is mandatory to model the heat release rate by empirical submodels that hardly contain any physical or chemical principles and instead attempt to reproduce the characteristic heat release rates obtained from experiments by simple mathematical equations with as few parameters as possible. The two most widely used approaches in this respect, the Wiebe function and the so-called polygon-hyperbola-combustion profile, will be discussed in the subsequent sections.

Wiebe Combustion Profile

The Wiebe function was originally developed by I.I. Vibe [27] and should therefore be correctly termed the Vibe function. However, since it is commonly referred to as the Wiebe function in the English literature, this spelling will be adopted throughout this text, too. The purpose of the function is to reproduce the typical S-shaped profile of the integrated heat release rate of SI engines. The ratio of the heat released until crank angle φ to the total amount of heat released by the end of combustion can be approximated by

$$\frac{Q_{chem}(\varphi)}{Q_{chem,tot}} = 1 - \exp\left[-a\left(\frac{\varphi - \varphi_{soc}}{\Delta\varphi_c}\right)^{m+1}\right], \tag{2.32}$$

where $Q_{chem,tot} = m_{fuel} \cdot LHV$, and φ_{soc} and $\Delta\varphi_c$ represent the start of combustion and the combustion duration, respectively. The constant m defines the shape of the integrated heat release curve and a is derived from the perception that only a certain fraction of the injected fuel has been burned at the end of combustion. Introducing the conversion efficiency η_{conv} at the end of combustion $\varphi = \varphi_{eoc}$,

$$\eta_{conv} = \left.\frac{Q_{chem}(\varphi)}{Q_{chem,tot}}\right|_{\varphi=\varphi_{eoc}} = 1 - e^{-a}, \tag{2.33}$$

the parameter a can be expressed as

$$a = -\ln(1 - \eta_{conv}). \tag{2.34}$$

$\eta = 0.9: a = 2.3$
Say: $\eta = 0.7: a = 0.35$
$\eta = 0.1: a \approx 0.1$

Thus, for a typically assumed conversion efficiency of 99.9% a becomes 6.908. The instantaneous heat release rate can be obtained by differentiating Eq. 2.32 with respect to the crank angle:

$$\frac{dQ_{chem}}{d\varphi} = a \cdot Q_{chem,tot} \cdot (m+1) \cdot \left(\frac{\varphi - \varphi_{soc}}{\Delta\varphi_c}\right)^m \cdot \exp\left[-a\left(\frac{\varphi - \varphi_{soc}}{\Delta\varphi_c}\right)^{m+1}\right]. \tag{2.35}$$

In Fig. 2.9 the instantaneous and integrated heat release rates as calculated by Eqs. 2.35 and 2.32, respectively, are plotted against the normalized combustion duration for various values of the shape parameter m. While the energy release rates of homogeneous charge SI engines can be reproduced with this Wiebe function reasonably well, it becomes obvious from Fig. 2.9 that the typical diesel combustion profile with a distinct premixed peak at the start of combustion (compare Fig. 2.10) can hardly be modeled with sufficient agreement. However, this problem can be diminished by superimposing two Wiebe functions, one for the premixed part of combustion and the other representing the later, diffusion controlled part of combustion. In that case, the two functions can have different form parameters as well as different combustion timings, and the total heat release is calculated as the sum of the two individual functions:

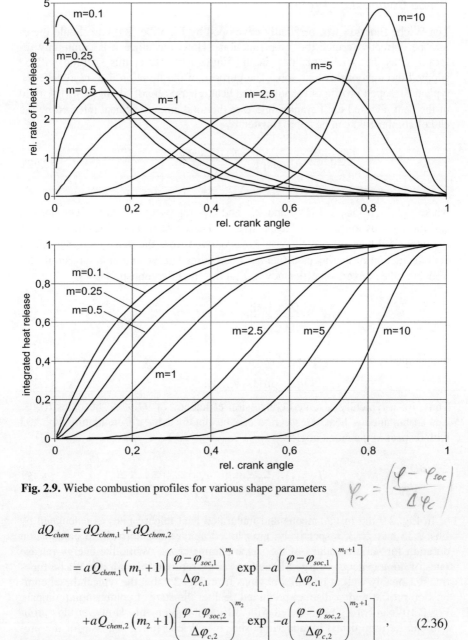

Fig. 2.9. Wiebe combustion profiles for various shape parameters $\varphi_r = \left(\dfrac{\varphi - \varphi_{soc}}{\Delta \varphi_c}\right)$

$$\begin{aligned}
dQ_{chem} &= dQ_{chem,1} + dQ_{chem,2} \\
&= aQ_{chem,1}(m_1+1)\left(\frac{\varphi - \varphi_{soc,1}}{\Delta\varphi_{c,1}}\right)^{m_1} \exp\left[-a\left(\frac{\varphi - \varphi_{soc,1}}{\Delta\varphi_{c,1}}\right)^{m_1+1}\right] \\
&\quad + aQ_{chem,2}(m_2+1)\left(\frac{\varphi - \varphi_{soc,2}}{\Delta\varphi_{c,2}}\right)^{m_2} \exp\left[-a\left(\frac{\varphi - \varphi_{soc,2}}{\Delta\varphi_{c,2}}\right)^{m_2+1}\right] ,
\end{aligned} \quad (2.36)$$

where $Q_{chem,1} + Q_{chem,2} = Q_{chem,tot}$. The left side of Fig. 2.10 shows a comparison between an experimentally obtained typical combustion profile of a direct injection diesel engine and *a* calculated profile by two combined Wiebe functions. While the main characteristics, i.e. the distinct premixed and diffusion parts of combustion, can be reproduced with this approach, it can also be seen that there usually remains a deviation in the decreasing part of the diffusion combustion that cannot be avoided.

Polygon-Hyperbola Combustion Profile

In order to be able to fit the typical diesel combustion profile yet better than with the double Wiebe function, especially during the late stages of diffusion combustion, Schreiner [25] presented a so-called polygon-hyperbola function. This function consists of an interconnection of four straight lines and one hyperbola. As displayed in the right diagram of Fig. 2.10, the first two straight lines (1-2-3) describe the premixed part of combustion, while the third and forth line (3-4-5) represent the increasing part and the plateau maximum of the diffusion combustion, respectively. The decreasing part of the diffusion combustion, where reaction rates are low because of both mixing constraints and low temperatures caused by the downward motion of the piston, is modeled by a hyperbola (5-6) that represents the late stages of combustion better than the Wiebe approach shown on the left diagram of Fig. 2.10. The heat release rates for the different stages of combustion can be expressed by the following set of equations:

Stage 1-2 ($\varphi_1 \leq \varphi \leq \varphi_2$):

$$\frac{dQ_{chem}}{d\varphi} = y_2 \frac{\varphi - \varphi_1}{\varphi_2 - \varphi_1} \qquad (2.37)$$

Stage 2-3 ($\varphi_2 \leq \varphi \leq \varphi_3$):

$$\frac{dQ_{chem}}{d\varphi} = y_2 \frac{\varphi_3 - \varphi}{\varphi_3 - \varphi_2} \qquad (2.38)$$

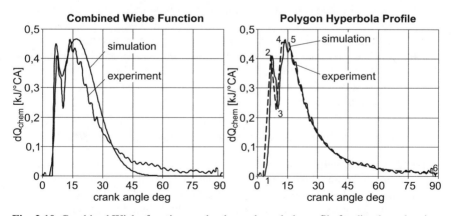

Fig. 2.10. Combined Wiebe functions and polygon-hyperbola profile for diesel combustion

Stage 3-4 ($\varphi_3 \leq \varphi \leq \varphi_4$):

$$\frac{dQ_{chem}}{d\varphi} = y_4 \frac{\varphi - \varphi_1}{\varphi_4 - \varphi_1} \qquad (2.39)$$

Stage 4-5 ($\varphi_4 \leq \varphi \leq \varphi_5$):

$$\frac{dQ_{chem}}{d\varphi} = y_4 \qquad (2.40)$$

Stage 5-6 ($\varphi_5 \leq \varphi \leq \varphi_6$):

$$\frac{dQ_{chem}}{d\varphi} = h_3 + h_1 (\varphi - \varphi_1)^{h_2}$$

$$= y_4 - \frac{y_4 - y_6}{1 - \left[\frac{\varphi_6 - \varphi_1}{\varphi_5 - \varphi_1}\right]^{h_2}}$$

$$+ \frac{y_4 - y_6}{(\varphi_5 - \varphi_1)^{h_2} - (\varphi_6 - \varphi_1)^{h_2}} (\varphi - \varphi_1)^{h_2} \qquad (2.41)$$

The constant y_2 defines the height of the premixed peak and the three parameters of the hyperbola h_1, h_2 and h_3 are derived by the constraint that the hyperbola runs through the states 5 and 6 of value $y_5=y_4$ and y_6, respectively. In addition the integral beneath the complete combustion profile has to equal the total amount of heat released $Q_{chem,tot}$. The height y_4 of the plateau phase of the diffusion combustion needs to be specified such that the center of gravity of the calculated heat release profile equals the center of gravity of the experimentally obtained profile and y_6 is chosen to model the reaction rate before the end of combustion.

Thus, a total of ten parameters needs to be specified in order to define the heat release rate profile given by Eqs. 2.37 to 2.41. These parameters are the start of combustion, the timing of the premixed peak, the combustion duration until the center of gravity of the profile, the combustion duration until the end of the diffusion plateau, the total combustion duration, the maximum heat release rates of the premixed and diffusion parts of combustion, the heat release rate prior to the end of combustion, the premixed fraction of the total heat release and finally the total amount of heat released by combustion. As suggested by Fig. 2.10 these ten parameters are usually sufficient to yield a very good agreement with experimentally obtained heat release rates.

Combustion Profile Variations for Varying Operating Conditions

It should be noted that both the Wiebe function and the polygon-hyperbola function rely on experimental data in order to adjust the model parameters. Therefore, precalculations of the heat release profile for varied boundary conditions (e.g. injection timing or pressure) or even for a completely different engine are generally not possible. However, there exist several approaches how to adjust the parameters within the Wiebe function [31] or the polygon hyperbola function [25] to different loads and speeds of the same engine, once the correct parameters are known for one particular operating condition.

These adjustments, that will not be discussed in detail in this text, can roughly predict the changes in the heat release rates for one specific engine configuration, but extrapolations to different boundary conditions of the combustion are impossible. Investigations that include changes in engine parameters other than load and speed are therefore subject to more sophisticated phenomenological or multidimensional combustion models as will be described in the following chapters. Nevertheless, the empirical heat release models described above are still useful and widely applied whenever computing time is more crucial than high accuracy and predictive capability of the models.

2.2.6 Ignition Delay

In order to apply the heat release models described above, the start of combustion has to be known as a boundary condition. While in spark ignition engines this combustion timing is controlled by the spark timing and therefore is a known parameter, in diesel engines only the start of injection can be controlled directly. As indicated by Fig. 2.11 there is generally a time lag between start of injection and auto-ignition in diesel engines during that physical and chemical subprocesses such as fuel atomization, evaporation, fuel-air-mixing and chemical pre-reactions take place. This time period is referred to as ignition delay τ_{id}.

Again, in thermodynamic cylinder models the subprocesses causing the ignition delay cannot be modeled in detail because of a lack of knowledge of spatially resolved fluid properties. Instead it has to be described by a simplified semi-empirical formulation in terms of the cylinder averaged values of temperature and pressure. Usually, the temperature and pressure dependencies are expressed by an Arrhenius type equation of the general form

$$\tau_{id} = C_1 p^{C_2} \exp\left(\frac{C_3}{T}\right). \tag{2.42}$$

This approach was first suggested by Wolfer [29] who carried out ignition experiments in a constant volume bomb with a fuel of cetane number greater than 50. For a temperature range between 590 K and 780 K and a pressure range between 8 and 48 atm the constants C_1 to C_3 were determined to be

$C_1 = 0.44$ ms, $\quad C_2 = -1.19,\quad C_3 = 4650$ K,

and the pressure has to be input in atm.

Sitkei [26] has expressed the ignition delay by

$$\tau_{id} = 0.5 + 0.133 p^{0.7} \exp\left(\frac{3,930}{T}\right) + 0.00463 p^{-1.8} \exp\left(\frac{3,930}{T}\right), \tag{2.43}$$

where the first term on the right hand side represents a physical delay and the second and third terms represent the chemical delays of the cold and blue flames.

Again, the units of τ_{id}, p and T are ms, atm and K, respectively. Hardenberg and Hase [10] proposed an Arrhenius equation for the ignition delay as well and suggested to calculated the activation energy E_A, i.e. the product of the molar gas constant R_m and the constant C_3 in Eq. 2.42, as a function of the fuel cetane number:

$$E_A = \frac{618{,}840}{CN + 25} \frac{J}{mol}. \tag{2.44}$$

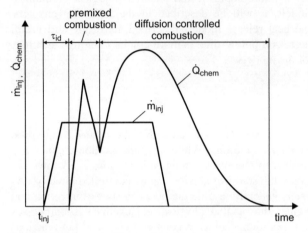

Fig. 2.11. Schematic illustration of injection, ignition delay and heat release rate in a DI diesel engine

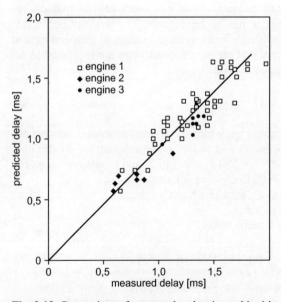

Fig. 2.12. Comparison of measured and estimated ignition delays [6]

To account for temporal changes in temperature and pressure during the ignition delay, the following empirical integral relation has to be satisfied:

$$\int_{t_{inj}}^{t_{inj}+\tau_{id}} \frac{1}{\tau_{id}} dt \geq 1. \tag{2.45}$$

An extensive overview over the many studies that have been performed on the subject of ignition delay – most of those are related to Eq. 2.42 – is given in ref. [23]. In general, a satisfying agreement between calculations and experiments can be achieved with this formulation, as it is indicated by the comparison in Fig. 2.12.

2.2.7 Internal Energy

In the foregoing sections all the components of the mass and energy balances (Eqs. 2.7 and 2.8) have been identified such that the change in internal energy of the control volume "combustion chamber" can be solved. However, for the combustion engineer the physical properties of temperature and pressure of the cylinder gases are more meaningful than a rather abstract value of the internal energy. Therefore, further thermodynamic correlations are necessary that link the change in internal energy to changes in temperature and pressure for a certain gas composition in the combustion chamber.

The total differential of the internal energy can be written as

$$\frac{dU}{dt} = \frac{d(mu)}{dt} = u\frac{dm}{dt} + m\frac{du}{dt}. \tag{2.46}$$

Since the specific internal energy u is a function of temperature, pressure and composition expressed by the overall equivalence ratio, $u = u(T,p,\phi)$, it becomes

$$\frac{dU}{dt} = u\frac{dm}{dt} + m\left(\frac{\partial u}{\partial T}\frac{dT}{dt} + \frac{\partial u}{\partial p}\frac{dp}{dt} + \frac{\partial u}{\partial \phi}\frac{d\phi}{dt}\right). \tag{2.47}$$

In Eq. 2.47 dm/dt and $d\phi/dt$ can be solved by conservation of mass principles and the temporal changes in temperature and pressure are related by the gas law in its differential form. For the simplified case of an ideal gas it reads

$$p\frac{dV}{dt} + V\frac{dp}{dt} = mR\frac{dT}{dt} + RT\frac{dm}{dt} + mT\frac{dR}{dt}. \tag{2.48}$$

Now only the partial derivatives of the specific internal energy with respect to temperature, pressure and composition remain unknown in Eq. 2.47. To estimate these terms several approaches have been suggested in the literature and the ones that are applied most often in thermodynamic combustion models will be described below.

Justi's Correlation

A simple and widely used empirical function established by Justi [16] assumes that the cylinder contents behave as an ideal gas. Therefore, the specific internal energy depends on temperature and composition only. For further simplification, the composition is not expressed in terms of the mass fractions of all species in the mixture but in terms of the reciprocal equivalence ratio $\lambda = 1/\phi$. This implies that the ratio of C- to H-atoms of the fuel must not be changed, that the global equivalence ratio has to be either stoichiometric or lean ($\phi \leq 1$) to assure complete combustion to CO_2 and H_2O and that the composition of the combustion air has to be constant. Thus, a varying water content of the intake air cannot be accounted for. However, since the C/H-ratio of gasoline and diesel fuel are similar adequate results have been obtained for both engines types with the following polynomial expression for the internal energy

$$u(T,\lambda) = 0.1445 \left[1356.8 + \left(489.6 + \frac{46.4}{\lambda^{0.93}} \right) (T - T_{ref}) 10^{-2} \right.$$

$$+ \left(7.768 + \frac{3.36}{\lambda^{0.8}} \right) (T - T_{ref})^2 \, 10^{-4}$$

$$\left. - \left(0.0975 + \frac{0.0485}{\lambda^{0.75}} \right) (T - T_{ref})^3 \, 10^{-6} \right] \quad \text{in} \left[\frac{kJ}{kg} \right], \quad (2.49)$$

where $T_{ref} = 273.15$ K. With this correlation the partial derivatives of u with respect to T, and λ in Eq. 2.47 can be determined analytically and the derivative with respect to pressure reduces to zero.

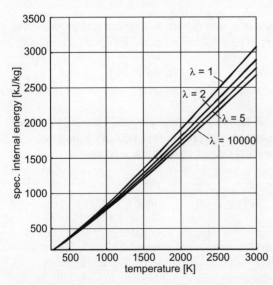

Fig. 2.13. Specific internal energy based on Justi's [16] model

Figure 2.13 shows the specific internal energy estimated by Eq. 2.49 for various air-fuel ratios as a function of temperature. It should be noted that for large values of λ, i.e. $\phi \rightarrow 0$, the specific internal energy of the mixture converges with the one of pure air.

Zacharias' Correlation

Zacharias [32] suggested a similar polynomial correlation for the internal energy, however, he additionally considered a pressure dependency to avoid the ideal gas assumption. Since the mixture composition is again approximated by the reciprocal equivalence ratio λ, the same limitations regarding the gas and fuel compositions as described in Justi's model apply.

$$u(T,p,\lambda) = R_0 T \left[-A \frac{\pi}{\vartheta^2} e^{\frac{D}{\vartheta^2}} \left(1 + 2\frac{D}{\vartheta}\right) + \sum_{i=0}^{6}\left[FA(i)\vartheta^i\right] - 1 \right] \text{ in } \left[\frac{\text{kJ}}{\text{kmol}}\right], \quad (2.50)$$

where:

$$R_0 = \frac{\tilde{R}}{28.89758 + 0.06021 \cdot r}$$

$$\pi = \frac{p}{0.980665 \text{ bar}}$$

$$\vartheta = \frac{T}{1000 \text{ K}}$$

$$r = \frac{\lambda - 1}{\lambda + \frac{1}{af_{stoic}}}$$

$$A = 2.77105 \cdot 10^{-4} - 9.00711 \cdot 10^{-5} \cdot r$$

$$D = 0.008868 - 0.006131 \cdot r$$

$FA(0) = 3.514956 - 0.005026 \cdot r$
$FA(1) = 0.131438 - 0.383504 \cdot r$
$FA(2) = 0.477182 - 0.185214 \cdot r$
$FA(3) = -0.287367 - 0.0694862 \cdot r$
$FA(4) = 0.0742561 + 0.01640411 \cdot r$
$FA(5) = -0.00916344 - 0.00204537 \cdot r$
$FA(6) = 0.000439896 - 0.00010161 \cdot r$

While the consideration of the internal energy's pressure dependence by Zacharias hints a better accuracy than Justi's ideal gas assumption, it also causes an increase in computational effort since an iterative solution of pressure and temperature becomes necessary. Because of the high temperatures generally encountered in combustion systems the numerically simpler approach by Justi is still often used in thermodynamic models. Only in engines with a high level of super- or turbocharging and with peak pressures in excess of 150 bar it is suggested to use a pressure dependent model as real gas effects do become important.

Thermodynamic Charts

Another possibility to specify the specific internal energy is to utilize thermodynamic charts. Since the ideal gas properties (specific heat, specific enthalpy, specific heat of formation, etc.) of many pure substances are tabulated as a function of temperature in the literature, e.g. in the NIST JANAF Tables [22], the specific internal energy of the mixture within the cylinder can be calculated by assuming an ideal mixture of ideal gases,

$$u = \sum_i \left(Y_i \frac{\tilde{u}_i}{MW_i} \right), \tag{2.51}$$

where Y_i is the mass fraction of species i and the superscript $\sim$ indicates molar properties.

If this approach is chosen it is no longer sufficient to describe the mixture composition in terms of the equivalence ratio ϕ or its reciprocal λ. Instead separate mass balances have to be solved for each species in the mixture. In order to do this the changes in species concentrations due to chemical reactions of the combustion process have to be accounted for in addition to the mass flows indicated in Eq. 2.7. This is usually done by assuming a single component reference fuel, e.g. iso-octane instead of gasoline, and determining the mass change of this fuel by dividing the heat release rate described in Sec. 2.2.5 by the appropriate lower heating value of the fuel. The estimation of the concentration changes of all other components in the mixture is then straight forward by assuming a constant air-fuel ratio in the flame and multiplying the concentration change of fuel by the respective stoichiometric coefficients of the other components involved in the chemical reaction.

The molar internal energies $\tilde{u}_i$ in Eq. 2.51 can easily by obtained from the tabulated values for the molar enthalpies $\tilde{h}_i$:

$$\tilde{u}_i = \tilde{h}_i - \tilde{R}T . \tag{2.52}$$

Even though with today's computer power the interpolation of temperature dependent data from tables is no constraint anymore, it is sometimes more convenient to express the property data by polynomial approximations. Heywood [12] suggests the relation

$$\tilde{h}_i = \tilde{R}T \left(a_{i1} + \frac{a_{i2}}{2}T + \frac{a_{i3}}{3}T^2 + \frac{a_{i4}}{4}T^3 + \frac{a_{i5}}{5}T^4 + \frac{a_{i6}}{T} \right) \tag{2.53}$$

for the specific enthalpy and specifies the parameters a_i for the most important species in combustion systems as given in Table 2.1.

Table 2.1. Coefficients for molar enthalpies of gases [12]

Species	T range [K]	a_{i1}	a_{i2}	a_{i3}	a_{i4}	a_{i5}	a_{i6}
CO_2	1000-5000	0.44608 e+1	0.30982 e-2	-0.12393 e-5	0.22741 e-9	-0.15526 e-13	-0.48961 e+5
	300-1000	0.24008 e+1	0.87351 e-2	-0.66071 e-5	0.20022 e-8	0.63274 e-15	-0.48378 e+5
H_2O	1000-5000	0.27168 e+1	0.29451 e-2	-0.80224 e-6	0.10227 e-9	-0.48472 e-14	-0.29906 e+5
	300-1000	0.40701 e+1	-0.11084 e-2	0.41521 e-5	-0.29637 e-8	0.80702 e-12	-0.30280 e+5
CO	1000-5000	0.29841 e+1	0.14891 e-2	-0.57900 e-6	0.10365 e-9	-0.69354 e-14	-0.14245 e+5
	300-1000	0.37101 e+1	-0.16191 e-2	0.36924 e-5	-0.20320 e-8	0.23953 e-12	-0.14356 e+5
H_2	1000-5000	0.31002 e+1	0.51119 e-3	0.52644 e-7	-0.34910 e-10	0.36945 e-14	-0.87738 e+3
	300-1000	0.30574 e+1	0.26765 e-2	-0.58099 e-5	0.55210 e-8	-0.18123 e-11	-0.98890 e+3
O_2	1000-5000	0.36220 e+1	0.73618 e-3	-0.19652 e-6	0.36202 e-10	-0.28946 e-14	-0.12020 e+4
	300-1000	0.36256 e+1	-0.18782 e-2	0.70555 e-5	-0.67635 e-8	0.21556 e-11	-0.10475 e+4
N_2	1000-5000	0.28963 e+1	0.15155 e-2	-0.57235 e-6	0.99807 e-10	-0.65224 e-14	-0.90586 e+3
	300-1000	0.36748 e+1	-0.12082 e-2	0.23240 e-5	-0.63218 e-9	-0.22577 e-12	-0.10612 e+4
OH	1000-5000	0.29106 e+1	0.95932 e-3	-0.19442 e-6	0.13757 e-10	0.14225 e-15	0.39354 e+4
NO	1000-5000	0.31890 e+1	0.13382 e-2	-0.52899 e-6	0.95919 e-10	-0.64848 e-14	0.98283 e+4
O	1000-5000	0.25421 e+1	-0.27551 e-4	-0.31028 e-8	0.45511 e-11	-0.43681 e-15	0.29231 e+5
H	1000-5000	0.25 e+1	0.0	0.0	0.0	0.0	0.25472 e+5

Table 2.2. Coefficients for molar enthalpies of vaporized fuels [12]

Fuel	A_{f1}	A_{f2}	A_{f3}	A_{f4}	A_{f5}	A_{f6}	A_{f8}
Methane CH_4	-0.29149	26.327	-10.61	1.5656	0.16573	-18.331	4.3
Propane C_3H_8	-1.4867	74.339	-39.065	8.0543	0.01219	-27.313	8.852
Hexane C_6H_{14}	-20.777	210.48	-164.125	52.832	0.56635	-39.836	15.611
Isooctane C_8H_{18}	-0.55313	181.62	-97.787	20.402	-0.03095	-60.751	20.232
Methanol CH_3OH	-2.7059	44.168	-27.501	7.2193	0.20299	-48.288	5.3375
Ethanol C_2H_5OH	6.990	39.741	-11.926	0	0	-60.214	7.6135
Gasoline $C_{8.26}H_{15.5}$	-24.078	256.63	-201.68	64.75	0.5808	-27.562	17.792
Gasoline $C_{7.76}H_{13.1}$	-22.501	227.99	-177.26	56.048	0.4845	-17.578	15.235
Diesel $C_{10.8}H_{18.7}$	-9.1063	246.97	-143.74	32.329	0.0518	-50.128	23.514

Units of A_{fi} such that $\tilde{h}_f$ is in kcal/mol

The enthalpy of vaporized fuels can also be estimated by a polynomial expression:

$$\tilde{h}_f = a_{f1}\vartheta + a_{f2}\frac{\vartheta^2}{2} + a_{f3}\frac{\vartheta^3}{3} + a_{f4}\frac{\vartheta^4}{4} - \frac{a_{f5}}{\vartheta} + a_{f6} + a_{f8}, \quad (2.54)$$

where $\vartheta = T/(1000\ K)$ and the respective parameters a_f are specified in Table 2.2 for various fuels.

2.3 Two-Stroke Scavenging Models

While the assumption of an ideally mixed combustion chamber in the single zone combustion model is acceptable for calculation of thermodynamic properties in four-stroke engines with typically small valve overlaps, the scavenging process of a two-stroke engine is more complicated. Only a very short time period is available for the gas exchange during that both inlet and exhaust ports are open at the same time. The purpose of the scavenging process is to replace the burned gases by fresh charge with as little mixing between the two gases as possible. Therefore, in modeling this process it is necessary to describe the cylinder gases with at least two zones to determine the gas composition after EPC, and the single zone cylinder model described above can no longer be used.

Three different port or valve configurations have been realized in two-stroke engines to prevent the burned and unburned gases from mixing and to avoid short-circuiting of the fresh charge directly into the exhaust port. Schematic illustrations of these cross-scavenged, loop-scavenged and uniflow-scavenged flow configuration are shown in Fig. 2.14. The quality of the scavenging process can be quantified by the delivery ratio, the trapping and scavenging efficiencies and the charge purity which are

$$\Lambda = \frac{\text{mass of delivered charge}}{\text{displaced volume} \times \text{ambient density}}, \qquad (2.55)$$

$$\eta_{tr} = \frac{\text{mass of delivered charge retained}}{\text{mass of delivered charge}}, \qquad (2.56)$$

$$\eta_{sc} = \frac{\text{mass of delivered charge retained}}{\text{mass of trapped cylinder charge}}, \qquad (2.57)$$

$$\text{Purity} = \frac{\text{mass of fresh charge in trapped cyl. charge}}{\text{mass of trapped cylinder charge}}, \qquad (2.58)$$

respectively. These values vary for the three flow configurations, but two limiting cases do apply for all scavenging systems.

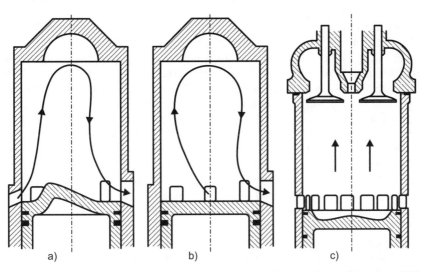

Fig. 2.14. a) Cross-, b) loop-, and c) uniflow-scavenged two-stroke configurations [12]

30   2 Thermodynamic Models

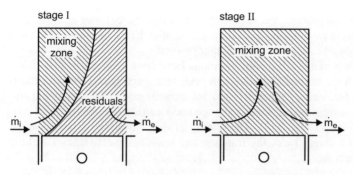

Fig. 2.15. Combined displacement/mixing scavenging model [19]

The poorest scavenging scenario is represented by the *complete mixing* model which is effectively a single zone cylinder model. It assumes that fresh charge that enters the cylinder mixes instantaneously and homogeneously with the burned cylinder contents. Hence, the charge exiting the cylinder consists of pure burned gases in the beginning but becomes more and more diluted as fresh charge continues to enter the cylinder. On the other hand, the ideal scavenging scenario is described by the *perfect displacement* model. Here it is assumed that fresh and burned charges are contained in two separate zones and do not mix at all. Consequently the overall composition of the combustion chamber changes proportionally to the mass flow rates entering and exiting the cylinder through the ports.

The real scavenging process shows a behavior somewhere in between these two limiting cases and a number of models have been proposed to describe the real process with better accuracy than the complete mixing or perfect displacement models. An extensive review of such scavenging models has been given by Merker and Gerstle [19].

In this text only a relatively simple combination of the two limiting models as presented by Eberle [7] or by Benson and Bradham [5] will be given. As indicated in Fig. 2.15 this combined model divides the scavenging process into two temporal stages. During the first stage displayed on the left hand side of Fig. 2.15 perfect displacement is assumed since only little fresh charge has yet entered the cylinder and is most likely not immediately located in the vicinity of the exhaust port or valve. This is taken into account by considering two spatial zones containing mixed charge and pure residuals, respectively. Prior to the scavenging process both zones have the same compositions, i.e. pure residuals, but once fresh charge starts to enter the cylinder it is added to the mixing zone only which therefore becomes more and more diluted. The residual or burned zone is not subject to any mixing and is perfectly displaced into the exhaust port by the growing mixing zone. Once the entire burned zone has been exhausted only the mixing zone is left during stage 2 (right hand side of Fig. 2.15) and complete mixing is assumed.

The model has one degree of freedom which is the so-called displacement ratio x. It is defined as the initial ratio of the residual zone volume to the entire cylinder volume and effectively determines the transition point between stage 1 and stage 2 of the model:

$$x = \frac{V_{res,0}}{V_{th}}. \tag{2.59}$$

For the purpose of determining the scavenging parameters it is assumed that the pressure and cylinder volume V_{th} remain approx. constant during the gas exchange process and that there is no heat transfer between either zone and the combustion chamber walls.

If z is defined as the instantaneous fraction of fresh charge in the entire cylinder, and V_i is the integral volume of fresh charge that has entered the cylinder since the start of scavenging, the change in fresh charge inside the cylinder $dz \cdot V_{th}$ is equal to dV_i during stage 1. During stage 2 it can be expressed as

$$dz \cdot V_{th} = dV_i - z dV_i = (1-z)\ dV_i. \tag{2.60}$$

At the transition point between stages 1 and 2 we can write $z = x$ and $V_i = x\, V_{th}$ and at the end of scavenging $z = \eta_{sc}$ and $V_i = \Lambda\, V_{th}$. Hence, Eq. 2.60 can be integrated,

$$\int_x^{\eta_{sc}} \frac{dz}{1-z} = \int_{xV_{th}}^{\Lambda V_{th}} \frac{dV_i}{V_{th}}, \tag{2.61}$$

and after some calculus yields the scavenging efficiency η_{sc}:

$$\eta_{sc} = \begin{cases} \Lambda & \text{for} \quad \Lambda \leq x \\ 1-(1-x)e^{(x-\Lambda)} & \text{for} \quad \Lambda > x \end{cases} \tag{2.62}$$

Figure 2.16 displays η_{sc} as a function of the delivery ratio Λ for various displacement ratios x.

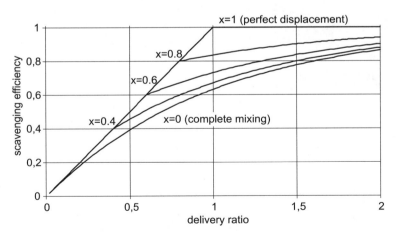

Fig. 2.16. Scavenging efficiencies calculated by the displacement/mixing model

In the described displacement/mixing model the value of the displacement ratio x has to be adjusted to fit experimental data of a specific scavenging configuration. Once this has been established the results obtained with the displacement/mixing model after the end of scavenging, i.e. at EPC, usually have sufficient accuracy to be used in thermodynamic combustion models. It should be noted though, that more comprehensive models for the scavenging process of two-stroke engines have been proposed (see ref. [19] for an extensive review) that do have a potential to describe the course of the characteristic scavenging parameters such as η_{sc} between EPO and EPC more realistically.

2.4 Empirical Two-Zone Combustion Model

Because of the exponential temperature dependence of the chemical reaction mechanism controlling the rate of nitrogen oxides (NO_x) formation in combustion engines (see Chap. 7) an estimation of tailpipe NO_x emissions is generally not possible with a single-zone cylinder model. Instead of the mass averaged cylinder temperature available in this model type local peak temperatures must be known in order to estimate realistic NO_x formation rates.

To reduce these shortcomings of the thermodynamic cylinder model, Heider et al. [11] presented a simple, empirical two-zone model effective during the combustion period of the engine's working process. It assumes that the cylinder gases are divided into two separate zones (Fig. 2.17) where the so-called reaction zone (index 1) contains burned products of high temperature, and the unburned zone (index 2) contains fresh air plus some residuals of a previous cycle in case of an exhaust gas recirculation. With this approach the NO_x formation rate can at least be approximated based on the hot temperatures in the reaction zone.

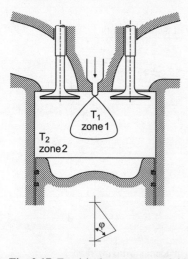

Fig. 2.17. Empirical two-zone model by Heider et al. [11]

2.4 Empirical Two-Zone Combustion Model

The calculation of the states within the two zones is set up on the solution of the zero-dimensional single-zone model and is further based on the following conditions and assumptions:

- Mass and volume conservation for both zones
- Ideal gas behavior for both zones
- The heat of combustion is completely released in the reaction zone 1. The mass of fuel in zone 1 is therefore given by the global heat release rate (see Sec. 2.2.5) and the heating value of the fuel.
- The equivalence ratio ϕ_1 in the reaction zone 1 is constant with time and approx. unity.
- The temperature difference between zones 1 and 2 has a maximum at the start of combustion and reduces to zero at EVO due to energy transfer between the zones.

Accordingly, one can write

$$m_1 + m_2 = m_{cyl}, \qquad (2.63)$$

$$V_1 + V_2 = V_{cyl}, \qquad (2.64)$$

$$p_1 = p_2 = p_{cyl}, \qquad (2.65)$$

where subscripts 1 and 2 denote the burned and unburned zones, respectively. The mass of zone 1 is calculated as

$$\begin{aligned} m_1(\varphi) &= m_{air,1}(\varphi) + m_{fuel}(\varphi) \\ &= m_{fuel}(\varphi) \cdot \left[\frac{af_{stoic}}{\phi_1} + 1\right] = \frac{Q_{chem}(\varphi)}{\text{LHV}} \cdot \left[\frac{af_{stoic}}{\phi_1} + 1\right], \end{aligned} \qquad (2.66)$$

and the zone volumes are related to the respective masses, temperatures and the cylinder pressure by the ideal gas law:

$$V_i = \frac{m_i R_i T_i}{p_{cyl}}; \qquad i = 1, 2. \qquad (2.67)$$

Furthermore the condition for the mean cylinder temperature has to be satisfied:

$$m_1 T_1 + m_2 T_2 = m_{cyl} T_{cyl}. \qquad (2.68)$$

As indicated above, the temperature difference between the two zones reduces during the course of combustion because of an energy transfer from zone 1 to zone 2. The rate of this transfer is empirically described in terms of the pressure difference between fired and motored engine operation $p_{cyl} - p_{mot}$, which was already used to model the turbulence increase because of combustion in Woschni's wall heat transfer model (see Sec. 2.2.4),

$$T_1(\varphi) - T_2(\varphi) = B(\varphi) \cdot A^*, \qquad (2.69)$$

where

$$B(\varphi) = \frac{\int_{\varphi_{soc}}^{\varphi_{EVO}} \left[p_{cyl}(\varphi) - p_{mot}(\varphi) \right] m_1(\varphi) d\varphi - \int_{\varphi_{soc}}^{\varphi} \left[p_{cyl}(\varphi) - p_{mot}(\varphi) \right] m_1(\varphi) d\varphi}{\int_{\varphi_{soc}}^{\varphi_{EVO}} \left[p_{cyl}(\varphi) - p_{mot}(\varphi) \right] m_1(\varphi) d\varphi}. \quad (2.70)$$

B in Eq. 2.70 reduces from 1 to 0 and implies that the temperature difference between zones 1 and 2 is A^* at the start of combustion and becomes zero at EVO. For small to medium sized diesel engines operated with an intake air swirl the equivalence ratio in the reaction zone is assumed to be stoichiometric, i.e. $\phi_1 = 1.0$, and A^* depends on the global equivalence ratio in the cylinder:

$$A^* = A \cdot \frac{\phi_{cyl}}{2.2} \cdot \left[1.2 + \left(\frac{1}{\phi_{cyl}} - 1.2 \right)^{0.15} \right]. \quad (2.71)$$

For large diesel engines without swirl ϕ_1 is assumed to be slightly lean, $\phi_1 = 0.971$, and A^* is independent of the engine load:

$$A^* = A = const. \quad (2.72)$$

A in Eqs. 2.71 and 2.72 is an engine specific parameter that has to be determined once for a particular engine. It typically ranges between 1500 and 1650 K.

With the above set of equations, the states in zones 1 and 2 are identified and the thermal NO_x formation can be estimated based on the hot temperature in zone 1 by the extended Zeldovich mechanism (see Chap. 7). Figure 2.18 displays typical histories of cylinder pressure as well as of the zone temperatures as calculated for a medium sized high-speed diesel engine.

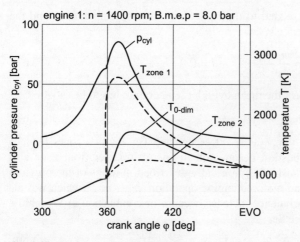

Fig. 2.18. Calculated pressure and zonal temperatures for a high-speed diesel engine [11]

2.5 Typical Applications

2.5.1 Heat Release Analysis

One standard application of the thermodynamic cylinder model described above is the heat release analysis of experimental data, see e.g. Schwarz and co-workers [24, 28]. Since the rate of energy released by the chemical reactions of fuel combustion cannot be measured directly, this information has to be derived from the temporal histories of the cylinder pressure which can be obtained relatively easily by pressure indication. This is achieved by solving the mass and energy balances (Eqs. 2.7 and 2.8) for the heat release rate dQ_{chem}/dt instead of the temporal change of internal energy dU_{cyl}/dt and Eq. 2.8 becomes

$$\frac{dQ_{chem}}{dt} = \frac{dU_{cyl}}{dt} - \frac{dQ_w}{dt} + p_{cyl}\frac{dV_{cyl}}{dt} - \frac{dm_{in}}{dt}h_{in} - \frac{dm_{exit}}{dt}h_{exit}$$
$$- \frac{dm_{fuel}}{dt}h_{fuel} - \frac{dm_{bb}}{dt}h_{bb} \quad . \tag{2.73}$$

The change in internal energy can be derived from the measured pressure trace by one of the relations given in Sec. 2.2.7 and all other terms present in the mass and energy balances are modeled exactly as described in the foregoing sections. Since the specific internal energy has a non-linear temperature and pressure dependence an error is generally encountered when a spatially uniform, mean temperature is assumed with the single-zone cylinder model. To diminish this error Kamimoto et al. [18] have suggested a two-zone model similar to the model discussed in Sect. 2.4, that contains one zone of stoichiometric combustion products and one of pure unburned air. They showed that the heat release rate derived from experimental pressure data based on the two-zone cylinder model exceeds the one based on the standard single-zone model by up to 10 percent.

Moreover, it should be noted that typically the wall heat transfer dQ_w is the dominant term on the right hand side of Eq. 2.73. Consequently, an uncertainty in the modeling of the wall heat transfer also has a significant effect on the outcome of the so-called "experimental" heat release rate which really is a calculated heat release rate based on an experimentally obtained pressure trace.

2.5.2 Analysis of Complete Power Systems

Thermodynamic combustion models allow for very short computation times but that they are characterized by a lack of insight into subscale spray and combustion phenomena inside the cylinder. Therefore, a typical field of application besides the above heat release rate analysis is hardly the optimization of the combustion process itself, but rather the design and tuning of complete power systems in that the engine represents only one of many interacting components. Because of the complexity of such systems consisting of the engine, turbocharger, governor, gearbox, propeller or generator etc., an experimental investigation of all system interdependencies during the project planning stage is almost impossible and simulation models become an invaluable tool.

Moreover, transient operation scenarios typically represent the greatest challenge to the system set up and its fine-tuning. Since the inertia of an entire propulsion system, e.g. the boat speed of a large vessel, is often much greater than the inertia of the engine, the changes in load and/or speed of the engine usually require multiple work cycles to reach stationary operating conditions again. Hence, a time efficient description of the in-cylinder processes is crucial and more comprehensive and time consuming combustion models are not an option for the simulation of a complete propulsion system.

To further reduce CPU-time requirements for transient simulations it is common practice to run cycle simulations with the thermodynamic cylinder model for a large number of engine parameters prior to the actual transient system simulation and store the results in characteristic engine maps for later interpolation. As an example of this procedure, Gerstle and Merker [9] have described the behavior of a medium-speed diesel engine by a set of n-dimensional maps that store the cycle averaged results (mass and enthalpy flows through the intake and exhaust, heat transfer to the coolant, overall equivalence ratio and cylinder torque) as a function of the seven input parameters fuel mass injected, start of injection, charging air pressure and temperature, exhaust back pressure, coolant temperature and engine speed. Figure 2.19 compares the estimated and measured temporal courses of the fuel pump control rod position, turbocharger speed, engine speed and exhaust manifold pressure for a load increase from 50% to 100% rated power in a constant speed generator application. It is obvious from the results that the level of agreement between simulation and experimental data is very good.

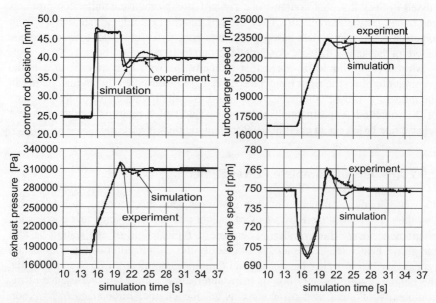

Fig. 2.19. Transient simulation of a 50% to 100% load increase [9]

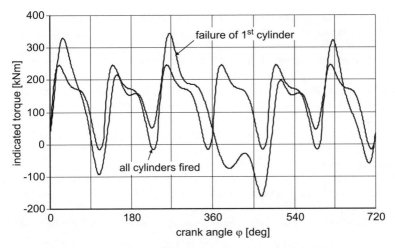

Fig. 2.20. Calculated cylinder torque for failure of one cylinder [9]

In a second example application, Gerstle and Merker [9] have worked with crank angle resolved data to model the effect of a failure of one cylinder of a 6-cylinder engine on engine torque, Fig. 2.20. The results could now be utilized as an input to investigate the additional stresses that can be caused by a cylinder failure on various system components such as crank shaft, bearings, gearbox etc..

These two simple examples clearly demonstrate the capability of time efficient thermodynamic cylinder models to describe various operational aspects of an engine and their influences on the behavior of a complete propulsion system.

References

[1] Annand WJ (1963) Heat Transfer in the Cylinders of Reciprocating Internal Combustion Engines. Proc Inst Mech Engineers, vol 177, no 36, pp 973–990
[2] Arcoumanis C, Fairbrother RJ (1992) Computer Simulation of Fuel Injection Systems for DI Diesel Engines. SAE Paper 922223
[3] Augugliaro G, Bella G, Rocco V, Baritaud T, Verhoeven D (1997) A Simulation Model for High Pressure Injection Systems. SAE Paper 971595
[4] Bargende M (1990) Ein Gleichungsansatz zur Berechnung der instationären Wandwärmeverluste im Hochdruckteil von Ottomotoren. Ph.D. Thesis, Technical University of Darmstadt, Germany
[5] Benson RS, Bradham PT (1969) A Method for Obtaining a Quantitative Assessment of the Influence of Charging Efficiency on Two-Stroke Engine Performance. Int J Mech Sci, vol 11, pp 303–312
[6] Dent JC, Mehta PS (1981) Phenomenological Combustion Model for a Quiescent Chamber Diesel Engine. SAE Paper 811235

[7] Eberle M (1968) Beitrag zur Berechnung des thermodynamischen Zusammenwirkens von Verbrennungsmotor und Abgasturbolader. Ph.D. Thesis, ETH Zurich, Switzerland
[8] Gerstle M, Merker GP (1998) Transient Simulation of Marine Diesel Engines. Proc 22nd CIMAC Cong, vol 2, pp 457–468, Copenhagen
[9] Gerstle M, Merker GP (1999) Transient Simulation of Marine Diesel Engine Systems by Improved Characteristic Cylinder Map Interpolation. IMarE Conf, vol 111, 2, pp 129–138
[10] Hardenberg HO, Hase FW (1979) An Empirical Formula for Computing the Pressure Rise Delay of a Fuel from its Cetane Number and from the Relevant Parameters of Direct-Injection Diesel Engines. SAE Paper 790493
[11] Heider G, Zeilinger K, Woschni G (1995) Two-Zone Calculation Model for the Prediction of NO Emissions from Diesel Engines. Proc 21st CIMAC Cong, Paper D52, Interlaken
[12] Heywood JB (1988) Internal Combustion Engine Fundamentals. McGraw-Hill, New York
[13] Hohenberg G (1979) Advanced Approaches for Heat Transfer Calculations. SAE Paper 790825
[14] Huber K (1990) Der Wärmeübergang schnellaufender, direkteinspritzender Dieselmotoren. Ph.D. Thesis, Technical University of Munich, Germany
[15] Incropera FP, DeWitt DP (1996) Introduction to Heat Transfer. 3rd edn, Wiley, New York
[16] Justi E (1938) Spezifische Wärme, Enthalpie, Entropie und Dissoziation technischer Gase. Springer, Berlin, Germany
[17] Kamel M, Watson N (1979) Heat Transfer in the Indirect Injection Diesel. SAE Paper 790826
[18] Kamimoto T, Minagawa T, Kobori S (1997) A Two-Zone Model Analysis of Heat Release Rate in Diesel Engines. SAE Paper 972959
[19] Merker GP, Gerstle M (1997) Evaluation on Two Stroke Engine Scavenging Models. SAE Paper 970358
[20] Merker GP, Schwartz C (2001) Technische Verbrennung – Simulation verbrennungsmotorischer Prozesse. B.G. Teubner, Stuttgart, Germany
[21] Namazian M, Heywood JB (1982) Flow in the Piston-Cylinder-Ring Crevices of a Spark-Ignition Engine: Effect on Hydrocarbon Emissions, Efficiency and power. SAE Paper 820088
[22] NIST (1993) JANAF Thermochemical Tables Database. Version 1.0, National Institute of Standards and Technology, Gaithersburg
[23] Ramos JI (1989) Internal Combustion Engine Modeling. Hemisphere, New York
[24] Reulein C, Schwarz C (2001) Gesamtprozessanalyse – Potenzial Grenzen und typische Anwendungen. Proc 4th Dresdner Motorenkolloquium, pp 257–266, Dresden, Germany
[25] Schreiner K (1995) Equivalent Combustion Rate with the Polygon-Hyperbola Function: Investigations into the Dependence of the Parameters in the Performance Map. Proc 5th Symp "The Working Process of the Internal Combustion Engine", pp 239–257, Technical University Graz, Austria
[26] Sitkei G (1964) Kraftstoffaufbereitung und Verbrennung bei Dieselmotoren. Springer, Berlin, Germany

[27] Vibe II (1962) Novoe o rabocem cikle dvigatelej: Skorost sgoranija i rabocij cikl dvigatelja. Masgiz, Moscow
[28] Witt A, Siersch W, Schwarz C (1999) New Methods in the Development of the Pressure Analysis for Modern SI Engines. 7th Symp "The Working Process of the Internal Combustion Engine", pp 53–67, Technical University Graz, Austria
[29] Wolfer HH (1938) Ignition Lag in Diesel Engines. VDI-Forschungsheft 392. Translated by Royal Aircraft Establishment, Farnborough Library no 358, UDC 621-436.047, August 1959
[30] Woschni G (1967) A Universally Applicable Equation for the Instantaneous Heat Transfer Coefficient in the Internal Combustion Engine. SAE Paper 670931
[31] Woschni G, Anisits F (1974) Experimental Investigation and Mathematical Presentation of Rate of Heat Release in Diesel Engines Dependent upon Engine Operating Conditions. SAE Paper 740086
[32] Zacharias F (1966) Analytical Description of the Thermal Properties of Combustion Gases (in German). Ph.D. Thesis, Technical University of Berlin, Germany

3 Phenomenological Models

3.1 Classification

While the thermodynamic combustion models described in Chap. 2 are relatively easy to handle and are characterized by a low computational effort, they are lacking the ability to make predictions of the effects of important engine parameters on combustion without prior measurements. The main reasons for this deficiency are that major subprocesses are either not modeled at all or described by solely empirical correlations and that the assumption of an ideally mixed combustion chamber makes it impossible to estimate pollutant formation rates that are strongly affected by local temperatures and mixture compositions. On the other hand, the multidimensional CFD models that are based on the locally resolved solutions of mass-, energy- and momentum-conservation and that include detailed submodels for spray and combustion phenomena, are computationally expensive, and they demand that the user has a much deeper understanding of the governing physical and chemical processes in order to correctly interpret the simulation results. Moreover, the predictive quality with respect to global quantities such as pressure traces and apparent heat release rates is not necessarily better than with simpler models. This is because the many subprocesses taking place inside a combustion chamber are often interacting with each other such that relatively small errors encountered within particular submodels may add up to a considerable error in the overall result of the computation.

Consequently, there is a need for a third model category that allows to execute time-efficient pre-calculations of heat release rates and exhaust emissions as a function of important engine parameters, e.g. engine compression ratio, boost temperature and pressure, injection timing and pressure, EGR rates or swirl ratios of the intake charge. This category is commonly referred to as phenomenological (or quasi-dimensional) models and can be classified in between the thermodynamic and the multidimensional models, Fig. 3.1. In these models the combustion chamber is typically divided into multiple zones which are characterized by different temperatures and compositions. The exact number of zones considered depends on the chosen model approach and can range between as few as two and as many as several hundreds. Because of this spatial resolution – albeit much coarser than in CFD-codes – the heat release rate does no longer have to be described by solely empirical correlations like the Wiebe-function. Instead, it can be pre-estimated based on physical and chemical submodels for local processes like spray formation, air-fuel-mixing, droplet evaporation, ignition and combustion including pollutant formation.

The advantage of phenomenological models compared to CFD-models is that the simplifying assumptions made have the effect that only ordinary differential equations with respect to time have to be solved. This is in contrast to the partial differential equations with respect to time and space that need to be solved in CFD-codes and, in addition to the reduced spatial resolution and the submodels that are typically not as detailed as the ones used in CFD-modeling, leads to a substantial reduction in computing time which can be as significant as three orders of magnitude, Fig. 3.2. However, the limitation of phenomenological models is that the turbulent flow field inside the combustion chamber is not resolved, and therefore the effects of geometrical changes of the combustion chamber can generally not be investigated with this model type.

It should be noted that phenomenological combustion models are most practical to describe diesel engine combustion. This is because in diesel engines the injection process, which can be described relatively well with the phenomenological approach, has the dominant effect on mixture formation and the subsequent course of combustion. In spark ignition engines, especially in engines with external mixture formation where the charge is already homogenously mixed at the ignition timing, combustion is controlled by the propagation of a premixed flame front rather than by mixing and diffusion phenomena. This flame front propagation is strongly affected by the turbulence structure and is therefore hard to describe with phenomenological models that do not resolve the three-dimensional turbulent flow field. For the above reasons, the majority of models discussed in the present chapter are diesel rather than SI engine combustion models.

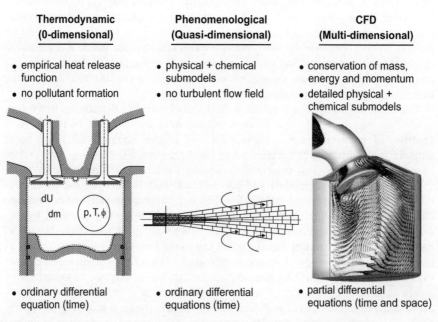

Fig. 3.1. Combustion model classification (Courtesy of DaimlerChrysler AG)

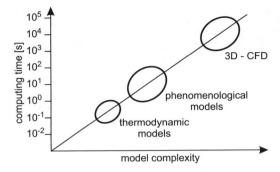

Fig. 3.2. Model depths and approximate computing requirements of combustion models

Various phenomenological spray and combustion models have been proposed in the literature that are characterized by a fairly different level of detail and hence by a different level of computing requirements [66]. However, the common purpose of all phenomenological models is to pre-calculate the heat release rate from a known injection rate profile and to establish a quasi-dimensional composition and temperature distribution inside the combustion chamber in order to allow an assessment of the formation rates of crucial diesel engine pollutants, i.e. nitrogen oxides and soot. While in some models, e.g. in the packet models discussed in Sect. 3.2.3, the calculation of gas mixing, heat release and emission formation is closely coupled, there are a number of other models where there are independent submodels for these processes which can be combined almost arbitrarily as desired by the user / programmer. For this reason, the following discussion of phenomenological models will be structured into heat release and gas mixing submodels as well as into wall heat transfer and pollutant formation submodels. The latter two apply to all approaches in a similar manner.

3.2 Heat Release in Diesel Engines

3.2.1 Zero-Dimensional Burning Rate Function

A relatively simple and easy to handle heat release model for direct injection diesel engines has been presented by Chmela and Orthaber [14]. It requires only slightly more computational effort than the empirical heat release functions discussed in Chap. 2 and could therefore also be applied in thermodynamic cylinder simulations. However, it does relate the fuel burning rate to characteristic engine parameters and is therefore viewed as a phenomenological model even though it does not require to subdivide the combustion chamber into zones of different composition and temperature.

The two parameters of paramount influence on the combustion rate are assumed to be the amount of fuel available for combustion at a particular point in time and the mixing rate between fuel and air which is further assumed to depend primarily on the local density of turbulent kinetic energy. Hence, the rate of heat release is expressed as the product of two functions f_1 and f_2, where f_1 depends on the available fuel mass $m_{fuel,av}$ and f_2 is related to the turbulent kinetic energy k:

$$\frac{dQ_{chem}}{dt} = C \cdot f_1(m_{fuel,av}) \cdot f_2(k). \tag{3.1}$$

The constant C scales the magnitude of the resulting heat release rate, and the fuel mass available for combustion can by derived as the difference between the injected fuel mass and the amount of fuel that has already been burned:

$$m_{fuel,av}(t) = m_{fuel}(t) - \frac{Q_{chem}(t)}{\text{LHV}}. \tag{3.2}$$

Equation 3.2 implies that the fuel evaporation process is assumed to be infinitely fast and all the fuel injected into the cylinder is immediately evaporated and available for combustion.

The formulation of the function f_2 in Eq. 3.1 is based on the general idea, that in a typical combustion engine three principal sources for production of turbulent kinetic energy can be identified: the kinetic energies associated with the inlet flow through the ports (i.e. a swirl or tumble motion) as well as with the squish flow during compression and, most importantly in DI diesel engines, the kinetic energy of the high pressure fuel injection. An assessment of the relative contributions of each of theses three sources has been executed for a single-cylinder research engine with a displacement volume of 2.0 liters, a compression ratio of $\varepsilon = 18$ and a common rail injection system with an injection pressure of 1200 bar. The respective results for the each kinetic energies that have been estimated based on simplified correlations for the three mechanisms are summarized in Table 3.1 for two different engine speeds. It is clearly visible that the kinetic energy associated with the fuel spray has by far the most dominant effect for both engine speeds. Therefore, the two other mechanisms will be neglected in the subsequent analysis.

The production rate of kinetic energy due to fuel injection can be expressed as

$$\frac{dE_{kin,prod}}{dt} = \frac{1}{2}\dot{m}_{fuel}(v_{inj})^2 = \frac{1}{2}\left[\frac{1}{\rho_{fuel}\ c_D A_{noz}}\right]^2 (\dot{m}_{fuel})^3, \tag{3.3}$$

where $\dot{m}_{fuel}$ is the fuel injection rate and $c_D \cdot A_{noz}$ is the effective cross sectional area of the nozzle holes. For simplicity, the dissipation rate of kinetic energy is assumed to be proportional to the total level of kinetic energy E_{kin}, such that the following differential balance equation is obtained:

$$\frac{dE_{kin}}{dt} = \frac{dE_{kin,prod}}{dt} - C_{diss} E_{kin}. \tag{3.4}$$

For the following considerations it is further assumed that not the total amount of kinetic energy inside the combustion chamber is linked to the fuel jet and available for the mixing process, but that a fraction of it will be transferred to the burned gases through the process of combustion. Hence, only the fraction of the kinetic energy that is proportional to the momentarily available fuel mass is available for the fuel-air mixing process:

$$E_{kin,mix} = E_{kin} \cdot \frac{m_{fuel} - Q_{chem}/LHV}{m_{fuel}}. \quad (3.5)$$

The specific turbulent kinetic energy is then approximated as the ratio of this available kinetic energy and the sum of fuel and air mass in the diffusion flame, which is assumed to be approximately stoichiometric:

$$k = C_{turb} \cdot \frac{E_{kin,mix}}{m_{fuel}(1 + af_{stoic})}. \quad (3.6)$$

The constant C_{turb} in Eq. 3.6 can be viewed as a "transfer efficiency" between kinetic and turbulent kinetic energy.

Following the general idea of the time scale combustion model originally suggested by Magnussen and Hjertager [43] (compare Sect. 3.2.4), which expresses the reaction rate as a function of a turbulent frequency (ε/k), it is now attempted to relate the function f_2 in Eq. 3.1 to the turbulence level of the cylinder contents. However, the difficulty in this approach is that within the zero-dimensional treatment of the combustion chamber the dissipation rate ε of the turbulent kinetic energy k cannot be approximated with simplified relations as it is done with k itself in the above equations. Therefore, the authors assume that the mixing rate, and thus the combustion rate in a turbulent diffusion flame, depend directly on the turbulent kinetic energy k:

$$f_2 = \exp\left[C_2 \frac{\sqrt{k}}{\sqrt[3]{V_{cyl}}}\right]. \quad (3.7)$$

Table 3.1. Assessment of contributions to kinetic energy inside the cylinder [14]

Engine Speed	[rpm]	1000	1800
Injection Quantity	[mm³/cycle]	255	253
Kinetic Energy of Squish Flow at TDC	[J]	0.15	0.40
Kinetic Energy of Swirl Motion at TDC	[J]	0.20	0.55
Kinetic Energy of Fuel Spray at End of Inj.	[J]	25.30	26.50

Single-cylinder research engine: $V_d = 2.0$ dm³, $\varepsilon = 18$, $r_S = 1.8$, $p_{inj} = 1200$ bar

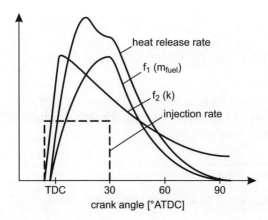

Fig. 3.3. Characteristic course of a diesel engine heat release rate as predicted by Chmela and Orthaber [14]

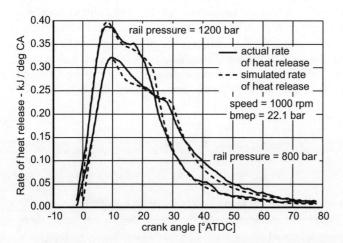

Fig. 3.4. Effect of rail pressure on the heat release rate of common rail diesel engine [14]

In order to obtain a frequency unit [1/s] the square root of k is divided by a characteristic length scale which is chosen as the cubic root of the momentary cylinder volume. An exponential formulation is selected to make sure that the combustion rate does not become zero for low levels of turbulent kinetic energy. Combining Eqs. 3.1, 3.2 and 3.7, the heat release rate becomes

$$\frac{dQ_{chem}}{dt} = C_1 \cdot \left(m_{fuel} - \frac{Q_{chem}}{\text{LHV}} \right) \cdot \exp\left[C_2 \frac{\sqrt{k}}{\sqrt[3]{V_{cyl}}} \right]. \tag{3.8}$$

Typical histories of the functions f_1 and f_2 as well as of the overall heat release rate obtained with the above model are displayed in Fig. 3.3 for an idealized rectangular injection shape. The effect of the two parameters *available fuel mass* and *turbulent kinetic energy density* on the calculated combustion rate is clearly visible. The peak in heat release rate occurs shortly after the start of injection and is caused by the early maximum in turbulent kinetic energy. Approximately at the end of injection a second distinct bend can be observed after which the heat release rate decreases rapidly. This bend is caused by the reduction of available fuel mass once the injection process has stopped.

Figure 3.4 compares experimental and simulated heat release rates for the research engine referenced in Tab. 3.1 as a function of the injection pressure. The influence of this parameter on the apparent heat release rate of the engine is predicted both qualitatively and quantitatively with good agreement. Similarly good accordance with experimental results can be obtained for variations in engine speed and load and in the nozzle geometry, i.e. the number and diameter of the nozzle holes [14].

It should be noted however, that with this simple model approach only the diffusion part of a typical diesel engine combustion can be simulated. The premixed part of combustion cannot be modeled with the above method since it is strongly affected by the evaporation of fuel during the ignition delay. Both the evaporation process and the duration of the ignition delay are not described in this model. Therefore, different methods have to be used to describe premixed combustion and ignition delay, e.g. the empirical correlations presented in Sects. 2.2.5 and 2.2.6, respectively.

3.2.2 Free Gas Jet Theory

A number of researchers have presented diesel engine heat release models that are based on the theory of undisturbed turbulent gas jets as proposed by Abramovich [1]. Examples are the studies by Shahed and co-workers [13, 60, 61], also referenced as the "Cummins engine model", or the ones by DeNeef [16] and Hohlbaum et al. [36, 44]. In all these models the diesel spray is treated as a quasi-steady gas jet penetrating into an also gaseous environment of combustion air. The result is a continuous profile of fuel vapor concentration ranging from very rich at the core of the spray to a very lean mixture at the periphery. Fresh air is continuously entrained into the spray and represents the main influence factor for the estimated heat release rate since the entire mixture between the rich and lean flammability rates may be converted to combustion products. This implies that the fuel is assumed to be fully evaporated at all positions within the spray where the rich flammability limit has been exceeded.

The Cummins Engine Model

In the Cummins engine model [13, 60, 61] the tip penetration of the spray which is schematically shown in Fig. 3.5 is calculated based on the empirical correlation

$$S = \frac{450\, d_{noz}^{0.5} \left(\rho_f / \rho_{ref}\right)^{0.4}}{\left(1 + \rho_{cyl} / \rho_{atm}\right)^{0.85}} \left(\frac{\rho_{cyl}}{\rho_{atm}}\right)^{0.5} \Delta p_{inj}^{0.25} \, t^{0.6}, \qquad (3.9)$$

where ρ_{ref}, ρ_f, ρ_{cyl} and ρ_{atm} are the densities of reference fuel, actual fuel, cylinder gases and atmospheric air, respectively. Δp_{inj} is the pressure drop across the nozzle hole and t the time increment since start of injection. In case of a swirl motion inside the cylinder, the spray penetration may be corrected by the equation

$$\frac{S - S_\omega}{S} = 0.35 \left(\frac{S_\omega Q_{cyl}}{d_{noz} Q_f}\right)^{0.44}, \qquad (3.10)$$

where S_ω is the tip penetration in case of swirl and Q_f and Q_{cyl} are the momenta of the fuel jet and the cylinder gases, respectively, that can be expressed in terms of the angular speed ω:

$$Q_{cyl} = \rho_{cyl} S_\omega^2 \omega^2, \qquad Q_f = \rho_f v_{inj}^2, \qquad v_{inj} = c_D \sqrt{\frac{2 \Delta p_{inj}}{\rho_{cyl}}}. \qquad (3.11)$$

The tangential displacement of the spray due to swirl is given as

$$S_t = d_{noz} \left(\frac{Q_{cyl}}{Q_f}\right) \left(\frac{r}{d_{noz}}\right)^{2.217}, \qquad (3.12)$$

where r is the radial coordinate in the combustion chamber.

The cross-sectional area of the spray is approximated by two ellipses that are combined to an equivalent circle of radius b,

$$b = \left(b_3 \frac{b_1 + b_2}{2}\right)^{1/2}, \qquad (3.13)$$

where the radius change in spray direction x, i.e. the spray angle, is

$$\frac{db_1}{dx} = 0.12 \left(1 + \frac{\rho_a}{\rho_{jet}}\right), \qquad (3.14)$$

and

$$b_2 = b_1 \left(1 + 0.0016 \cdot \mathrm{Re}^{0.66}\right), \qquad b_3 = b_1 + 0.11 x \left(\frac{\rho_a}{\rho_{jet}}\right). \qquad (3.15)$$

The Reynolds number in Eq. 3.15 is defined in terms of the angular swirl speed ω:

$$\mathrm{Re} = \frac{2\pi r \omega d_{noz}}{v_a}. \qquad (3.16)$$

The distribution of the fuel concentration c across the equivalent circular spray slice is given by

$$c = c_m \left[1 - \left(\frac{y}{b}\right)^{3/2} \right], \qquad (3.17)$$

where c is defined as the mass fraction of fuel in the mixture, c_m denotes the fuel mass fraction on the axis and y is the coordinate in radial direction of the spray cross-section. The centerline fuel concentration c_m has a hyperbolical profile along the spray coordinate x,

$$c_m = (\alpha(t) x + 1)^{-1} \quad \text{for } S_L \leq x \leq S, \qquad (3.18)$$

$$c_m = 0 \quad \text{otherwise}, \qquad (3.19)$$

and $\alpha(t)$ can be solved implicitly from a mass balance, stating that the integrated amount of fuel within the spray has to match the fuel mass that has been injected until time t,

$$\int_0^t \dot{m}_{inj} \, dt = 2\pi \int_{S_L}^{S} \int_0^b c \rho y \, dy \, dx \,, \qquad (3.20)$$

where

$$\rho = \frac{p_{cyl}}{T_{cyl} \left[(1-c) R_a + c R_f \right]}, \qquad (3.21)$$

and S_L is the tail position of the spray after the end of injection which is calculated by a correlation similar to Eq. 3.9 [13]. Note that $\alpha(t)$ reduces to a constant in case of a steady jet.

In a subsequent study Kuo et al. [42] adjusted the model by replacing Eq. 3.17 by the relation

$$c = c_m \exp\left[-0.693 \left(\frac{y}{R_{1/2}}\right)^{5/2} \right], \qquad (3.22)$$

which has been established by evaluating results obtained from a three-dimensional spray model. $R_{1/2}$ in Eq. 3.22 is defined as the radial location in the spray where the fuel concentration c reaches a value of $c_m/2$.

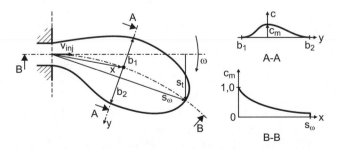

Fig. 3.5. Schematic spray geometry and its concentration distribution profiles [13]

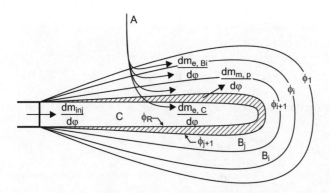

Fig. 3.6. Progressive evolution of combustion zones and entrainment rates [61]

In order to estimate the rates of combustion and pollutant formation, a set of progressively evolving discrete combustion zones is superimposed on the continuous fuel-air distribution calculated above. Figure 3.6 displays this concept in which the cylinder is considered to be divided into $(n+2)$ zones at incipient ignition. Zone A corresponds to the medium into that the fuel is injected, zone C to the rich core of the spray and the n zones B_i to combustible mixture between the lean and rich flammability limits. Since the mixture is already stratified at the time of ignition, which needs to be estimated with a separate submodel, more than one mixture zone B_i may be present initially. During the course of combustion the fuel mass in each zone B_i remains constant and the additional fuel that is diluted beyond the rich flammability limit is assigned to an additional combustion zone B_j. In contrast to these constant fuel masses in zones B_i, air from zone A is continuously entrained into the zones B_i and C.

The mixture zones B_i are bounded by the rich and lean flammability limits ϕ_R and ϕ_L and by intermediate equivalence ratios ϕ_i as follows:

$$\phi_R \geq \phi_i \geq \phi_L, \tag{3.23}$$

$$m_{f,Bi} = 2\pi \int_{S_L}^{S} \int_{y(\phi_{i+1})}^{y(\phi_i)} c\rho y\, dy\, dx, \tag{3.24}$$

$$m_{a,Bi} = 2\pi \int_{S_L}^{S} \int_{y(\phi_{i+1})}^{y(\phi_i)} (1-c)\rho y\, dy\, dx, \tag{3.25}$$

where $m_{f,Bi}$ and $m_{a,Bi}$ are the masses of fuel and air in zone B_i, respectively and the iso-equivalence ratio boundary $y(\phi_i)$ is calculated from Eqs. 3.18–3.21. The fuel and air masses in zone C are computed in analogy to Eqs. 3.24 and 3.25 by considering the integration limits from $y(\phi = \infty)$ to $y(\phi_R)$. The fuel beyond the lean limit ϕ_L cannot be burned and is therefore added to zone A.

Based on the concept that all the fuel crossing the rich flammability limit ϕ_R is added to a new combustion zone B_j while the fuel masses in all other combustion zones B_i are fixed, a new set of boundaries ϕ_i can be calculated at each computational step after an updated spray geometry is obtained. The boundaries are calculated starting from the leanest zone A by using Eq. 3.24 proceeding to progressively richer zones. In these calculations the only unknown is the lower limit of integration in y-direction, i.e. $y(\phi_{i+1})$.

Once the zone boundaries are determined, the combustible mixture preparation and air entrainment for each zone can be calculated. The entrainment rate into a zone B_i is simply equal to the change of mass of this zone, because the fuel mass remains constant:

$$\frac{dm_{e,Bi}}{d\varphi} = \frac{dm_{Bi}}{d\varphi}. \tag{3.26}$$

The air entrainment into the core zone C that results in mixture preparation for zone B_j is

$$\frac{dm_{e,C}}{d\varphi} = 2\pi \frac{d}{d\varphi} \int_{S_L}^{S} \int_{0}^{y(\phi_{j+1})} (1-c)\rho y \, dy \, dx, \tag{3.27}$$

where ϕ_{j+1} is the boundary of the richest combustion zone B_j. The mixture prepared for combustion is defined as the fresh mixture crossing the rich limit of combustion ϕ_R and can be written as

$$\frac{dm_{f,p}}{d\varphi} = 2\pi \frac{d}{d\varphi} \int_{S_L}^{S} \int_{y(\phi_R)}^{y(\phi_{j+1})} c\rho y \, dy \, dx, \tag{3.28}$$

$$\frac{dm_{a,p}}{d\varphi} = 2\pi \frac{d}{d\varphi} \int_{S_L}^{S} \int_{y(\phi_R)}^{y(\phi_{j+1})} (1-c)\rho y \, dy \, dx, \tag{3.29}$$

$$\frac{dm_{m,p}}{d\varphi} = \frac{dm_{f,p}}{d\varphi} + \frac{dm_{a,p}}{d\varphi}, \tag{3.30}$$

where the subscripts a, f, m and p stand for air, fuel, total mixture and 'prepared for combustion', respectively.

With the above equations the mass and energy balances for the various zones can be written as follows:

$$\frac{dm_C}{d\varphi} = \frac{dm_{inj}}{d\varphi} + \frac{dm_{e,C}}{d\varphi} - \frac{dm_{m,p}}{d\varphi}, \tag{3.31}$$

$$\frac{dm_A}{d\varphi} = \frac{dm_{inj}}{d\varphi} - \frac{dm_{e,C}}{d\varphi} - \sum_{i=1}^{j} \frac{dm_{e,Bi}}{d\varphi}, \tag{3.32}$$

$$\frac{dm_{Bi}}{d\varphi} = \frac{dm_{e,Bi}}{d\varphi} + \delta_{ij} \frac{dm_{m,p}}{d\varphi}, \quad i = 1,...j \tag{3.33}$$

where δ_{ij} is the Kronecker delta, i.e. $\delta_{ij} = 1$ if $i = j$ and $\delta_{ij} = 0$ if $i \neq j$.

$$\frac{d(m_C u_C)}{d\varphi} = -p\frac{dV_C}{d\varphi} + \frac{dQ_C}{d\varphi} + \frac{dm_{inj}}{d\varphi}h_f + \frac{dm_{e,C}}{d\varphi}h_A - \frac{dm_{m,p}}{d\varphi}h_m, \quad (3.34)$$

$$\frac{d(m_A u_A)}{d\varphi} = -p\frac{dV_A}{d\varphi} + \frac{dQ_A}{d\varphi} - \frac{dm_A}{d\varphi}h_A, \quad (3.35)$$

$$\frac{d(m_{Bi} u_{Bi})}{d\varphi} = -p\frac{dV_{Bi}}{d\varphi} + \frac{dQ_{Bi}}{d\varphi} + \frac{dm_{e,Bi}}{d\varphi}h_A + \delta_{ij}\frac{dm_{m,p}}{d\varphi}h_m, \quad i = 1,...j \quad (3.36)$$

In Eqs. 3.34–3.36 the internal energies and enthalpies are absolute quantities including the enthalpy of formation. Thus, an additional term for the chemical heat release by combustion is not necessary. Each of the zones is considered to be homogenously mixed and its composition is calculated by solving for the chemical equilibrium of the major combustion species.

The contribution of each zone to the total wall heat transfer is based on the product of zone mass and temperature,

$$\frac{dQ_{w,z}}{d\varphi} = \frac{m_z T_z}{m_A T_A + m_C T_C + \sum_{i=1}^{j} m_{Bi} T_{Bi}} \cdot \frac{dQ_{w,tot}}{d\varphi}, \quad (3.37)$$

where the zonal index z applies to all zones within the cylinder, i.e. A, B_i, B_j and C.

The total wall heat transfer is calculated by Annand's correlation which combines both convection and radiation effects [2]:

$$\frac{dQ_{w,tot}}{d\varphi} = -\frac{A}{6\ \mathrm{rpm}}\left[0.45\frac{k}{B}\mathrm{Re}^{0.7}(T - T_w) + 7.8\cdot 10^{-6}\left(T^4 - T_w^4\right)\right]. \quad (3.38)$$

In Eq. 3.38 A is the combustion chamber surface area, B the cylinder bore and k the gas thermal conductivity. The Reynolds number is defined in terms of the bore and the piston velocity, $\mathrm{Re} = \rho v_{pis} B/\mu$, and T is the mass averaged gas temperature in the cylinder:

$$T = \frac{m_A T_A + m_C T_C + \sum_{i=1}^{j} m_{Bi} T_{Bi}}{m_A + m_C + \sum_{i=1}^{j} m_{Bi}}. \quad (3.39)$$

In addition to the zonal mass and energy balances in Eqs. 3.31–3.36 the ideal gas law has to be satisfied in all the zones,

$$pV_z = m_z R_z T_z, \quad (3.40)$$

and the sum of all zonal volumes has to equal the momentary cylinder volume:

$$V_{cyl} = V_A + V_C + \sum_{i=1}^{j} V_{Bi}. \quad (3.41)$$

The above set of differential equations is implicit in T but can be solved numerically in order to obtain the time dependent zonal temperatures, masses, compositions and the cylinder pressure which is assumed to be equal in all zones. Details on the numerical solution scheme are given in ref. [61]. The properties of the various combustion zones can then be used as boundary conditions in order to calculate the kinetics of pollutant formation as will be discussed in Chap. 7.

Analytical Description of Spray Geometry and Mixture Formation

In contrast to the empirical expressions for the spray penetration in the Cummins engine model, DeNeef [16] and Hohlbaum [36] suggested a calculation of the spray movement that is based on momentum conservation principles. In order to apply the respective conservation equations, the spray is averaged over its circular cross-section and reduced to its centerline, such that a "slice" of the spray can be treated as a moving mass point. Figure 3.7 shows a schematic illustration of a fuel spray penetrating into a swirling gas in a cylindrical coordinate system (r, φ, z). Accordingly, momentum conservation for the radial, tangential and vertical directions of the cylindrical coordinate system can be written as

$$\frac{d}{dt}\left(dm_{jet}\dot{r}\right) = dF_r, \qquad (3.42)$$

$$\frac{1}{r}\frac{d}{dt}\left(dm_{jet}r^2\dot{\varphi}\right) = \frac{d}{dt}\left(dm_a\right)r\omega + dF_t, \qquad (3.43)$$

$$\frac{d}{dt}\left(dm_{jet}\dot{z}\right) = 0, \qquad (3.44)$$

respectively, where dm_{jet} is the mass of a spray slice of thickness dx. dF_r and dF_t are the radial and tangential forces acting upon that spray slice, respectively, and subscript a denotes the unburned air surrounding the jet. The radial force is caused by a radial pressure gradient in the cylinder due to the swirl motion,

$$dF_r = -dV\frac{dp}{dr} = -\frac{dm_{jet}}{\bar{\rho}}\rho_a r\omega^2, \qquad (3.45)$$

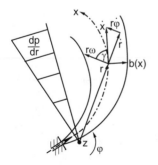

Fig. 3.7. Schematic illustration of a swirl deflected jet [16]

and the tangential drag force can be approximated as

$$dF_t = 0.1 \frac{1}{\overline{c}} \frac{v_{inj}}{b} r(\omega - \dot{\varphi}) \, dm_f, \qquad (3.46)$$

with the radius $b = b(x)$ of the circular spray slice [36]. $\overline{\rho}$ and $\overline{c}$ denote mass averaged quantities over the cross-section of the spray.

The above relations lead to the motion equations of the spray front:

$$\ddot{r} + \overline{c} \frac{d}{dt}\left(\frac{1}{\overline{c}}\right) \dot{r} = r\left[\dot{\varphi}^2 - (1-\overline{c})\omega^2\right], \qquad (3.47)$$

$$\ddot{\varphi} + 2\frac{\dot{r}}{r}\dot{\varphi} = \left[\overline{c}\frac{d}{dt}\left(\frac{1}{\overline{c}}\right) + 0.1\frac{v_{inj}}{b}\right](\omega - \dot{\varphi}), \qquad (3.48)$$

$$\ddot{z} + \overline{c}\frac{d}{dt}\left(\frac{1}{\overline{c}}\right)\dot{z} = 0. \qquad (3.49)$$

The spray velocity $\dot{x}$ and the spray tip penetration S now become

$$\dot{x} = \sqrt{\dot{r}^2 + (r\dot{\varphi})^2 + \dot{z}^2}, \qquad S = x = \int_0^t \dot{x} \, dt. \qquad (3.50)$$

The spray angle, i.e. the change in the spray radius b with respect to the axial spray coordinate x, significantly influences the calculated rate of air-fuel mixing. It is suggested to choose a standard value of approximately

$$(db/dx)_{\omega=0} = 0.16 \qquad (3.51)$$

for non-swirl cases. This value may have to be adjusted slightly in order to match realistic spray angles that are actually influenced by injection pressure, nozzle geometry and fluid properties. As shown in refs. [1] and [33], the spray angle diverges from the above value if the fuel is injected into an air swirl. For this case DeNeef [16] has derived the correction

$$\frac{db}{dx} = \frac{1-C\,(r\omega/v_{inj})}{1+C\,(r\omega/v_{inj})} \cdot \left(\frac{db}{dx}\right)_{\omega=0}, \qquad (3.52)$$

$$C = \frac{r\dot{\varphi}}{\dot{x}} - \frac{1}{2}\sqrt{2}\frac{\dot{r}}{\dot{x}}. \qquad (3.53)$$

Once the spray tip penetration and the outer spray contour are known, the mixture distribution within the jet needs to be determined. For this purpose the mean fuel mass fraction $\overline{c}$ along the spray coordinate x can be calculated on the basis of mass conservation. Considering that the fuel mass contained within one slice of the spray of thickness dx has to remain constant ($dm_{jet} \cdot \overline{c} = $ const.) and further assuming that the mean spray density $\overline{\rho}$ within this slice is very small compared to the liquid fuel density ρ_f, the change in the mean fuel concentration can be expressed in terms of the spray angle (db/dx) as follows:

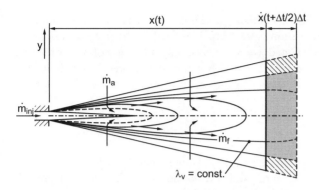

Fig. 3.8. Equivalence ratio distribution and mixture preparation rate in a steady spray [16]

$$\frac{d}{dt}\left(\frac{1}{\bar{c}}\right) = \frac{4}{d_{noz}^2 v_{inj}} \frac{\rho_a}{\rho_f} \left[2\left(\frac{db}{dx}\right) b \dot{x}^2 + b^2 \ddot{x}\right]. \tag{3.54}$$

With the mean fuel concentration $\bar{c}(x)$ known, the local concentrations $c(x,y)$ can be solved by the same distribution function that is used in the Cummins engine model (Eq. 3.17).

As opposed to the studies by Shahed et al. [13, 60, 61], DeNeef [16] relates the combustion rate to the fuel mass prepared with air in a stoichiometric ratio per unit time. This quantity can be derived as follows. Since the fuel concentration has been solved for at every position within the spray, iso-contours of the air-fuel equivalence ratio $\lambda = 1/\phi$ can be easily calculated as indicated in Fig. 3.8. The dimensionless radius y/b of any specific air-fuel equivalence ratio λ_v at an axial spray position x can be expressed as

$$\frac{y}{b}(\lambda_v, x) = \left[1 - \frac{c(\lambda_v)}{c_m(x)}\right]^{2/3}. \tag{3.55}$$

Because of the assumption of a steady jet, the equivalence ratio distribution within the spray does not vary with time. The only change is that in each computational time step Δt a new 'slice' is added to the spray front, i.e. the hatched and shaded areas on the right hand side of Fig. 3.8. Because of mass conservation this new spray front has to contain the exact amount of fuel injected during that time increment. Hence, the mass of fuel that is crossing a contour of constant air-fuel equivalence ratio λ_v within one time step (hatched area) must be equal to the difference between the injected fuel mass $\dot{m}_{inj} \cdot \Delta t$ and the fuel contained within the spray front in the region richer than λ_v (shaded area):

$$\Delta m_{f,\lambda_v} = \dot{m}_{inj}\Delta t - \pi y^2(\lambda_v)\rho_a c_m \left[1 - \frac{4}{7}\left(\frac{y(\lambda_v)}{b}\right)^{3/2}\right]\dot{x}\Delta t. \tag{3.56}$$

In order to determine the total amount of fuel that is prepared in a stoichiometric ratio with air within the entire spray, Eq. 3.56 needs to be integrated between the rich flammability limit λ_R and $\lambda = 1.0$. Since only a fraction of $d\lambda_v$ of the fuel that crosses from $(\lambda = \lambda_v)$ to $(\lambda = \lambda_v + d\lambda_v)$ becomes newly prepared with air (the remaining fraction has already been prepared in previous time steps), the relation

$$\Delta m_{f,stoic} = \lambda_{v,R} \dot{m}_{f,\lambda_{v,R}} \Delta t + \int_{\lambda_{v,R}}^{\lambda_v=1} \dot{m}_{f,\lambda_v} d\lambda_v \cdot \Delta t \qquad (3.57)$$

is obtained.

After the end of the injection period, it is assumed that the spray parts close to the nozzle are not existent anymore, while the regions further downstream are still in steady state and not affected by the stop in injection. This principle is implemented by simply calculating the formation of a second spray which is then subtracted from the original one. A similar procedure is applied in the Cummins engine model as well.

The combustion rate is expressed with a quasi-kinetic approach that describes the burned fraction of the stoichiometrically prepared fuel,

$$X = \frac{m_{f,b}}{m_{f,stoic}}, \qquad (3.58)$$

by the Arrhenius type differential equation

$$dX = A \rho_{jet} T_{jet}^\beta \frac{af_{stoic}(1-X)^2}{af_{stoic}-1} \exp\left[-\frac{E_A}{R_m T_{jet}}\right] dt, \qquad (3.59)$$

where T_{jet} and ρ_{jet} are the mean temperature and density within the entire spray and the Arrhenius constants A, β and E_A have to adjusted to match experimentally obtained heat release rates.

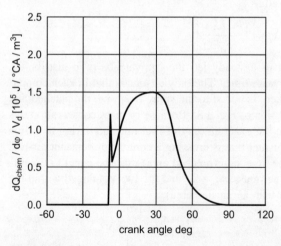

Fig. 3.9. Simulated heat release rate for a heavy-duty diesel engine at full load [36]

Since the sub-processes of fuel atomization and droplet evaporation are not modeled explicitly in this model type, it is hardly feasible to estimate an ignition delay based on chemical kinetics depending on local fuel vapor and oxygen concentrations as well as on local temperatures. Instead, it is assumed that the ignition starts, when the centerline fuel concentration c_m at the spray tip becomes leaner than the rich flammability limit for the first time. However, by this timing the local equivalence ratio at the outer spray regions has already been further diluted than the rich limit. Hence, a certain amount of fuel is already stoichiometrically prepared for combustion and can react rapidly. This causes the typical premixed peak in the heat release history of diesel engine combustion. Figure 3.9 displays such an exemplary heat release rate which has been estimated for a full load case of a high-speed heavy-duty diesel engine [36].

The above procedure of determining the ignition delay means that the rich flammability limit ϕ_R may be adjusted in the calculation in order to match realistic ignition delay periods. This implies however, that the adjustment of the combustion start and the amount of the predicted premixed combustion has a direct influence on the rate of the later diffusion combustion as well. The reason is that the rich combustion limit enters the calculation of the stoichiometrically prepared fuel mass in Eq. 3.57. In a realistic engine scenario however, there is not such a tight coupling between the ignition delay and the rate of diffusion combustion, since the former is primarily governed by fuel evaporation and reaction kinetics, whereas the latter is most importantly affected by turbulence and the mixing rate between fuel vapor and oxygen.

It should be noted, that both in the Cummins engine model and in this modified gas jet analysis the spray angle that needs to be specified as a boundary condition has a significant influence on the mixing rate and thus, on the combustion rate. Moreover, for obvious reasons the analogy between the real fuel spray and an undisturbed turbulent gas jet is not valid in cases where the spray impinges on a cylinder wall. Therefore, this model type seems to be most appropriate for large diesel engines with significant air swirl ratios.

3.2.3 Packet Models

In the packet model that was originally proposed by Hiroyasu et al. [34, 35] and later applied and extended by several other authors, e.g. [6, 26, 40, 65], the fuel jet is described by numerous discrete so-called packets in an attempt to model both the global geometry of the penetrating spray and detailed local subprocesses such as fuel atomization and evaporation, fuel-air mixing, ignition, combustion and pollutant formation.

The general model concept is such that during the compression stroke there is only one zone that is treated as ideally mixed within the combustion chamber. It contains fresh air and, in case of an internal or external exhaust gas recirculation, homogeneously mixed residuals. Once fuel injection has started, there are continuously formed more spray packets that initially contain liquid fuel only. The spray packets are arranged in a way that the spray is discretized both in axial and in radial direction, Fig. 3.10. During injection new axial spray slices are formed in equal time increments and therefore the amount of fuel within each packet depends on the instantaneous injection rate at the time of packet formation. The total number of packets in axial spray direction depends on the injection duration. There is a constant mass distribution as a function of the radial packet position. After spray breakup fresh gases from the surrounding air zone are entrained into the spray packets. Both fuel droplet evaporation and combustion, i.e. heat release, take place within the borders and under the local conditions of the each packets. Thus, there can be a mixture of liquid fuel, vaporized fuel, fresh air and combustion products within each spray packet. It is assumed that there is no mass or energy exchange other than the air entrainment between the various packets. However, there occurs heat transfer between each packet and the combustion chamber walls. Mass and energy balances can then be solved for the air zone and for all spray packets, separately. Therefore, each zone has its own temperature and composition history, whilst the pressure is assumed to be uniform over the entire cylinder volume and varies only with time.

Spray Penetration and Air-Fuel Mixing

Initially the injected fuel within the spray packets is treated as a continuous liquid phase as indicated by the shaded packets in the vicinity of the nozzle in Fig. 3.10. It travels at an initially constant velocity through the combustion chamber,

$$v_{inj} = 0.39 \sqrt{\frac{2 \Delta p_{inj}}{\rho_f}}, \quad (3.60)$$

until a characteristic breakup time that is given as

$$t_{bu} = 28.65 \frac{\rho_f d_{noz}}{\sqrt{\rho_a \Delta p_{inj}}} \quad (3.61)$$

is reached. At this point the liquid phase within the packet disperses into many small droplets and it is decelerated because of aerodynamic interactions with the surrounding gas phase. The spray tip velocity after breakup is expressed as a function of time elapsed since start of injection:

$$v(t) = 1.48 \cdot \left(\frac{\Delta p_{inj} d_{noz}^2}{\rho_a} \right)^{1/4} \cdot \frac{1}{\sqrt{t}}. \quad (3.62)$$

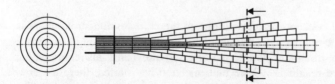

Fig. 3.10. Schematic illustration of the packet spray model

Because there are stronger interactions between air and fuel at the periphery of the jet than at its core an exponential decrease of the packet velocity in radial direction is assumed. For a spatial resolution of 5 packets in radial direction, Stiesch and Merker [64] found good results with the empirical function

$$v_k = v_1 \cdot \exp\left[-3.86 \cdot 10^{-2}(k-1)^2\right], \qquad k = 1...5, \tag{3.63}$$

where k is the radial packet index ($k=1$ at centerline, $k=5$ at spray perimeter) and v_1 is obtained from Eq. 3.62. Moreover, Nishida and Hiroyasu [53] suggested that the breakup occurs earlier at the spray outside than on the centerline and assumed a linear decrease in breakup time towards the spray perimeter. This effect is indicated by the shaded packets in Fig. 3.10 representing packets prior to breakup,

$$t_{bu,k} = t_{bu,1} \cdot \frac{6-k}{5}, \qquad \text{for } k = 1...5, \tag{3.64}$$

where $t_{bu,1}$ is obtained from Eq. 3.61.

Because spray packets that are injected early typically face higher drag forces and are therefore decelerated more quickly than packets injected later during the injection duration, Stiesch and Merker [64] suggested to implement a correction for the spray velocity equation 3.62 not only in radial direction (Eq. 3.63) but also in axial direction. This axial dependence is expressed as

$$v_i = C_1 \cdot v \cdot \left[1 + \left(\frac{i-1}{i_{max}-1}\right)^{C_2} \cdot \frac{\Delta t_{inj}}{C_3}\right], \tag{3.65}$$

where $i=1$ represents the first and $i=i_{max}$ the last injected slice of packets. The constant C_1 is slightly less than unity. It indicates that the first packets may actually penetrate slower than the visible spray tip since they are overtaken by faster packets injected at a later timing. C_2 is in the range of 0.5 and C_3 determines the velocity difference between the first and last packets and may be adjusted to account for air swirl.

As opposed to the gas jet analogy discussed in Sec. 3.2.2 where a continuous fuel concentration profile is modeled in order to determine the rate of air fuel mixing, the entrainment rate of unburned gases into the spray packets is now calculated on the basis of momentum conservation for each packet. This means that the total packet momentum, which can easily be estimated as the product of the injected fuel mass per packet $m_{f,0}$ and its initial velocity v_{inj} specified by Eq. 3.60, does not change with time:

$$v_{i,k}(t) \cdot m_{i,k}(t) = const. = m_{f,0} \cdot v_{inj}. \tag{3.66}$$

Thus, the instantaneous packet mass after breakup $m_{i,k}(t)$ is directly coupled to the decreasing packet velocity $v_{i,k}(t)$ determined by Eqs. 3.62, 3.63 and 3.65 and the resulting temporal increase in packet mass is due to entrainment of unburned air (and residuals in case of EGR) into the packet. This implies that the packets located at the jet periphery are characterized by greater entrainment rates and leaner equivalence ratios because their velocity is decelerated more rapidly.

An investigation of two different-sized diesel engines with displacement volumes of 2.0 and 0.52 liters per cylinder revealed, that after the start of combustion the entrainment rates calculated by Eq. 3.66 ought to be reduced by factors of 0.7 and 0.4 for the large and the small engine, respectively [41]. However, in the same study it was also found that after wall impingement the entrainment rate into the spray packets should be enhanced by a factor of approx. 1.5 in both engines.

In a similar attempt to describe the influence of spray wall impingement on the air fuel mixing rate more realistically, Nishida and Hiroyasu [53] proposed to alter the time dependence of Eq. 3.62 after impingement to $t^{-0.75}$ instead of $t^{-0.5}$. This causes the packets to decelerate more quickly after they have hit a wall which will result in an increased entrainment rate as well.

Kouremenos et al. [40] recommended to estimate the air entrainment rate into spray packets based on the volume change of each zone with respect to time. In order to determine this quantity they utilized the spray angle correlation [3]

$$\alpha = 0.05 \left(\frac{d_{noz}^2 \rho_a \Delta p_{inj}}{\mu_a^2} \right)^{0.25}, \qquad (3.67)$$

and subdivided the total spray angle into equal increments for the each packets in radial spray direction. The result is an increased entrainment rate by approx. 20% compared to the momentum conservation method, which fitted available experimental data better. However, in a more recent study [38], the same group of authors utilized the momentum conservation principle in combination with a velocity adjustment after impingement as suggested in ref. [53], too, stating that it yields more stable results that require less re-adjustments of model parameters when operating conditions are varied.

Fuel Atomization and Droplet Evaporation

Fuel atomization for a particular packet is assumed to occur instantaneously at the breakup time of that packet determined by Eqs. 3.61 and 3.64. Thereafter all droplets within the packet are represented by a Sauter mean diameter (SMD) which characterizes a single droplet with the same volume to surface area ratio as the ratio of the respective quantities integrated over the whole droplet size distribution present in a real spray:

$$\mathrm{SMD} = \frac{\sum_{i=1}^{N_{drops}} d_i^3}{\sum_{i=1}^{N_{drops}} d_i^2}. \qquad (3.68)$$

The initial value of the SMD after breakup is estimated by an empirical correlation fitted from experimental data. In the original study [34], the equation found by Hiroyasu and Kadota [32] was used,

$$\mathrm{SMD} = 23.9 \cdot 10^{-6} \Delta p_{inj}^{-0.135} \rho_a^{0.12} B^{0.131}, \qquad (3.69)$$

where SMD is in mm, Δp_{inj} in kPa, ρ_a in kg/m³ and B is the fuel mass injected in mg/stroke. Other frequently applied correlations have been presented by Elkotb [24],

$$\text{SMD} = 6156 \cdot 10^{-6} \cdot v_f^{0.385} \cdot \rho_f^{0.737} \cdot \rho_a^{0.06} \cdot \Delta p_{inj}^{-0.54}, \quad (3.70)$$

where the unit of the liquid fuel viscosity v_f is m²/s, and by Varde et al. [72] who related the SMD to the diameter of the injection nozzle:

$$\frac{\text{SMD}}{d_{noz}} = 8.7 \left(\text{Re}_f \, \text{We}_f \right)^{-0.28}. \quad (3.71)$$

The Reynolds and Weber numbers refer to the injection velocity, the nozzle diameter and the liquid fuel properties.

After breakup hot air entrained into the packets starts to heat up the fuel droplets that thereupon begin to evaporate. In most studies based on phenomenological spray and combustion models, e.g. [18, 34, 40, 65, 79] the droplet evaporation model by Borman and Johnson [8] is applied. In this model the temperature change of a liquid droplet is due to convective heat transfer to and mass diffusion from the droplet and can be obtained from an energy balance

$$\frac{dT_f}{dt} = \frac{1}{m_f c_{p,f}} \left(\frac{dQ_f}{dt} + \frac{dm_f}{dt} \Delta h_{evap} \right), \quad (3.72)$$

where Δh_{evap} is the latent heat of evaporation and

$$\frac{dQ_f}{dt} = \pi \cdot \text{SMD} \cdot k_s \cdot \left(T_g - T_f \right) \cdot \frac{z}{e^z - 1} \cdot \text{Nu}, \quad (3.73)$$

$$\frac{dm_f}{dt} = -\pi \cdot \text{SMD} \cdot D \cdot \rho_s \cdot \ln \left(\frac{p_{cyl}}{p_{cyl} - p_v} \right) \cdot \text{Sh}. \quad (3.74)$$

In the above equations T_g is the gas temperature within the packet, p_v is the saturated fuel vapor pressure typically determined by the Clausius-Clapeyron equation, D indicates the binary diffusion coefficient and index s refers to the surface conditions, that are approximated as the arithmetic mean of saturated vapor and the gas phase within the packet. The variable z is a dimensionless correction factor taking account of the reduced heat transfer in the presence of a simultaneous mass transfer:

$$z = \frac{c_{p,v} \cdot \left(dm_f / dt \right)}{\pi \cdot \text{SMD} \cdot k_s \cdot \text{Nu}}. \quad (3.75)$$

The appropriate Nusselt and Sherwood numbers have been proposed by Ranz and Marshall [58]:

$$\text{Nu} = 2.0 + 0.6 \cdot \text{Re}^{1/2} \cdot \text{Pr}^{1/3}, \quad (3.76)$$

$$\text{Sh} = 2.0 + 0.6 \cdot \text{Re}^{1/2} \cdot \text{Sc}^{1/3}, \quad (3.77)$$

where a value of approx. 30% of the instantaneous packet velocity ought to be used as a relative velocity between gas and droplets within the packet in order to determine the Reynolds number [53].

Figure 3.11 shows exemplary mass and temperature histories of an n-$C_{14}H_{30}$ droplet evaporating under constant boundary conditions as predicted by the above model. The initial and boundary conditions are specified within the figure. It can be seen that the droplet temperature increases rapidly in the beginning due to the convective heat transfer from the hot gas phase. This causes an increased saturated vapor pressure at the droplet surface such that the diffusion rate is enhanced. However, because of this enhanced diffusion rate a greater amount of latent heat of vaporization needs to be provided by the droplet and therefore a quasi-steady state is reached where the droplet temperature stays almost constant until the droplet is fully evaporated.

Ignition and Combustion

As all the other subprocesses the ignition delay is calculated separately for each spray packet in terms of the respective local properties. An Arrhenius-type equation is used [34],

$$\tau_{id} = 4.0 \cdot 10^{-3} \left(\frac{p}{p_{ref}} \right)^{-2.5} \phi_g^{-1.04} \exp\left(\frac{6000}{T} \right), \quad (3.78)$$

where the constants may have to be slightly adjusted for different engines and operating conditions in order to fit experimental data. In order to account for temporal changes in the packet properties during the ignition delay, the ignition integral

$$\int_0^t \frac{1}{\tau_{id}} dt \geq 1 \quad (3.79)$$

has to exceed unity as a condition for combustion start within a packet.

The heat release rate per packet is determined by assuming a stoichiometric single-step reaction from fuel vapor to CO_2 and H_2O. The fuel burning rate per computational time step is limited by the more stringent of two criteria. Firstly, only the vaporous fraction of fuel may be burned, and the second constraint is represented by the amount of air entrained into the packet. The energy released by combustion is obtained by the product of the fuel mass burning rate and the lower heating value of the fuel:

$$\frac{dQ_{chem}}{dt} = \text{LHV} \cdot \min\left(\frac{m_v}{\Delta t}, \frac{m_a / af_{stoic}}{\Delta t} \right). \quad (3.80)$$

Nishida and Hiroyasu [53] have included a third constraint on the maximum combustion rate which describes a maximum chemical reaction rate for premixed charge. Based on a work of Edelman and Harsha [20], they suggested an Arrhenius equation that expresses the rate of change in fuel vapor density due to combustion as a function of temperature and the mass fractions of oxygen and fuel vapor. The exponents of the fuel and oxygen mass fractions are chosen such that a maximum burning rate is obtained for stoichiometric mixture ($y_{v,stoic} = 0.22$):

$$\frac{d\rho_{v,b}}{dt} = -5 \cdot 10^{10} \frac{m^3}{kg\ s} \cdot \rho_g^2\ y_v^1\ y_{O2}^5 \exp\left(-\frac{12000\,K}{T}\right). \quad (3.81)$$

By executing parametric studies with a packet combustion model accounting for all three criteria, Stiesch and Merker [64] found that the evaporation rate typically represents the limiting factor for the premixed part of combustion, since a significant amount of air is predicted to be entrained directly at the breakup time of a packet. However, the available fuel vapor is burned rapidly and the entrainment rate of fresh air into a packet becomes the limiting factor during the subsequent diffusion portion of the diesel combustion. Finally, the chemical constraint as modeled by Eq. 3.81 becomes a factor when either temperatures are low towards the end of combustion or the equivalence ratio within a spray packet is extremely lean due to rapid air entrainment.

Zonal Thermodynamics

It is assumed that each spray packet as well as the zone of unburned air contributes to the overall wall heat transfer of the combustion chamber by a fraction proportional to the product of mass and temperature of the respective zone:

$$\frac{dQ_{w,z}}{dt} = \frac{m_z T_z}{\sum_z (m_z T_z)} \cdot \frac{dQ_{w,tot}}{dt}. \quad (3.82)$$

The total heat transfer $dQ_{w,tot}$ between the cylinder contents and the walls is calculated by either Woschni's [78] or Annand's [2] global expressions, Sect. 2.2.4 or Eq. 3.38, respectively.

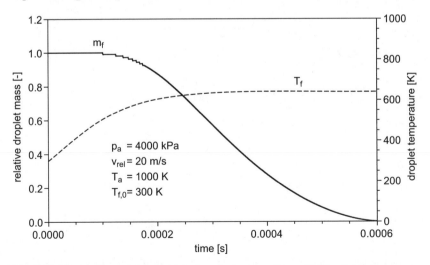

Fig. 3.11. Temperature and mass histories of an evaporating n-$C_{14}H_{30}$ droplet

The changes in temperature and volume of each zone can be obtained from an energy balance and an equation of state. Typically the ideal gas law is applied. However, it should be noted that for modern diesel engines with peak pressures of up to 20 MPa better results may be achieved when real gas effects are incorporated. For the simplified case of an ideal gas, the temperature changes of the spray packets (index sp) and the unburned air zone (index a) become

$$dT_{sp} m_{sp} c_{p,sp} = dQ_{w,sp} + dQ_{chem,sp} - dQ_{f,sp} - p_{cyl} V_{sp} \left(\frac{dm_{sp}}{m_{sp}} + \frac{dR_{sp}}{R_{sp}} - \frac{dp_{cyl}}{p_{cyl}} \right)$$
$$- dm_{f,sp} h_{v,sat} + dm_{entr,sp} h_a - dm_{sp} h_{sp} \ , \qquad (3.83)$$

and

$$dT_a m_a c_{p,a} = dQ_{w,a} - p_{cyl} V_a \left(\frac{dm_a}{m_a} - \frac{dp_{cyl}}{p_{cyl}} \right) + dm_a \ R_a \ T_a \ , \qquad (3.84)$$

respectively [65].

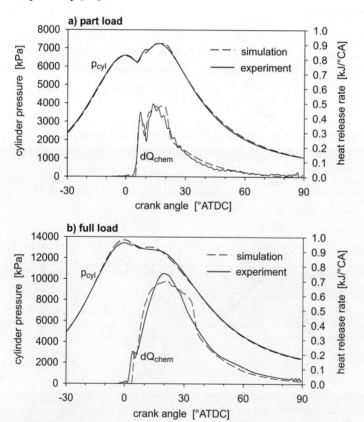

Fig. 3.12. Heat release rates and pressure traces for a heavy-duty high speed diesel engine. V_d=4.0 liter/cyl, n=1500 rpm. a) imep=980 kPa, b) imep=2220 kPa [65]

The only unknown left in Eqs. 3.83 and 3.84 is the rate of pressure change dp_{cyl}. It is initially taken from the previous time step and is then updated by an iteration under the constraint, that the sum of the volume changes of all zones has to match the geometrical change in cylinder volume during that time step which can be determined from Eqs. 2.17–2.20:

$$\sum_z dV_z = dV_{cyl}. \qquad (3.85)$$

Figure 3.12 compares calculated and measured heat release rates and pressure traces for two operating conditions of a heavy-duty high speed diesel engine with a displacement volume of 4.0 liters per cylinder. Diagram *a* displays medium load conditions which have been utilized to adjust all model constants and diagram *b* represents a full load case of that engine that has been pre-calculated with the same set of constants. It is apparent that both heat release rates and resulting pressure traces can be simulated with good accuracy once the model has been correctly adjusted to a particular engine. Especially the decreasing amount of the premixed fraction in the heat release rate for greater engine loads is predicted very well.

It should be noted that while the spray angle was shown to be of paramount influence on the combustion rate prediction in the gas jet models described in Sect. 3.2.2, the relations for the spray penetration, Eqs. 3.62 ff, that govern the air-fuel mixing rate have the most significant effect on the model outcome in the packet approach.

Pre-Injection

Thoma et al. [71] have extended the packet model approach to describe DI diesel engines with pre-injections. Based on spray experiments carried out by Stegemann et al. [63] they concluded that the spray penetration equation 3.62 that has been established for continuously injected sprays is not valid for small amounts of pre-injection. Instead, the fuel injected during the short pre-injection pulse is decelerated and thus mixed with air more quickly. Therefore, a time dependence of $1/t$ has been suggested in Eq. 3.62 in agreement with experimental data from a truck diesel engine operated with a pre-injection quantity of 6 mg. A further adjustment of this exponent for varying amounts of pre-injection might be necessary however.

Once the main injection pulse has started the packets of the pre-injection pulse are combined to a single pre-injection zone with mass averaged properties from the former spray packets, Fig. 3.13. The packets of the main injection pulse will penetrate into the pre-injection zone because of the reduced spray velocity of the small pre-injection quantity and the entrainment into the main injection packets will be from the pre-injection zone instead of from the fresh air zone while there is an spatial overlap between the zones. The entrainment of hot gases into the main spray packets causes a reduced ignition delay within the main packets such that the typical premixed peak in the combustion profile of a single injection case is diminished or even eliminated. Figure 3.14 displays that this behavior can be predicted by the model. Note, that only about 50% of the fuel injected during the first pulse is actually burned during the first peak in the heat release history. The remaining fraction is diluted beyond the lean flammability limits and can only be burned after mixing with the richer spray packets of the main injection. This phenomenon can be predicted by considering a maximum chemical reaction rate in terms of the local equivalence ratio similar to Eq. 3.81.

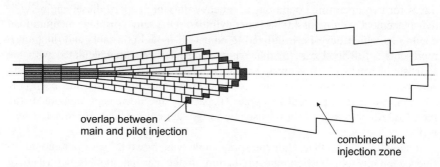

Fig. 3.13. Mixing of pre- and main-injection pulses [71]

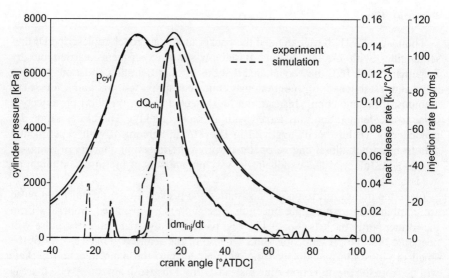

Fig. 3.14. Experimental and predicted heat release and pressure histories of a DI diesel engine with pre-injection. $V_d = 1.2$ liter/cyl, $n = 1500$ rpm, imep = 920 kPa [71]

3.2.4 Time Scale Models

Single Pulse Injection

Boulouchos and co-workers [9, 74] presented a phenomenological diesel combustion model which is based on characteristic time scales for the heat release rates similar to the eddy-breakup models frequently applied in CFD-codes. Different time scales are considered for the premixed and diffusion fractions of combustion since it is assumed that the former process is primarily governed by chemical kinetics whereas turbulent mixing represents the major constraint for the latter.

The fuel atomization and evaporation is modeled similarly to the procedure for the packet models described in the previous section. However, in this application the spray is discretized in axial direction only, and the tip penetration is described by the formula developed by Dent [17]:

$$S = 3.07 \left(\frac{\Delta p_{inj}}{\rho_g} \right)^{1/4} \cdot (t \, d_{noz})^{1/2} \cdot \left(\frac{294}{T_g} \right)^{1/4}. \tag{3.86}$$

It is now assumed that all the fuel evaporated prior to the first occurrence of ignition in the entire cylinder is burned in the characteristic premixed peak, while the remaining fuel is burned in the mixing controlled diffusion combustion. The ignition delay is determined by solving the ignition integral, Eq. 3.79, where τ_{id} is calculated as a function of temperature and pressure for a stoichiometric air-fuel mixture [67].

The time scale characteristic for the kinetically controlled premixed fraction of combustion is assumed to be proportional to the ignition delay such that the fuel mass burning rate can be written as

$$\frac{dm_{prem}}{dt} = C_{prem} \frac{1}{\tau_{id}} f_{prep} m_{prem,av}, \tag{3.87}$$

where $m_{prem,av}$ is the total fuel mass assigned to the premixed combustion and the factor f_{prep} is due to the fact that only the fraction f_{prep} of $m_{prem,av}$ that has exceeded the ignition integral of unity has yet been prepared for combustion.

An analogous formulation to Eq. 3.87 is applied for the diffusion combustion, where the characteristic time scale is controlled by turbulent mixing phenomena rather than by chemical kinetics:

$$\frac{dm_{diff}}{dt} = C_{diff} \frac{1}{\tau_{turb}} f_{A,turb} m_{diff,av}. \tag{3.88}$$

However, in contrast to CFD models that include turbulence submodels which can be utilized in order to determine a characteristic mixing frequency $1/\tau_{turb}$, a different approach has to be found in order to specify this quantity within the framework of a phenomenological cylinder model. Boulouchos and Eberle [9] suggested to express the frequency by the ratio of the turbulent diffusivity, which can be shown to be proportional to the turbulent viscosity and the second power of a typical dimension of the problem:

$$\frac{1}{\tau_{turb}} = \frac{u'l_I}{X_{char}^2}. \tag{3.89}$$

In order to determine the turbulent viscosity a simplified formulation with two turbulence sources is chosen. The first is the motion of the charge air, for that the turbulence intensity u' is proportional to the mean piston speed and the length scale is proportional to the clearance height. The second contribution to turbulence production is due to injection where u' and l_I can be solved based on conservation equations [30]. The initial values depend on the injection velocity and the nozzle diameter, respectively:

$$u'l_I = \left(u'l_I\right)_{charge} + \left(u'l_I\right)_{inj}. \tag{3.90}$$

The characteristic dimension of the diffusion process is approximated by the instantaneous cylinder volume, the global equivalence ratio and the number of nozzle holes [74]:

$$X_{char} = \left(\frac{\phi \cdot V_{cyl}}{N_{noz}}\right)^{1/3}. \tag{3.91}$$

The factor $f_{A,turb}$ takes account of the increase in flame surface density because of turbulent wrinkling and is written as the ratio of turbulent to kinematic viscosity:

$$f_{A,turb} = \frac{u'l_I}{v}. \tag{3.92}$$

Thus, Eq. 3.88 becomes

$$\frac{dm_{diff}}{dt} = C_{diff} \frac{u'l_I}{X_{char}^2} \frac{u'l_I}{v} m_{diff,av}. \tag{3.93}$$

Finally, the available fuel mass for both combustion mechanisms, i.e. premixed and diffusion combustion, results from an integration of the evaporation rate and the burning rate:

$$m_{i,av} = \int_{t_{i,0}}^{t} \left(\frac{dm_{i,evap}}{dt} - \frac{dm_i}{dt}\right) dt. \tag{3.94}$$

In an application to a 9.5 liter/cyl marine diesel engine it was shown that the above model is capable of predicting heat release rates for medium and high load operating conditions very well [74]. However, for conditions with 25% load or less, deviations between predictions and experimental data were observed. They are most likely attributed to an overprediction of the evaporation rate during ignition delay which causes too much fuel to be burned during the early premixed combustion.

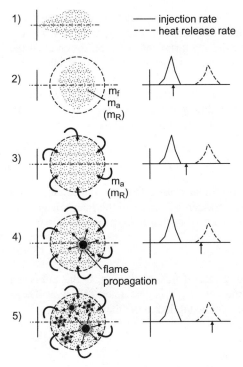

Fig. 3.15. Temporal sequence of the pre-combustion model. 1) Gaseous fuel of one jet forms one premixed zone 2) An initial amount of air and residual gas is entrained 3) After the end of injection air and residuals continue to be entrained due to turbulent diffusion 4) A turbulent flame propagates from one ignition source (ascending heat release trace) 5) flame propagation from multiple ignition sources (descending heat release trace) [4]

Pre-Injection

In a later study Barba et al. [4] used a very similar diffusion combustion model as described in Eqs. 3.88–394 and added a new submodel for premixed combustion of a pre-injection pulse in order to simulate modern passenger car diesel engines equipped with common rail injection systems. In this model the pre-combustion resulting from the pilot injection pulse is completely decoupled from the combustion of the main injection pulse. Only the ignition delay of the main combustion and therefore the portion of premixed combustion within the main combustion is affected by the heat release of pre-injection combustion. The phenomenology of the pre-combustion model is schematically shown in Fig. 3.15.

The fuel injected during the typically short pilot injection pulse forms a spherical zone with an initial amount of air and residuals entrained. This zone is considered to be homogeneously mixed and the entrainment rate during the injection period is assumed to be proportional to the fuel evaporation rate. The exact air-fuel ratio is empirically chosen to be 11.6, such that a slightly rich mixture of equivalence ratio $\phi = 1.25$ is obtained. After the end of pilot injection air and residuals continue to be entrained into the premixed zone. The entrainment rate is now modeled by a necessarily simplified model for turbulent diffusion in terms of the Reynolds number as a characteristic measure of turbulence. Details about this mixing model are provided in ref. [4].

The ignition delay is calculated based on the Arrhenius equation 3.78 and the ignition integral, Eq. 3.79, however with a different set of constants to account for the typical conditions at the early timing of the pre-injection pulse [4]:

$$\tau_{id,pre} = 0.04\, \phi^{-0.2} \left(\frac{p}{p_{ref}}\right)^{-1.2} \exp\left(\frac{6000}{T}\right). \tag{3.95}$$

After ignition it is assumed that a turbulent flame starts to propagate from a single ignition location, which leads to an ascending trace in the heat release rate because of the growth in flame surface area A_F (burning mode 1 in Fig. 3.16):

$$\frac{dm_{mix}}{dt} = \rho_u s_t A_F. \tag{3.96}$$

In Eq. 3.96 dm_{mix}/dt is the mass burning rate of the mixture, ρ_u is the unburned mixture density, and the turbulent flame velocity s_t is calculated following Damkoehler's relation [15], which has been widely applied in the literature [31]:

$$s_t = s_l \left(1 + c_1 \left[\frac{u'}{s_l}\right]^{c_2}\right). \tag{3.97}$$

The constants c_1 and c_2 are chosen as 1.6 and 0.8, respectively, and the turbulence intensity is assumed to be proportional to the mean piston velocity. The laminar burning speed s_l is derived according to Metghalchi, Keck and Rhodes [45, 46, 59], with a slightly modified set of parameters in order to allow combustion up to fuel-air equivalence ratios of about 0.3:

$$s_l = s_{l,0} \left(\frac{T}{T_0}\right)^{\gamma} \left(\frac{p}{p_0}\right)^{\delta} (1 - 2.1 f_R), \tag{3.98}$$

$$s_{l,0} = 0.276 - 0.47\, (\phi - 1.1)^2, \tag{3.99}$$

$$\gamma = 2.18 - 0.8(\phi - 1), \tag{3.100}$$

$$\delta = -0.16 + 0.22(\phi - 1). \tag{3.101}$$

The reference state is defined by $T_0 = 298$ K, $p_0 = 98$ kPa, and f_R indicates the residual mass fraction.

Once the combustion has started the temperature in the pre-injection zone increases in the flame is assumed to spread out from multiple new ignition sources. Now the available fuel mass m_f within the pre-injection zone becomes the limiting factor for the combustion rate (burning mode 2 in Fig. 3.16). This is expressed by the frequency approach

$$\frac{dm_f}{dt} = cg\frac{1}{3}\frac{s_t}{r_{pre}}m_f, \tag{3.102}$$

where the frequency factor is modeled as the ratio of turbulent flame speed s_t and the radius of the pre-injection zone r_{pre}. The constant c is of order one, and g is an empirical function that delays combustion due to dilution with air and residuals:

$$g = \frac{1}{\Lambda^2}, \quad \Lambda = \frac{m_a + m_R}{m_f}. \tag{3.103}$$

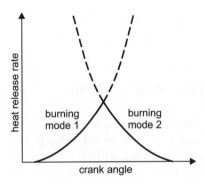

Fig. 3.16. Ascending and descending branches in the pre-injection combustion model [4]

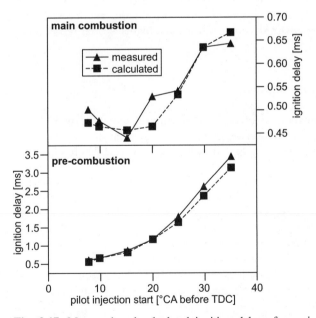

Fig. 3.17. Measured and calculated ignition delays for various pilot injection timings. $V_d = 0.44$ liter/cyl, $n = 1000$ rpm, bmep $= 200$ kPa [4]

As indicated in Fig. 3.16 the reaction rates obtained for both burning modes 1 and 2, Eqs. 3.96 and 3.102 respectively, are estimated at any time and the lesser of the two is taken as the actual mass burning rate.

The above premixed combustion model for pilot injections has been combined with a diffusion combustion model as described in Eqs. 3.88–3.94 and applied to simulate modern passenger car diesel engines with common rail injection systems [4]. As an example Fig. 3.17 displays the effect of pre-injection timing on the ignition delays of pre- and main combustion. An early pre-injection timing is characterized by a long ignition delay of the pre-combustion because temperature and pressure in the combustion chamber are still fairly low. Therefore, the pre-injection zone becomes more diluted with air such that the pre-combustion rate is reduced and a longer ignition delay of the main-combustion results as well.

3.3 Gas Composition and Mixing in Diesel Engines

Many of the governing reactions for the formation of exhaust emissions, and especially the formation of nitrogen oxides, are characterized by an exponential temperature dependence, compare Chap. 7. Therefore, it is generally not feasible to calculate reaction kinetics based on the mass averaged cylinder temperature. Instead it is necessary to subdivide the combustion chamber into multiple zones in order to resolve local peak temperatures at least approximately and calculate emission formation rates individually for each zone as a function of the local conditions. At least two different zones for burned and burned gases are necessary, but the numerical accuracy is obviously improved when a higher spatial resolution is chosen.

3.3.1 Two-Zone Cylinder Models

A very simple and computationally efficient approach to divide the cylinder gases into two zones containing cold unburned mixture and hot combustion products, respectively, is the empirical model suggested by Heider et al. [29] that has been described in Sect. 2.4.

Merker et al. [44] used a different two-zone model where the composition of the two zones is directly related to the combustion rate that is calculated following Sect. 3.2.2. Figure 3.18 displays a schematic diagram of this approach where zone 1 contains unburned mixture, i.e. air, unburned fuel and the residual fraction, and zone 2 describes the burned products that may be diluted with fresh air entrained from zone 1 as well. The flame front is modeled as an infinitely thin segregation between zones 1 and 2 and does not contain any mass itself. It is however assumed, that the flame front is characterized by a spatially and temporally constant equivalence ratio.

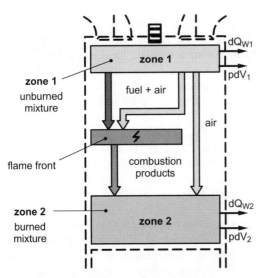

Fig. 3.18. Schematic illustration of the two-zone cylinder model Merker et al. [44]

Before the start of combustion all cylinder contents are contained within the unburned zone 1. Thereafter mass is subtracted from zone 1 and added to zone 2 both via combustion through the flame front and by turbulent mixing which is assumed to result in a direct entrainment of unburned air into the product zone 2. Thus, the mass balances for the two zones can be written as

$$dm_1 = dm_{inj} - dm_{FF} - dm_{12}, \qquad (3.104)$$

$$dm_2 = dm_{FF} + dm_{12}, \qquad (3.105)$$

where the amounts of products that are added to zone 2 through the flame front dm_{FF} are determined by the fuel mass burning rate $dm_{f,b}$ and the flame front air-fuel equivalence ratio, which is assumed to be slightly rich ($\lambda_{FF} = 0.7$):

$$dm_{FF} = dm_{f,b} \cdot (1 + \lambda_{FF} \cdot af_{stoic}). \qquad (3.106)$$

The mixing of unburned mixture into zone 2 that bypasses the flame front is calculated based on the concept that the instantaneous air-fuel equivalence ratio of the combined mass flows into zone 2 (both dm_{FF} and dm_{12}), that is referred to as the mixing stoichiometry λ^*, increases linearly with time:

$$\lambda^* = \frac{dm_{a,FF} + dm_{a,12}}{af_{stoic} \cdot dm_{f,b}}. \qquad (3.107)$$

Since the equivalence ratio of the mass flow dm_{FF} is constant with time, this increase of λ^* takes into account the fact that the bypass mass flow dm_{12} is very small in the beginning when zone 2 is still small and increases with time since the mass of zone 2 and thus its surface area in contact with zone 1 is growing. Consequently, the initial value of the combined mixing stoichiometry at ignition and during premixed combustion is $\lambda^*_0 = \lambda_{FF}$. The exact gradient of the linear increase in λ^* during diffusion combustion depends on the turbulence level within the cylinder. In general it should be determined such that the local equivalence ratio within zone 2 reaches the integral equivalence ratio in the entire cylinder by about the time the exhaust valve opens. Characteristic histories of the air-fuel equivalence ratios in the flame front and in zone 2 as well as of the mixing stoichiometry λ^* are displayed in Fig. 3.19.

Energy conservation equations can be set up for the two zones by the assumption that the chemically released energy of combustion is completely added to the hot products zone 2. The total wall heat transfer of the cylinder, which is calculated by the Woschni model (Sect. 2.2.4), is distributed between the two zones with a weighting factor equal to the product of zone mass and zone temperature, Eq. 3.82.

A similar approach to the one above has been proposed by Shi et al. [62], albeit with the consideration of four different zones instead of two: an unburned air zone, a spray mixing zone, a product zone and a burning zone that, in contrast to the flame front in the model by Merker et al. [44], is associated with its own mass and spatial extension. Turbulent mixing between different zones is calculated in this model as well. However, here the mixing rates between two neighboring zones i and j are determined based on the product of the respective zone masses as an approximate measure of the contact area between the zones:

$$dm_{ij} \sim \frac{m_i \cdot m_j}{m_i + m_j}. \tag{3.108}$$

Thus, the mixing rate has a maximum when the two zones have equal masses. Additionally, the mixing rate contains a proportionality factor which is a turbulent frequency determined from a simplified turbulence model.

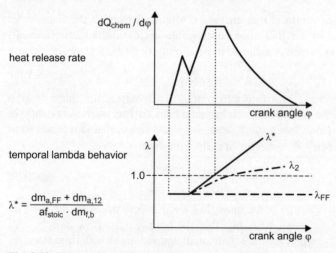

Fig. 3.19. Temporal histories of the air-fuel equivalence ratios [44]

3.3.2 N-Zone Cylinder Models

In this category cylinder models are summarized that consider a number of different product zones generated in sequential order during the course of combustion. An example is the work by Papadopoulos [55] where, similar as in the two-zone models, the unburned mixture is treated separately from the combustion products. However, after ignition a new additional product zone is formed during each computational time step such that a differentiated temperature distribution can be calculated even within the burned mixture. As indicated in Fig. 3.20, a newly build product zone initially consists of combustion products directly from the flame front such that they are characterized by stoichiometric equivalence ratio and adiabatic flame temperature. Only in subsequent time steps unburned air is entrained into the already existing product zones such that their respective masses and volumes increase while the local equivalence ratio becomes leaner.

The initial mass of a product zone can be estimated in a straight forward manner as the product of the burning rate function and the computational time increment. The entrainment rate of fresh air into a product zone $dm_{a,p}$ during the subsequent time steps after the initiation of the zone is derived following the concept of the mixing surface area between two zones (Eq. 3.108) with a characteristic mixing time τ_m:

$$\frac{dm_{a,p}}{dt} = \frac{1}{\tau_m} \cdot \frac{m_a \cdot m_p}{m_a + m_p} . \tag{3.109}$$

For simplicity, this time scale is related to the ratio of an air or mixing velocity v_a to the clearance height c:

$$\frac{1}{\tau_m} = C_m \frac{v_a}{c} . \tag{3.110}$$

C_m is a tuning parameter and the air velocity is determined from the swirl ratio in the cylinder.

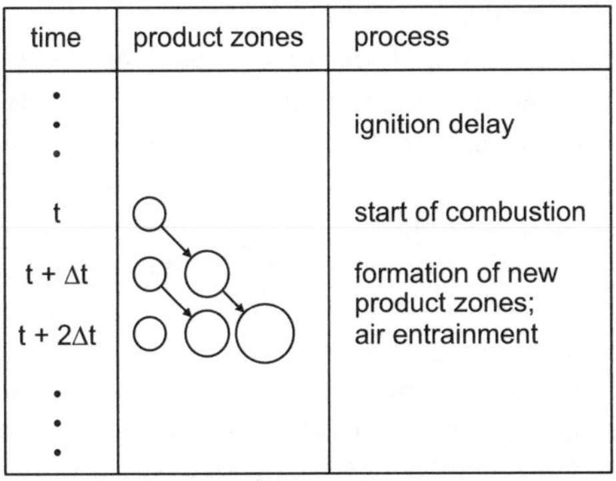

time	product zones	process
⋮		
		ignition delay
t	○	start of combustion
t + Δt	○○	formation of new product zones;
t + 2Δt	○○○	air entrainment
⋮		

Fig. 3.20. Sequential generation and air entrainment into product zones [55]

It should be noted however, that in modern diesel engines with high injection pressures turbulent mixing is driven by the kinetic energy of the fuel jet rather than by gas swirl. Thus, Eq. 3.110 ought to be adjusted accordingly, which has been done in the work by Weisser and Boulouchos [74]. They have chosen a very similar cylinder model with successively generated product zones. Here the mixing time scale τ_m is evaluated by the same relation that the authors have utilized as a turbulent time scale in their combustion model described in Sect. 3.2.4 (Eq. 3.89). This time scale includes both effects of gas motion and injection (Eq. 3.90).

Additional variations by Weisser and Boulouchos [74] to the model of Papadopoulos [55] are that in an attempt to reduce computing requirements a total number of product zones N_p can be specified such that a new zone is not formed with every numerical time step but only after a specific fraction of the total fuel injected has been burned. Moreover, it is assumed that no product zone can be diluted beyond the global equivalence ratio in the cylinder. Therefore, the air mass substituted into Eq. 3.109 is not the total mass of the unburned zone but only that fraction of air, that corresponds to the fuel mass allocated to one product zone:

$$m_a = af_{cyl} \frac{m_{f,cyl}}{N_p}. \tag{3.111}$$

3.3.3 Packet Models

When the heat release rate is simulated with a packet model as described in Sect. 3.2.3 there is no need to superimpose an additional zonal cylinder model in order to establish a temperature and composition distribution within the combustion chamber. The spray packets reproducing the spray shape as well as the unburned zone are all characterized by individual temperatures and compositions calculated by solving mass and energy balances. The resulting distributions can directly be used in order to estimate local pollutant formation rates, e.g. [6, 34, 40].

However, dependent on the injection duration of the investigated operating conditions and the numerical time increment of the spray model which typically ranges around 20 µs, the total number of packets in the spray can easily reach about 1000, which will cause significant computing costs if the reaction kinetics are solved within each packet. This is also due to the fact that the numerical integration of the chemical reaction rates typically requires smaller numerical time steps such that a subcycling becomes necessary. To at least diminish this problem Stiesch and Merker [65] suggested to combine various spray packets that are similar with respect to temperature for the purpose of calculating exhaust emissions. This property is predominantly governing the nitrogen oxides formation. It was shown that with a number of approx. 20 combined so-called exhaust zones a result for the NO_x-emissions can be obtained that is identical to the one obtained when reaction kinetics are solved separately for each of the approx. 1000 spray packets. Thus, a significant reduction in computing time is achieved.

The 20 exhaust zones are defined in a way that the coldest zone combines all packets with a temperature below a cutoff temperature of about 1400 K. This rather coarse resolution is possible because reaction rates are very slow in this low temperature range. The remaining 19 exhaust zones combine spray packets with temperatures in equal increments between 1400 K and 3000 K. The latter is in the range of the highest possible adiabatic flame temperature in diesel engines. If the soot formation is to be calculated as well, a slightly greater number of exhaust zones ought to be chosen in order to obtain a numerically accurate result, i.e. a result that does not deviate from the one where reaction rates are calculated separately for each spray packet. However, with respect to soot emission predictions, the best results will be obtained if the combination to exhaust zones is not only based on similar packet temperatures but based on both temperatures and equivalence ratios.

3.4 Advanced Heat Transfer Models

3.4.1 Heat Transfer Mechanisms

In general, the relatively simple and easy to handle heat transfer models of Annand [2] or Woschni [78], that solve for a time dependent but spatially averaged heat transfer coefficient (see Sect. 2.2.4), can be applied in phenomenological combustion models, provided that an assumption is made about how much a particular zone contributes to the total heat transfer rate between the cylinder gases and the combustion chamber walls, e.g. Eq. 3.82. In fact, today Woschni's heat transfer model is still probably the most widely used within a phenomenological framework, and it is sometimes applied even in multidimensional computations [54].

However, it is well known that especially in diesel engines radiative heat transfer may have a significant contribution in addition to convective heat transfer. Heywood [30] suggested a relative importance of radiation of 20 to 35 percent. Other authors proposed even greater contributions to the total heat flux that range up to 50 percent at full load operating conditions [10]. The radiative heat transfer in diesel engines is caused by both radiation of hot gases and by radiation of soot particles within the diffusion flame. It is well agreed in the literature that the latter has a significantly greater impact on the radiative heat flux, and thus most heat transfer models concentrate on the radiation of soot only.

Since soot radiation strongly depends on the soot concentration and temperature whereas convection is primarily governed by the turbulence level in the cylinder, the radiative heat flux typically peaks about 10 degrees of crank angle later than the convective heat flux. This and the fact, that convection is driven by the simple temperature difference between gas and wall,

$$\dot{q}_{conv} \sim \left(T_g - T_w\right), \tag{3.112}$$

as opposed to the difference between the fourth powers of soot particle and wall temperatures that drive radiation,

$$\dot{q}_{rad} \sim \left(T_s^4 - T_w^4\right), \tag{3.113}$$

suggest, that it is advantageous to model the convective and radiative heat fluxes separately. Within the framework of a phenomenological combustion model, this is possible because the spatial resolutions of composition and temperature in the combustion chamber can be utilized to estimate the time dependent soot concentrations. It should be noted though, that a general difficulty in the evaluation of soot radiation exists in that the prediction of the soot concentration itself is typically subject to significant uncertainties.

A number of researchers have addressed the problem of combined convective and radiative heat transfer in diesel engines. Morel and Keribar [47] developed a model where both the convective heat transfer coefficient as well as the gas and wall temperatures are both space and time dependent. The effective gas velocity is derived from the mean flow velocities supplied by a newly developed flow model and from a turbulent kinetic energy analysis. In ref. [48] the same authors presented a compatible radiation model which derives the decisive soot temperature in the combustion chamber from a two-zone model and determines the soot concentration on the basis of a kinetic model for soot formation and oxidation. Boulouchos and co-workers [10, 11] treated convection and radiation separately as well. Soot concentration distributions are approximated by means of typical particulate concentrations within a simplified spray model, and radiant heat fluxes impacting on an incremental cylinder surface area can be estimated by spatial integration under consideration of the respective view factors. Both Suhre and Foster [68] and Wolff et al. [75] have modeled a particulate layer adsorbed on the cylinder wall in order to take into consideration its insulation effect. Adsorption of soot on the wall is due to thermophoresis – within the boundary layer, particulates move in the opposite direction of the temperature gradient – and soot oxidation off the wall is described by an Arrhenius equation.

3.4.2 Convective and Radiative Heat Transfer Model

Eiglmeier et al. [22, 23] presented a detailed convective and radiative heat transfer model for diesel engines. As it incorporates the most important findings of the above references and represents a thorough overview of the current status of phenomenological heat transfer modeling it will be summarized in the following sections.

Geometric Assumptions

The spatial radiation heat flux distribution and flow inside the cylinder are highly dependent on the combustion chamber geometry. Within a phenomenological or quasi-dimensional combustion model only a coarse spatial resolution is performed and hence, several assumption have to be made. Here it is assumed that the piston has a cylindrical bowl located at its center and that the combustion chamber is surrounded by the six isothermal sub-surfaces indicated in Fig. 3.21 [48].

It is further assumed that the soot is distributed uniformly within the burned zone of a two-zone cylinder model. As indicated in Fig. 3.22 this product zone is represented by two cylindrical clouds located centrally in the cylinder. The expansion of the burned zone is broken down into three phases:

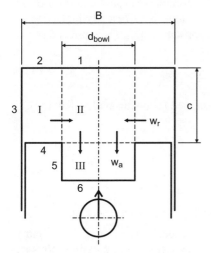

Fig. 3.21. Assumed isothermal sub-surfaces of the combustion chamber: (1) cylinder head above piston bowl (2) cylinder head above piston crown (3) liner (4) piston crown (5) piston bowl side wall (6) piston bowl bottom

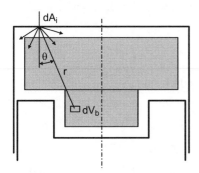

Fig. 3.22. Geometry of the soot cloud [23]

1. expansion within the piston bowl during premixed combustion (one cloud)
2. expansion into the squish region (the second cloud is generated)
3. expansion in all directions during diffusion combustion

Typically the burned zone starts to expand into the squish region at a crank angle between 5° and 10°ATDC. These results have been confirmed by experimental data from Gibson et al. [27].

Soot Layer Insulation

Soot particles formed during combustion can reach the thermal boundary layer near the combustion chamber wall by turbulent gas motion. Depositions of soot particles on the wall are then formed due to thermophoresis and a portion of this layer may also be reoxidized if the wall temperature is sufficiently high. Hence, the net change of the soot layer thickness $\delta_{s,w,i}$ on sub-surface i is expressed as

$$\frac{d\delta_{s,w,i}}{dt} = v_{d,i} - v_{ox,i}, \qquad (3.114)$$

where the soot deposition velocity $v_{d,i}$ is determined following Wolff et al. [75]:

$$v_{d,i} = f_v \cdot \frac{V_b}{V_{cyl}} \cdot \frac{2 C_s C_c v_g \left[\frac{k_g}{k_s} + C_t \mathrm{Kn} \right]}{(1 + 3 C_m \mathrm{Kn}) \left[1 + 2\frac{k_g}{k_s} + 2 C_t \mathrm{Kn} \right]} \cdot \frac{1}{T_{th,i}} \cdot \frac{T_{cyl} - T_{ss,i}}{\delta_{th,i}}. \qquad (3.115)$$

The soot volume fraction in the burned zone is defined as $f_v = m_s/(\rho_s V_b)$ and V_b and V_{cyl} denote the volumes of the burned zone and the entire cylinder, respectively. k_g and k_s are the gas and soot thermal conductivities, Kn is the Knudsen number which is assumed to be unity and T_{cyl}, T_{ss} and T_{th} indicate the mass averaged cylinder temperature, the temperature of the soot layer surface and the mean temperature within the thermal boundary layer, respectively. The constants in Eq. 3.115 are chosen as $C_m=1.14$, $C_s=1.17$, $C_t=2.18$ and C_c is the Cunningham error correction.

Table 3.2. Soot oxidation parameters

	A_j [mol, cm, s]	E_j [J/mol]
$k_A = A_A \exp(-E_A/RT_{ss})$ g cm^{-2} s^{-1} bar^{-1}	20	125,600
$k_B = A_B \exp(-E_B/RT_{ss})$ g cm^{-2} s^{-1} bar^{-1}	4.46·10^{-3}	63,640
$k_T = A_T \exp(-E_T/RT_{ss})$ g cm^{-2} s^{-1}	1.51·10^5	406,100
$k_Z = A_Z \exp(-E_Z/RT_{ss})$ g cm^{-2} s^{-1} bar^{-1}	21.3	-17,200

The soot oxidation velocity is based on the relation by Nagle and Strickland-Constable [52]

$$v_{ox,i} = \frac{M_C f_A}{\rho_s}\left[\left(\frac{k_{A,i} p_{O_2}}{1+k_{z,i} p_{O_2}}\right)x_i + k_{B,i} p_{O_2}(1-x_i)\right], \qquad (3.116)$$

where

$$x_i = \frac{1}{1+\dfrac{k_{T,i}}{k_{B,i} p_{O_2}}}, \qquad (3.117)$$

and the reaction parameters k_A, k_B, k_T and k_Z are given by the Arrhenius equations specified in Table 3.2. M_C is the molecular weight of carbon (12 kg/kmol), the density of the soot layer is assumed to be approx. $\rho_s = 170$ kg/m^3 and the factor $f_A \approx 12$ accounts for the fact that the roughness of the soot layer increases the effective surface area for oxidation.

It turns out that the net rate of change in soot layer thickness is relatively small such that a constant thickness can be assumed during the entire cycle when the engine is operated under steady conditions. However, the steady thickness may vary for wall sub-surfaces of different temperature and also when engine load and speed are changed.

Because of the insulating effect of the soot layer, the soot surface temperature T_{ss} can no longer be assumed to be independent of time. Therefore, the transient heat conduction within the soot layer and the combustion chamber wall needs to be calculated by Fourier's one-dimensional heat conduction equation assuming a plane geometry and constant physical properties. The resulting equations,

$$\rho_w c_w \frac{\partial T_{w,i}}{\partial t} = \frac{\partial}{\partial x}\left(k_w \frac{\partial T_{w,i}}{\partial x}\right), \qquad 0 \leq x \leq \delta_{w,i} \qquad (3.118)$$

$$\rho_s c_s \frac{\partial T_{s,i}}{\partial t} = \frac{\partial}{\partial x}\left(k_s \frac{\partial T_{s,i}}{\partial x}\right), \qquad \delta_{w,i} \leq x \leq \delta_{w,i} + \delta_{s,w,i}(t) \qquad (3.119)$$

where $\delta_{w,i}$ is the wall thickness and x denotes the coordinate perpendicular to the wall, can be solved numerically with the Crank-Nicholson scheme.

Convection

The convective heat transfer in combustion engines is caused by the turbulent in-cylinder gas flow. In contrast to the analogy to turbulent flow through a circular tube that is often applied for determining the convective heat transfer coefficient, e.g. Eq. 2.22, in this model an effective flow rate averaged over each of the six sub-surfaces is estimated. This effective velocity U_{eff} is composed of the two velocity components parallel to the sub-surface under consideration, i.e. squish and swirl flow, as well as the turbulence intensity:

$$U_{eff,i} = \sqrt{U_{sq,i}^2 + U_{sw,i}^2 + \frac{2}{3}k}, \qquad (3.120)$$

where k is the specific turbulent kinetic energy based on the entire mass of cylinder gases.

The squish flow from the annular region above the piston crown into the piston bowl during the compression stroke (and vice versa during expansion) is modeled by dividing the combustion chamber into three idealized control volumes indicated in Fig. 3.21 and solving mass balances for these zones. The radial and axial velocity components become

$$w_r = \frac{1}{\pi d_{bowl} c} \frac{dV_I}{dt} - \frac{V_I}{V_{cyl}} \frac{dV_{cyl}}{dt}, \qquad (3.121)$$

$$w_a = -\frac{c}{V_{cyl}} \frac{dV_{cyl}}{dt}, \qquad (3.122)$$

respectively, with the instantaneous clearance height c and the piston bowl diameter d_{bowl}. Spatial averaging of the velocity profiles in Eqs. 3.121 and 3.122 over the each sub-surfaces provides the squish flow component $U_{sq,i}$ [5].

For simple calculation of the swirl flow, the combustion chamber is divided into two sub-volumes with solid body swirl, one above the piston bowl and one above the piston crown, Fig. 3.23. The swirl intensities in both sub-volumes are described by the corresponding constant angular velocities. Murakami et al. [51] have confirmed the applicability of this assumption with LDV measurements.

Further assuming rotationally symmetrical and axially uniform tangential velocity profiles, it is possible to calculate the angular velocities ω_1 and ω_2 using the corresponding approaches for wall friction at the adjacent combustion chamber surfaces and for the angular momentum exchange between the two sub-volumes:

$$\frac{d}{dt}(J_1 \omega_1) = M_{fr,pbb} + M_{fr,chi} + M_{fr,pbs} + M_{fr21,shear} + M_{fr21,squish}, \qquad (3.123)$$

$$\frac{d}{dt}(J_2 \omega_2) = M_{fr,pc} + M_{fr,cho} + M_{fr,lin} - M_{fr21,shear} - M_{fr21,squish}. \qquad (3.124)$$

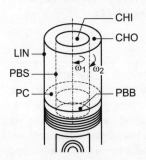

Fig. 3.23. Modeling of the swirl flow [23]

The frictional moments $M_{fr,i}$, where the indices i denote the sub-surfaces indicated in Fig. 3.23, are determined based on proximity equations for disk and plate surfaces according to Eiglmeier [21]. The momentum exchange between volumes 1 and 2 due to squish flow $M_{fr21,squish}$ and shear forces $M_{fr21,shear}$ has been derived in ref. [21] as well. The calculation and spatial averaging of the tangential velocities at the corresponding sub-surfaces of the cylinder walls then lead to the swirl flow component $U_{sw,i}$.

The influence of turbulence on the convective heat transfer enters the model through the specific turbulent kinetic energy k in Eq. 3.120. Within the framework of a phenomenological cylinder model that, by definition, does not resolve the turbulent flow field, a value for k has to be approximated by a simplified approach. This is done by a zero-dimensional energy cascade. The total kinetic energy E_{kin}, whose initial value at IVC is approximated as the kinetic energy of the intake flow, is converted into turbulent kinetic energy k by turbulent dissipation expressed through the production rate P:

$$\frac{dE_{kin}}{dt} = -P . \qquad (3.125)$$

Following Poulos and Heywood [56], the production rate can be written as

$$P = 0.3307 \cdot C \cdot \frac{E_{kin}}{(4 \ V_{cyl} / \pi \ B^2)} \cdot \sqrt{k} , \qquad (3.126)$$

where the constant C has been chosen as $C = 1.5$ in ref. [23]. The rate of change in specific turbulent kinetic energy k becomes

$$\frac{dk}{dt} = \frac{P}{m_{cyl}} - \varepsilon + \frac{2}{3} \frac{k}{\rho_{cyl}} \frac{d\rho_{cyl}}{dt} + \frac{\dot{m}_{inj} v_{inj}^2}{m_{cyl}} , \qquad (3.127)$$

where the index cyl denotes the gases trapped within the combustion chamber. The last term in Eq. 3.127 is based on the assumption that the entire kinetic energy of the fuel jet is instantaneously converted into turbulent kinetic energy, and the second to last term accounts for the increase in k due to the increase in gas density by compression and combustion. The viscous dissipation rate ε of the turbulent kinetic energy is approximated as [56]

$$\varepsilon = \frac{\pi}{4} \frac{B^2}{V_{cyl}} \left(\frac{2}{3} k \right)^{3/2} . \qquad (3.128)$$

The convective heat transfer coefficient is now evaluated based on the empirical equation for flow over a flat, isothermal plate, where the 1/7-power-law has been assumed for the velocity profile within the hydrodynamic boundary layer [12]:

$$h_i = 0.0153 \frac{\rho_{cyl} c_p U_{eff,i}}{\left(\frac{U_{eff,i} \delta_{th,i}}{v_{cyl}} \right)^{0.25}} \left(\frac{T_{conv,i}}{T_{ss,i}} \right)^{0.4} . \qquad (3.129)$$

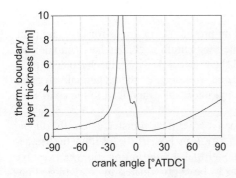

Fig. 3.24. Typical history of thermal boundary layer thickness in a DI diesel engine [23]

The gas temperature relevant for convective heat transfer is determined as

$$T_{conv,i} = T_u \left(1 - \frac{m_{b,i}}{m_{tot,i}}\right) + T_b \frac{m_{b,i}}{m_{tot,i}}, \qquad (3.130)$$

which utilizes the burned mass fraction within the geometrical zones I, II and III (specified in Fig. 3.21) for the respective sub-surfaces i in order to estimate to what extend the each sub-surfaces are in contact with fresh gas and the burned zone. The convective heat flux is now written as

$$\dot{q}_{conv,i} = h_i \left(T_{conv,i} - T_{ss,i}\right). \qquad (3.131)$$

The only unknown remaining in the above set of equations is the thermal boundary layer thickness δ_{th} that appears in Eqs. 3.115 and 3.129. For the compressible flow in a combustion engine cylinder it is affected not only by the convective heat transfer but also by the pressure and density changes of the cylinder gases. Formulation of a one-dimensional energy equation for these relations leads, after several simplifying assumptions and a number of mathematical rearrangements, to the following implicit differential equation in terms of the mean density of the thermal boundary layer $\rho_{m,i}$ and the isentropic exponent γ [21]:

$$\frac{d\delta_{th,i}}{dt} = \frac{\rho_{cyl}}{\rho_{m,i}} \frac{dp}{dt} \frac{1}{p\gamma} \delta_{th,i} - \frac{1}{\rho_{m,i}} \frac{d\rho_{m,i}}{dt} \delta_{th,i} + \frac{\dot{q}_{conv,i}}{\rho_{m,i} c_p T_{conv,i}}. \qquad (3.132)$$

Figure 3.24 displays a typical history of the thermal boundary layer thickness in a DI diesel engine as calculated by Eq. 3.132. During the compression stroke the numerical solution for the thickness approaches infinity as the gas and wall temperatures become equal. At this point the resulting convective heat flux is obviously zero.

Radiation

In order to estimate radiant heat fluxes in an engine cylinder the spatial and temporal soot concentration and temperature distributions have to be known. In this phenomenological model the simplifying geometric assumptions for the combustion chamber and the soot cloud that are indicated in Figs. 3.21 and 3.22 are made. A homogeneous soot concentration is present within the burned zone consisting of two ideal cylinders located in the bowl and the squish regions as described above. The soot formation and oxidation is modeled by a two-equation approach where the formation rate is based on the Hiroyasu model and the oxidation rate based on the relations by Nagle and Strickland-Constable, see Chap. 7.

The radiation transport equation between the soot cloud and the soot layer surface on the walls is solved by a simplified zonal approximation according to Hottel and Sarofim [37]. It is assumed that the dispersion of the temperature radiation due to liquid fuel droplets is negligible. The total absorbed heat flux of the soot covered sub-surface A_i can be represented as the difference between the incident and emitted heat fluxes. The emittance of the deposited soot is close to unity, i.e. the soot layer behaves as a blackbody, which means that the incident radiation heat flux is almost completely absorbed. Thus, the radiation from the surfaces A_i is emitted exclusively at the respective surface temperatures $T_{ss,i}$.

The incident heat flux per sub-surface is generally composed of both the temperature radiation of the soot cloud and of the incident radiation of the other optically visible combustion chamber sub-surfaces. Due to the fact that during combustion the soot layer surface temperatures $T_{ss,i}$ are relatively low in comparison to the radiation temperature of the particles in the soot cloud T_{rad}, the fraction of incident wall radiation is negligible. Thus, the radiant heat flux to the wall can be written as

$$\dot{q}_i = \frac{\overline{gs_i}}{A_i}\sigma T_{rad}^4 - \varepsilon_i \sigma T_{ss,i}^4 . \tag{3.133}$$

In order to calculate the so-called direct exchange area $\overline{gs_i}$ the combustion chamber surfaces are subdivided into incremental surfaces dA and the soot cloud into incremental volumes dV, Fig. 3.22. Following ref. [37], the direct exchange area becomes

$$\overline{gs_i} = \frac{1}{\pi}\int_{V_b}\int_{A_i}\frac{a \cdot \cos\theta_i \cdot \tau(r)}{r^2} dA_i\, dV_b , \tag{3.134}$$

where a is the absorption coefficient of the soot cloud and r the distance between dA and dV. Its dependence on the soot volume fraction f_v within the burned zone and the radiation temperature T_{rad} was defined by Morel and Keribar [48]:

$$a = 1575 \cdot f_v \cdot T_{rad} . \tag{3.135}$$

The transmittance $\tau(r)$ specifies for each sub-volume dV the fraction of the emitted radiation actually reaching the wall. The remaining portion of the radiation is reabsorbed by the surrounding optical dense soot cloud before it can reach the wall:

$$\tau(r) = \exp\left[-\int a\, dr\right]. \tag{3.136}$$

In the literature various controversial recommendations have been presented for the definition of the effective soot radiation temperature. The importance of this definition is obvious because of its influence on the radiation transport equation (Eq. 3.133) at the fourth power. However, it is generally accepted that the use of the burned temperature in a two-zone model leads to an underestimation of the radiant heat flux. This can be explained when considering that the soot just formed is initially close to adiabatic flame temperature, and only later assumes the temperature of the diluted combustion products within the burned zone of the two-zone cylinder model. Therefore, the radiation temperature is here calculated as an average between adiabatic flame temperature and the temperature of the burned zone:

$$T_{rad} = \left(1 - \frac{m_b}{m_{cyl}}\right) T_{ad} + \frac{m_b}{m_{cyl}} T_b \, . \tag{3.137}$$

It should be noted that the above radiative heat transfer model can be applied not only to a two-zone but also to more detailed cylinder models, e.g. to the packet models presented in Sect. 3.2.3. In that case a more realistic prediction of the soot cloud geometry and the soot temperature can be expected. However, the evaluation of the direct exchange area by numerical integration of Eq. 3.134 will become more complex.

Exemplary Results

In Fig. 3.25 simulated heat fluxes (combined convection and radiation) are compared with experimental data obtained from a 1.8 liter single-cylinder diesel engine at five of the six sub-surfaces specified in Fig. 3.21 [23]. The respective engine specifications and operating conditions are summarized in Table 3.3. It can be seen that both qualitative and quantitative agreement between predictions and experimental data are very good. Hence, the convective and radiative heat transfer model is not only suitable to describe the influence on in-cylinder processes but also to investigate thermal loads on engine components.

Table 3.3. Engine specifications and operating conditions corresponding to Fig. 3.25 [23]

Engine parameter	Quantity	Operating parameter	Quantity
Bore [mm]	128	Engine speed [rpm]	1000
Stroke [mm]	142	imep [kPa]	800
Compression ratio	16.25	Engine torque [Nm]	111
Swirl ratio	2.15	Boost pressure [kPa]	0
Piston bowl diameter [mm]	67	Coolant temperature [°C]	79
Piston bowl depth [mm]	27	Oil temperature [°C]	82

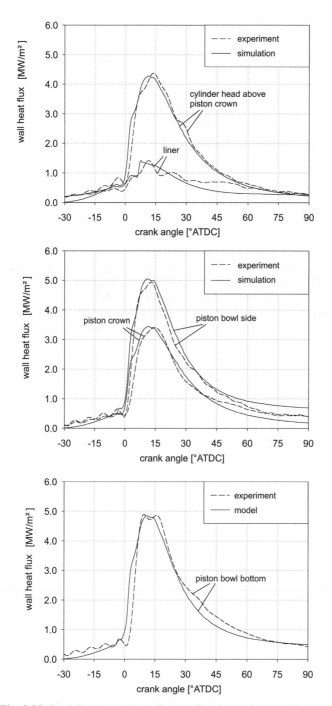

Fig. 3.25. Spatially resolved heat flux profiles for engine specified in Table 3.3 [23]

3.5 SI Engine Combustion

3.5.1 Burning Rate Calculation

In spark ignition engines with external mixture formation by fuel injection into the intake manifold it is a reasonable assumption that air and fuel vapor are homogeneously mixed by the time combustion is initiated. This is in strong contrast to the conditions in direct injection diesel engines, and consequently the driving mechanisms and flame structure in SI engines differ significantly from the ones observed in CI engines. Therefore a different modeling approach becomes necessary.

In SI engines the homogeneous mixture is inflamed by spark discharge at the spark plug location. From there a premixed flame front starts to spread out into the combustion chamber. On a microscale the thin flame front wherein almost all of the chemically bound energy is released is subject to turbulent wrinkling. However, on a macroscale it may be assumed that – in the absence of strong swirl or tumble – the leading edge of the flame is represented by a portion of the surface of a sphere. Thus, the flame propagation can be estimated based on geometric considerations when the approximate design of a combustion chamber is known, Fig 3.26. An example of flame geometry calculations by Poulos and Heywood [56] is shown in Fig. 3.27. The effective flame area is plotted versus the flame radius, i.e. the distance from the spark plug, for two different combustion chamber shapes and two spark plug locations. It is obvious from the results that a compact chamber geometry with a central plug position is beneficial in order to achieve short burning durations that enable a greater thermal efficiency and reduce the knocking tendency of the engine.

The mass burning rate can generally be written as

$$\frac{dm_b}{dt} = \rho_u A_f s_t, \qquad (3.138)$$

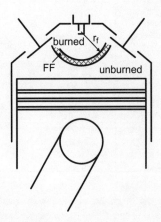

Fig. 3.26. Schematic illustration of premixed flame propagation in a homogeneously operated SI engine

where the unburned mixture density ρ_u can be obtained from a thermodynamic two-zone analysis as described in Sect. 3.5.2. The flame area A_f depends on the combustion chamber geometry and the position of the spark plug. Thus, the burning rate calculation becomes a problem of determining the turbulent flame speed s_t, which is affected by the gas composition, temperature, pressure and turbulence level.

The laminar burning velocity in the absence of turbulence has been investigated by Keck and co-workers [45, 46, 59] for a number of hydrocarbon fuels. A general temperature and pressure dependency of the form

$$s_l = s_{l,0} \left(\frac{T_u}{T_0}\right)^\alpha \left(\frac{p}{p_0}\right)^\beta (1 - 2.1 f_R) \tag{3.139}$$

was found, where f_R is the residual mass fraction and the reference temperature and pressure are $T_0 = 298$ K and $p_0 = 101.3$ kPa, respectively. The exponents α and β as well as the burning velocity at atmospheric conditions $s_{l,0}$ are fuel dependent quantities. For propane, isooctane and methanol they are given as

$$\alpha = 2.18 - 0.8(\phi - 1), \tag{3.140}$$

$$\beta = -0.16 + 0.22(\phi - 1), \tag{3.141}$$

$$s_{l,0} = B_m + B_\phi (\phi - \phi_m)^2, \tag{3.142}$$

where ϕ_m is the equivalence ratio at which $s_{l,0}$ has its maximum of value B_m. The respective parameters are specified in Table 3.4.

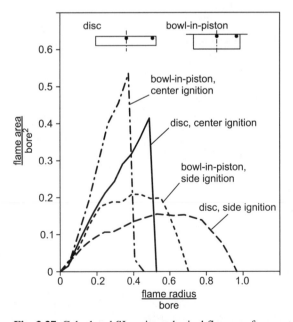

Fig. 3.27. Calculated SI engine spherical flame surface areas [56]

Table 3.4. Parameters for Eq. 3.142

Fuel	ϕ_m	B_m [cm/s]	B_ϕ [cm/s]	Ref.
Methanol	1.11	36.9	-140.5	[46]
Propane	1.08	34.2	-138.7	[46]
Isooctane	1.13	26.3	-84.7	[46]
Gasoline	1.21	30.5	-54.9	[59]

The influence of turbulence on the burning velocity, usually expressed as the turbulent to laminar flame speed ratio s_t/s_l, is hard to evaluate in phenomenological cylinder models. This is because the turbulent flow field is not explicitly solved as it is done in CFD-models. Instead a simplified analysis has to be made that approximates the turbulence level as a function of as few characteristic parameters as possible.

Often a correlation of the form initially suggested by Damköhler [15],

$$\frac{s_t}{s_l} = 1 + C\frac{u'}{s_l}, \qquad (3.143)$$

has been applied, where u' is the turbulence intensity, i.e. the mean velocity fluctuation of the turbulent gas flow, and C is a constant or involves a limited number of scaling parameters. For example, Groff [28] used the relation

$$\frac{s_t}{s_l} = 2.273 + 1.681\frac{u'}{s_l}, \qquad (3.144)$$

and Tabaczynski et al. [69, 70] derived the function

$$\frac{s_t}{s_l} = C\left(\frac{u'}{s_l}\right)^{1/3}\left(\frac{\rho_u}{\rho_{u,0}}\right)^{1/9}\left(\frac{u'L}{\nu}\right)^{1/3}, \qquad (3.145)$$

where L is a length scale proportional to the chamber height.

Keck [39] found that the turbulence intensity is approximately proportional to the square root of the unburned mixture density at the spark timing,

$$u' = 0.08\bar{u}_i\left(\frac{\rho_u}{\rho_i}\right)^{1/2}, \qquad (3.146)$$

where index i denotes the inlet conditions and $\bar{u}_i$ is the mean inlet gas speed:

$$\bar{u}_i = \eta_v \frac{A_p}{A_{iv}} c_m. \qquad (3.147)$$

In Eq. 3.147 η_v is the volumetric efficiency, A_p the piston area, A_{iv} the maximum open area of the intake valve and c_m the mean piston speed.

In a more sophisticated approach the turbulence intensity is modeled by an analysis of the specific turbulent kinetic energy k. For isotropic turbulence one can write

$$k = \frac{1}{2}\left(u'^2_x + u'^2_y + u'^2_z\right) = \frac{3}{2}u'^2 \quad \Leftrightarrow \quad u' = \sqrt{\frac{2}{3}k}, \qquad (3.148)$$

where k can be derived by Eqs. 3.125–3.128. The only difference between the SI engine case here and the diesel application in Sect. 3.4.2 is that the last term on the right hand side of Eq. 3.127 representing the kinetic energy of the diesel jet now reduces to zero.

Blizard and Keck [7] developed a mass burning rate formulation that is slightly different than the general formulation in Eq. 3.138. In their widely applied so-called entrainment model they assume that the large eddies entrain fresh mixture into the turbulent flame brush whereas the small eddies burn in a laminar manner. The mass burning rate is written as

$$\frac{dm_b}{dt} = \rho_u A_f s_l + \frac{\mu}{\tau_b} \qquad (3.149)$$

where μ is a parametric mass that is interpreted as the mass entrained within the flame region that has yet to burn:

$$\mu = m_e - m_b, \qquad (3.150)$$

$$\frac{d\mu}{dt} = \rho_u A_f u'\left(1 - \exp[-t/\tau_b]\right) - \frac{\mu}{\tau_b}. \qquad (3.151)$$

The characteristic burning time τ_b is the ratio of a length scale L and the laminar flame speed s_l

$$\tau_b = L/s_l, \qquad (3.152)$$

where the length scale L is approximated in terms of the intake valve lift L_{iv}:

$$L = 0.8 L_{iv} \left(\frac{\rho_i}{\rho_u}\right)^{3/4}. \qquad (3.153)$$

3.5.2 Gas Composition

The cylinder contents are often modeled by defining two different zones within the combustion chamber, one containing unburned mixture and the other containing burned combustion products. Each zone is considered to be ideally mixed at any time and a spatially independent pressure is assumed over the entire cylinder. The first law of thermodynamics can be applied for both the unburned and burned zones, denoted by subscripts u and b, respectively. Neglecting flow into crevices and blowby we obtain

$$\frac{d}{dt}(m_u u_u) = \frac{dQ_{w,u}}{dt} - p\frac{dV_u}{dt} - h_u\frac{dm_b}{dt}, \qquad (3.154)$$

$$\frac{d}{dt}(m_b u_b) = \frac{dQ_{w,b}}{dt} + \frac{dm_b}{dt}\text{LHV} - p\frac{dV_b}{dt} + h_u\frac{dm_b}{dt}. \qquad (3.155)$$

The two energy balances are coupled through equations of state, e.g. the ideal gas law, and continuity considerations:

$$pV_u = m_u R_u T_u, \qquad (3.156)$$

$$pV_b = m_b R_b T_b, \qquad (3.157)$$

$$V_u + V_b = V_{cyl}, \qquad (3.158)$$

$$m_u + m_b = m_{cyl}. \qquad (3.159)$$

The above set of first-order ordinary differential equations can be solved when the mass burning rate dm_b/dt is specified. This can either be done by an empirical approach such as the Wiebe function discussed in Chap. 2, or on the basis of the flame speed analysis described in Sect. 3.5.1. If the latter method is chosen additional geometric assumptions have to be made. Often a spherical flame surface with an origin at the spark plug as indicated in Fig. 3.26 is assumed in order to calculate the effective flame surface area that determines the integral mass burning rate by Eq. 3.138. It should be noted however, that a sphere represents the smallest possible ratio of flame surface area to burned volume. Therefore, the effective flame area and the mass burning rate may be significantly underpredicted as the real flame shape can be distorted due to turbulence effects and interactions with the chamber walls.

Furthermore, it should be noted that the assumption of an ideally mixed burned zone may cause errors in the results. While the unburned zone is fairly well reproduced by this uniform temperature concept, significant temperature gradients may exist in the burned gases due to the differences between first burning and then compressing the burned gas, as compared to first compressing and then burning the fresh mixture [57]. This problem could be diminished by considering multiple, subsequently generated product zones as described in the diesel engine models in Sect. 3.3.2.

3.5.3 Engine Knock

Fundamentals

Engine knock is an SI engine phenomenon where the compressed end gas, i.e. the unburned portion of the air-fuel mixture within the cylinder, ignites before it is reached by the propagating flame front, Fig. 3.28. This ignition can be caused by either surface ignition at hot spots on the combustion chamber walls or by spontaneous autoignition within the unburned mixture. Primary sources for surface ignition are hot exhaust valves, spark plugs and metal asperities such as edges of head cavities or piston bowls [30]. They have to be avoided by the engine designer, e.g. by not placing these hot spots towards the end of the regular flame path. Autoignition may generally occur because of the pressure and temperature increase in the unburned mixture that is caused by the expansion of the hot combustion products within the burned region of the combustion chamber.

In both cases the essentially instantaneous release of a significant quantity of energy causes a shock wave to propagate from the end gas through the combustion chamber. This shock wave, an accompanying expansion wave, and the reflections of these waves by the cylinder walls result in substantial pressure oscillations that are characteristic for engine knock. A typical pressure trace of an engine experiencing knock is displayed in Fig. 3.29. In case of knock the pressure is no longer uniform throughout the combustion chamber but steep spatial gradients exist. The significant amplitude of the pressure fluctuations is clearly audible and can cause severe damage to the engine structure. Thus, its occurrence must be excluded by designing compact combustion chambers with central spark plug locations and short flame paths or by retarding the spark timing. However, the latter option will reduce the thermal efficiency of the engine.

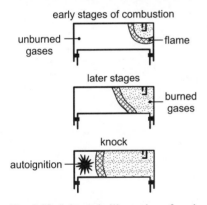

Fig. 3.28. Schematic illustration of engine knock [57]

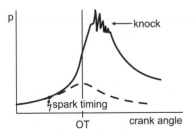

Fig. 3.29. Pressure trace for an engine experiencing knock

The chemical kinetics governing autoignition in hydrocarbon-air mixtures are very complex, e.g. [30, 73]. However, it is widely agreed that in iso-paraffinic fuels that exhibit two-stage ignition behavior (a cool flame followed by a hot flame) hydroperoxides are important intermediate species found in increasing concentrations within the end gas of a combustion chamber. Hydrogen peroxide which is relatively stable at temperatures below 800 K is produced by the chain propagation reaction

$$\mathrm{H\dot{O}_2 + RH \rightarrow H_2O_2 + \dot{R}}, \qquad (3.160)$$

where R denotes an organic radical. At higher temperatures H_2O_2 quickly decomposes into two hydroxyl radicals, representing the major chain branching step in autoignition:

$$\mathrm{H_2O_2 + M \rightarrow 2\dot{O}H + M}. \qquad (3.161)$$

Müller and co-workers [49, 50] have developed reduced reaction mechanisms for autoignition of n-heptane and mixtures of n-heptane and isooctane, respectively. In ref. [49] a three-step mechanism was proposed that describes the reaction path from fuel F via an intermediate species I to combustion products P:

$$\begin{array}{ll} F + \alpha O_2 & \rightarrow \quad P \\ F + 2O_2 & \leftrightarrow \quad I \\ I + (\alpha - 2)O_2 & \rightarrow \quad P \end{array} \qquad (3.162)$$

The species are defined as

$$F = \frac{\mathrm{ON}}{100} \cdot \mathrm{C_8H_{18}} + \left(1 - \frac{\mathrm{ON}}{100}\right) \cdot \mathrm{C_7H_{16}}$$

$$I = \frac{\mathrm{ON}}{100} \cdot I_8 + \left(1 - \frac{\mathrm{ON}}{100}\right) \cdot I_7 + \mathrm{H_2O} \qquad (3.163)$$

$$P = \left[8 \cdot \frac{\mathrm{ON}}{100} + 7 \cdot \left(1 - \frac{\mathrm{ON}}{100}\right)\right] \cdot \mathrm{CO_2} + \left[9 \cdot \frac{\mathrm{ON}}{100} + 8 \cdot \left(1 - \frac{\mathrm{ON}}{100}\right)\right] \cdot \mathrm{H_2O}$$

where ON is the fuel octane number, α is the stoichiometric coefficient of oxygen

$$\alpha = 12.5 \cdot \frac{\mathrm{ON}}{100} + 11 \cdot \left(1 - \frac{\mathrm{ON}}{100}\right), \qquad (3.164)$$

and I_7 and I_8 denote the peroxides $\mathrm{OC_7H_{13}OOH}$ and $\mathrm{OC_8H_{15}OOH}$, respectively.

Modeling

The simplest, yet least predictive method to model knock in SI engines is to utilize the ignition or knock integral as it is used in phenomenological diesel combustion models (Eq. 3.79) and to apply it to the unburned zone of end-gases,

$$I_k = \int_0^{t_k} \frac{1}{\tau} dt = 1, \qquad (3.165)$$

where t is the elapsed time from the start of the end-gas compression process and t_k is the time of autoignition. The induction time τ is modeled by an Arrhenius equation. Several suggestions have been proposed for the respective constants. Probably the most widely used relation is the one by Douaud and Eyzat [19], that relates the induction time to the octane number of the fuel:

$$\tau = 18.69 \left(\frac{ON}{100}\right)^{3.4017} p^{-1.7} \exp\left(\frac{3800\,K}{T}\right). \qquad (3.166)$$

It should be noted though that Eqs. 3.165 and 3.166 represent a significant abstraction of the real problem, and consequently it is often not possible to accurately describe the onset of knocking over an entire engine map. A readjustment of the parameters in Eq. 3.166 may therefore be necessary in order to predict knock for varied operating conditions or different engine types.

In order to yield a formulation of greater applicability Franzke [25] proposed that the ignition or knocking integral expressed by Eq. 3.165 has to exceed unity prior to a specific crank angle φ_k in order to result in knocking engine performance. This crank angle φ_k is related to the combustion duration by the relation

$$K = \frac{\varphi_k - \varphi_{1\%}}{\varphi_{95\%} - \varphi_{1\%}}, \qquad (3.167)$$

where $\varphi_{1\%}$ and $\varphi_{95\%}$ denote the crank angle positions of 1% and 95% mass conversion, respectively. The factor K needs to be determined for one reference case per engine and thereafter φ_k can be calculated by rearranging Eq. 3.167 for other operating conditions as well.

Worret et al. [76, 77] investigated a database of 1700 operating conditions of a Mercedes-Benz 6-cylinder inline gasoline engine, including 142 data points directly at the knock limit. The engine parameters varied in the study were engine speed, equivalence ratio, residual mass fraction, inlet temperature and pressure as well as compression ratio. They suggested to replace Eq. 3.166 by

$$\tau = 2.37 \cdot p^{-1.299} \cdot \exp\left(\frac{4617\,K}{T}\right), \qquad (3.168)$$

but realized that Eq. 3.167 was still not generally applicable since the variance in the K-factor was found to be as large as ±30% for the 142 data points at the knock limit. To overcome these shortcomings they defined the K-factor in terms of the 75%-mass-conversion timing which is typically characterized by a greater repeatability than the 95%-timing. Furthermore, they found that both the critical quantity of the ignition integral I_k and the K-factor need to be related to the air-fuel equivalence ratio λ and the 50%- and 75%-mass-conversion timings for varying operating conditions:

$$\frac{I_k}{I_{k,ref}} = \left(\frac{\varphi_{75\%}+6}{\varphi_{75\%,ref}+6}\right)^{-0.557}, \qquad (3.169)$$

$$\frac{K}{K_{ref}} = \left(\frac{\varphi_{50\%}+8}{\varphi_{50\%,ref}+8}\right)^{-0.236} \cdot \left(\frac{1.292-0.251\cdot\lambda}{1.292-0.251\cdot\lambda_{ref}}\right). \qquad (3.170)$$

The subscript *ref* refers to one global reference case at the knock limit which is defined by $n = 2000$ rpm, $\lambda = 1$, $T_{int} = 298$ K, $p_{int} = 99.5$ kPa and an ignition timing at $-17.25°$ATDC for the investigated engine. The induction time τ was computed by Eq. 3.168 which results in a value of 1.0 for the reference knock integral $I_{k,ref}$.

References

[1] Abramovich, GN (1963) The Theory of Turbulent Jets. MIT Press, Cambridge
[2] Annand WJ (1963) Heat Transfer in the Cylinders of Reciprocating Internal Combustion Engines. Proc Inst Mech Engineers, vol 176, no 36, pp 973–990
[3] Arai M, Tabata M, Hiroyasu H, Shimizu M (1984) Desintegrating Process and Spray Characterization of Fuel Jet Injected by a Diesel Nozzle. SAE Paper 840275
[4] Barba C, Burkhardt C, Boulouchos K, Bargende M (2000) A Phenomenological Combustion Model for Heat Release Rate Prediction in High-Speed DI Diesel Engines with Common Rail Injection. SAE Paper 2000-01-2933
[5] Bargende M (1990) Ein Gleichungsansatz zur Berechnung der instationären Wandwärmeverluste im Hochdruckteil von Ottomotoren. PhD Thesis, Technical University of Darmstadt, Germany
[6] Bazari Z (1992) A DI Diesel Combustion and Emission Predictive Capability for Use in Cycle Simulation. SAE Paper 920462
[7] Blizard NC, Keck JC (1974) Experimental and Theoretical Investigation of Turbulent Burning Model for Internal Combustion Engines. SAE Paper 740191
[8] Borman GL, Johnson JH (1962) Unsteady Vaporization Histories and Trajectories of Fuel Drops Injected into Swirling Air. SAE Paper 598 C
[9] Boulouchos K, Eberle MK (1991) Aufgabenstellungen der Motorthermodynamik heute – Beispiele und Lösungsansätze. MTZ, vol 52, no 11, pp 574–583
[10] Boulouchos K, Hannoschöck N (1986) Der Wärmetransport zwischen Arbeitsmedium und Brennraumwand. MTZ, vol 47, no 9
[11] Boulouchos K, Isch R (1990) Modeling of Heat Transfer During Combustion: A Quasi-Dimensional Approach with Emphasis on Large Low-Speed Diesel Engines. Int Symp COMODIA 90, pp 321–328
[12] Cebeci T, Bradshaw P (1988) Physical and Computational Aspects of Convective Heat Transfer. Springer, Berlin, New York
[13] Chiu WS, Shahed SM, Lyn WT (1976) A Transient Spay Mixing Model of Diesel Combustion. SAE Paper 760128

[14] Chmela FG, Orthaber GC (1999) Rate of Heat Release Prediction for Direct Injection Diesel Engines Based on Purely Mixing Controlled Combustion. SAE Paper 1999-01-0186
[15] Damkoehler G (1940) Der Einfluss der Turbulenz auf die Flammgeschwindigkeit in Gasgemischen. Z Elektrochem, vol 46, pp 601–626
[16] DeNeef AT (1987) Untersuchung der Voreinspritzung am schnellaufenden, direkteinspritzenden Dieselmotor. PhD Thesis, ETH Zurich, Switzerland
[17] Dent JC (1971) Basis for the Comparison of Various Experimental Methods for Studying Spray Penetration. SAE Paper 710571
[18] Dent JC, Mehta PS (1981) Phenomenological Combustion Model for a Quiescent Chamber Diesel Engine. SAE Paper 811235
[19] Douaud AM, Eyzat P (1978) Four-Octane-Number Method for Predicting the Anti-Knock Behavior of Fuels and Engines. SAE Paper 780080
[20] Edelman RB, Harsha PT (1978) Laminar and Turbulent Gas Dynamics in Combustors – Current Status. Prog Energy Combust Sci, vol 4, no 1
[21] Eiglmeier C (2000) Phänomenologische Modellbildung des gasseitigen Wandwärmeüberganges in Dieselmotoren. PhD Thesis, University of Hanover, Germany
[22] Eiglmeier C, Merker GP (2000) New Approaches to the Phenomenological Modeling of the Gas-Side Wall Heat Transfer in Diesel Engines. MTZ Worldwide, no 5/2000, pp 17–24
[23] Eiglmeier C, Lettmann H, Stiesch G, Merker GP (2001) A Detailed Phenomenological Model for Wall Heat Transfer Prediction in Diesel Engines. SAE Paper 2001-01-3265
[24] Elkotb MM (1982) Fuel Atomization for Spray Modeling. Prog Energy Combust Sci, vol 8, pp 61–91
[25] Franzke DE (1981) Beitrag zur Ermittlung eines Klopfkriteriums der ottomotorischen Verbrennung und zur Vorausberechnung der Klopfgrenze. Ph.D. Thesis, Technical University of Munich, Germany
[26] Gao Z, Schreiber W (2001) The Effects of EGR and Split Fuel Injection on Diesel Engine Emisssion. Int J Automotive Technology, vol 2, no 4, pp 123–135
[27] Gibson DH, Mahaffey WA, Mukerjee T (1990) In-Cylinder Flow and Combustion Modeling of 1.7L Caterpillar Engine. SAE Paper 900253
[28] Groff EG (1987) An Experimental Evaluation of an Entrainment Flame-Propagation Model. Combustion and Flame, vol 67, pp 153–162
[29] Heider G, Zeilinger K, Woschni G (1995) Two-Zone Calculation Model for the Prediction of NO Emissions from Diesel Engines. Proc 21st CIMAC Cong, Interlaken, Paper D52
[30] Heywood JB (1988) Internal Combustion Engine Fundamentals. McGraw-Hill, New York
[31] Heywood JB (1994) Combustion and its Modeling in Spark-Ignition Engines. Int Symp COMODIA 94, pp 1–15
[32] Hiroyasu H, Kadota T (1974) Fuel Droplet Size Distribution in Diesel Combustion Chamber. SAE Paper 740725
[33] Hiroyasu H, Kadota T, Arai M (1980) Combustion Modeling in Reciprocating Engines. Symp at GM-Research-Laboratories 1978, Plenum Press, New York, London, pp 349–405

[34] Hiroyasu H, Kadota T, Arai M (1983a) Development and Use of a Spray Combustion Model to Predict Diesel Engine Efficiency and Pollutant Emissions, Part 1: Combustion Modeling. Bull JSME, vol 26, no 214, pp 569–575

[35] Hiroyasu H, Kadota T, Arai M (1983b) Development and Use of a Spray Combustion Model to Predict Diesel Engine Efficiency and Pollutant Emissions, Part 2: Computational Procedure and Parametric Study. Bull JSME, vol 26, no 214, pp 576–583

[36] Hohlbaum B (1992) Beitrag zur rechnerischen Untersuchungen der Stickoxid-Bildung schnellaufender Hochleistungsdieselmotoren. PhD Thesis, University of Karlsruhe, Germany

[37] Hottel HC, Sarofim AF (1967) Radiative Transfer. Mc-Graw-Hill, New York

[38] Hountalas DT, Kouremenos DA, Pariotis EG, Schwarz V, Binder KB (2002) Using a Phenomenological Model to Investigate the Effect of Injection Rate Shaping on Performance and Pollutants of a DI Heavy Duty Diesel Engine. SAE Paper 2002-01-0074

[39] Keck JC (1982) Turbulent Flame Structure and Speed in Spark-Ignition Engines. Proc 19th Symp (Int) Combustion, pp 1451–1466, The Combustion Institute, Pittsburgh, PA

[40] Kouremenos DA, Rakopoulos CD, Hountalas DT (1997) Multi-Zone Modeling for the Prediction of Pollutant Emissions and Performance of DI Diesel Engines. SAE Paper 970635

[41] Kuo TW (1987) Evaluation of a Phenomenological Spray-Combustion Model for Two Open-Chamber Diesel Engines. SAE Paper 872057

[42] Kuo TW, Yu RC, Shahed SM (1983) A Numerical Study of the Transient Evaporating Spray Mixing Process in the Diesel Environment. SAE Paper 831735

[43] Magnussen BF, Hjertager BH (1976) On Mathematical Modeling of Turbulent Combustion with Special Emphasis on Soot Formation and Combustion. 16th Symp (Int) Combust, pp 719–729, The Combustion Institute, Pittsburgh, PA

[44] Merker GP, Hohlbaum B, Rauscher M (1993) Two-Zone Model for Calculations of Nitrogen-Oxide Formation in Direct-Injection Diesel Engines. SAE Paper 932454

[45] Metghalchi M, Keck JC (1980) Laminar Burning Velocity of Propane-Air Mixtures at Hight Temperature and Pressure. Combust Flame, vol 38, pp 143–154

[46] Metghalchi M, Keck JC (1982) Burning Velocities of Mixtures of Air with Methanol, Iso-octane, and Indolene at High Pressure and Temperature. Combust Flame, vol 48, pp 191–210

[47] Morel T, Keribar R (1985) A Model for Predicting Spatially and Time Resoved Convective Heat Transfer in Bowl-in Piston Combustion Chambers. SAE Paper 850204

[48] Morel T, Keribar R (1986) Heat Radiation in DI Diesel Engines. SAE Paper 860445

[49] Müller UC (1993) Reduzierte Reaktionsmechanismen für die Zündung von n-Heptan und iso-Oktan unter motorrelevanten Bedingungen. Ph.D. Thesis, RWTH Aachen, Germany

[50] Müller UC, Peters N, Linan A (1992) Global Kinetics for n-Heptane Ignition at High Pressures. 24th Symp (Int) Combust, pp 777–784, The Combustion Institute, Pittsburgh, PA

[51] Murakami A, Arai M, Hiroyasu H (1988) Swirl Measurements and Modeling in Direct Injection Diesel Engines. SAE Paper 880385

[52] Nagle J, Strickland-Constable RF (1962) Oxidation of Carbon between 1000–2000 C. Proc 5th Carbon Conf, vol 1, Pergamon Press, London

[53] Nishida K, Hiroyasu H (1989) Simplified Three-Dimensional Modeling of Mixture Formation and Combustion in a DI Diesel Engine. SAE Paper 890269
[54] Otto F, Dittrich P, Wirbeleit F (1998) Status of 3D-Simulation of Diesel Combustion. Proc 3rd Int Indicating Symp, Mainz, pp 289–308
[55] Papadopoulos S (1987) Reduktion der Stickoxidemissionen des direkteinspritzenden Dieselmotors durch Dieseloelwasseremulsionen bzw. Wassereinspritzung. Ph.D. Thesis, ETH Zurich, Switzerland
[56] Poulos SG, Heywood JB (1983) The Effect of Chamber Geometry on Spark-Ignition Engine Combustion. SAE Paper 830334
[57] Ramos JI (1989) Internal Combustion Engine Modeling. Hemisphere, New York
[58] Ranz WE, Marshall WR (1952) Evaporation from Drops. Chem Eng Prog, vol 48, no 3, pp 141–146 and 173–180
[59] Rhodes DB, Keck JC (1985) Laminar Burning Speed Measurements of Indolene-Air-Diluent Mixtures at High Pressures and Temperature. SAE Paper 850047
[60] Shahed SM, Chiu WS, Yumlu VS (1973) A Preliminary Model for the Formation of Nitric Oxide in Direct Injection Diesel Engines and Its Application in Parametric Studies. SAE Paper 730083
[61] Shahed SM, Chiu WS, Lyn WT (1975) A Mathematical Model of Diesel Combustion. Proc Inst Mech Engineers, C94/75, pp119–128
[62] Shi SX, Su WH, Zhao KH, Yue Y (1993) Development of a Four-Zone Analytical Combustion Model for a DI Compression-Ignition Engine. Proc 20th CIMAC Cong, London, Paper D13
[63] Stegemann J, Seebode J, Baumgarten C, Merker GP (2002) Influence of Throttle Effects at the Needle Seat on the Spray Characteristics of a Multihole Injection Nozzle. Proc 18th ILASS-Europe Conf, pp 31–36, Zaragoza, Spain
[64] Stiesch G, Merker GP (1998) A Phenomenological Heat Release Model for Direct Injection Diesel Engines. 22nd CIMAC Int Congr Combust Engines, vol 2, pp 423–430
[65] Stiesch G, Merker GP (1999) A Phenomenological Model for Accurate and Time Efficient Prediction of Heat Release and Exhaust Emissions in Direct-Injection Diesel Engines. SAE Paper 1999-01-1535
[66] Stiesch G, Eiglmeier C, Merker GP, Wirbeleit F (1999) Possibilities and Application of Phenomenological Combustion Models in Diesel Engines. MTZ worldwide, no 4/99, pp 19–24
[67] Stringer FW, Clarke AE, Clarke JS (1969) The Spontaneous Ignition of Hydrocarbon Fuels in a Flowing System. Proc IMechE 184
[68] Suhre B, Foster D (1992) In-Cylinder Soot Deposition Rates Due to Thermophoresis in a Direct Injection Engine. SAE Paper 921629
[69] Tabaczynski RJ, Ferguson CR, Radhakrishnan K (1977) A Turbulent Entrainment Model for Spark-Ignition Engine Combustion. SAE Paper 770647
[70] Tabaczynski RJ, Trinker FH, Shannon BAS (1980) Further Refinement and Validation of a Turbulent Flame Propagation Model for Spark-Ignition Engines. Combustion and Flame, vol 39, pp 111–121
[71] Thoma M, Stiesch G, Merker GP (2002) A Phenomenological Spray and Combustion Model for Diesel Engines with Pre-Injection. 5th Int Symp Internal Combust Diagnostics, pp 90–101, Baden-Baden, Germany
[72] Varde KS, Popa DM, Varde LK (1984) Spdray Angle and Atomization in Diesel Sprays. SAE Paper 841055

[73] Warnatz J, Maas U, Dibble RW (2001) Combustion: Physical and Chemical Fundamentals, Modeling and Simulation, Experiments, Pollutant Formation. 3rd ed, Springer, Berlin
[74] Weisser G, Boulouchos K (1995) NOEMI – A Tool for the Precalculation of Nitric Oxide Emissions of DI Diesel Engines. Proc 5th Symp "The Working Process of the Combustion Engine", pp 23–50, Technical University Graz, Austria
[75] Wolff A, Boulouchos K, Mueller R (1997) Computational Investigation of Unsteady Heat Flux Through an IC Engine Wall Including Soot Layer Dynamics. SAE Paper 970063
[76] Worret R (2002) Entwicklung eines Kriteriums zur Vorausberechnung der Klopfgrenze. FVV-Report no 700, Forschungsvereinigung Verbrennungskraftmaschinen, Frankfurt, Germany
[77] Worret R, Bernhardt S, Schwarz F, Spicher U (2002) Application of Different Cylinder Pressure Based Knock Detection Methods in Spark Ignition Engines. SAE Paper 2002-01-1668
[78] Woschni G (1967) A Universally Applicable Equation for the Instantaneous Heat Transfer Coefficient in the Internal Combustion Engine. SAE Paper 670931
[79] Zhang Y (1992) A Simplified Model for Predicting Evaporating Spray Mixing Process in DI Diesel Engines. SAE Paper 922228

4 Fundamentals of Multidimensional CFD-Codes

4.1 Conservation Equations

The abbreviation CFD stands for computational fluid dynamics which indicates the numerical solution of multidimensional flow problems that may be of unsteady and turbulent nature. In general, multidimensional flow problems are governed by conservation principles for mass energy and momentum. The application of these principles results in a set of partial differential equations in terms of time and space that need to be integrated numerically as they are too complex to be solved analytically.

For the sake of simplicity and clearness the subsequent analysis is based on single-component, single-phase flows. However, it should be noted that in engine combustion chambers the gas phase usually consists of multiple components that are subject to chemical reactions. Furthermore, in direct injection engines two-phase flows with evaporating fuel droplets are encountered. In order to take account of these effects, additional transport- or source-terms have to be added to the conservation equations as it is indicated in Sect. 4.5.

Mass Conservation

The continuity equation based on mass conservation can be derived for an infinitesimal volume element $dV = dx_1\, dx_2\, dx_3$ as indicated in Fig. 4.1. In a Eulerian approach, i.e. with a coordinate system fixed in space, the control volume is fixed in space as well. It can be passed by the flow without resistance and the mass within the control volume will increase if the inflow exceeds the outflow. In the opposite case it will decrease. Either case is associated with a density change in the control volume. The mass balance reads

$$\frac{\partial}{\partial t}\left(dx_1\, dx_2\, dx_3\, \rho\right) = d\dot{m}_{x_1} + d\dot{m}_{x_2} + d\dot{m}_{x_3}\,, \tag{4.1}$$

where

$$d\dot{m}_{x_1} = \left(\dot{m}_{x_1}\right)_{x_1} - \left(\dot{m}_{x_1}\right)_{x_1+dx_1} = dx_2\, dx_3\left(\rho v_1\right)_{x_1} - dx_2\, dx_3\left(\rho v_1\right)_{x_1+dx_1}\,, \tag{4.2}$$

and the second and third terms on the right hand side of Eq. 4.1 are treated accordingly. Furthermore, the second term on the right hand side of Eq. 4.2 can be expressed by a Taylor series,

$$(\rho v_1)_{x_1+dx_1} = (\rho v_1)_{x_1} + \frac{\partial}{\partial x_1}(\rho v_1)_{x_1} dx_1, \tag{4.3}$$

and we obtain

$$\dot{m}_{x_1} = -dx_2\, dx_3\, \frac{\partial}{\partial x_1}(\rho v_1)_{x_1} dx_1. \tag{4.4}$$

The terms $\dot{m}_{x_2}$ and $\dot{m}_{x_3}$ are again treated accordingly. Since the size of the control volume is constant with time, the dx_i in Eq. 4.1 can be taken out of the differential operator. Combining Eqs. 4.1 and 4.4 then yields the general form of mass conservation:

$$\frac{\partial \rho}{\partial t} + \frac{\partial}{\partial x_i}(\rho v_i) = 0. \tag{4.5}$$

In Eq. 4.5 and throughout this chapter the Einstein convention will be utilized. It states that whenever the same index appears twice in any term, summation over the range of that index is implied, i.e.

$$\frac{\partial \rho}{\partial t} + \frac{\partial}{\partial x_i}(\rho v_i) = \frac{\partial \rho}{\partial t} + \frac{\partial(\rho v_1)}{\partial x_1} + \frac{\partial(\rho v_2)}{\partial x_2} + \frac{\partial(\rho v_3)}{\partial x_3} = 0. \tag{4.6}$$

In many applications the fluid may be treated as incompressible. This is true not only for flows of liquids, whose compressibility may indeed be neglected, but also for gases if the Mach number is below approx. 0.3 [3]. Applying the chain rule, Eq. 4.5 can be written as

$$\frac{\partial \rho}{\partial t} + v_i \frac{\partial \rho}{\partial x_i} + \rho \frac{\partial v_i}{\partial x_i} = 0, \tag{4.7}$$

and for incompressible flows (ρ = const.) it reduces to

$$\mathrm{div}(\vec{v}) \equiv \frac{\partial v_i}{\partial x_i} = 0. \tag{4.8}$$

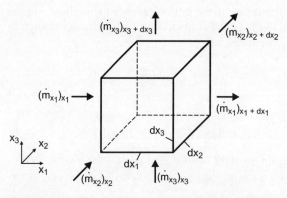

Fig. 4.1. Mass fluxes entering and exiting the control volume

Momentum Conservation

The Navier-Stokes equations for conservation of momentum can be derived in a similar manner as the mass conservation equation in the above section. This procedure is carried out in detail in many textbooks about fluid mechanics, e.g. [3, 15], and therefore only a short summary and the resulting equations will be given at this point.

As opposed to the mass conservation equation, Eq. 4.5, there is not only one equation for conservation of momentum, but one for each of the three Cartesian dimensions x_1, x_2 and x_3. Based on the principle that the temporal change of the momentum equals the sum of all external forces acting on the infinitesimal control volume, the following equation is obtained for each of the three dimensions x_j:

$$\frac{\partial(\rho v_j)}{\partial t} + v_i \frac{\partial(\rho v_j)}{\partial x_i} = -\frac{\partial p}{\partial x_j} + \frac{\partial \tau_{ij}}{\partial x_i} + \rho F_j, \quad j = 1, 2, 3 \quad (4.9)$$

The first term on the right hand side of Eq. 4.9 denotes the pressure gradient in direction x_j, the second term includes the stress tensor

$$\tau_{ij} = \begin{pmatrix} \tau_{x_1 x_1} & \tau_{x_1 x_2} & \tau_{x_1 x_3} \\ \tau_{x_2 x_1} & \tau_{x_2 x_2} & \tau_{x_2 x_3} \\ \tau_{x_3 x_1} & \tau_{x_3 x_2} & \tau_{x_3 x_3} \end{pmatrix}, \quad (4.10)$$

and the external forces F_j are typically zero in x_1- and x_2-direction and equal to the gravitational force in vertical x_3-direction:

$$F_1 = F_2 = 0, \quad F_3 = -g. \quad (4.11)$$

The viscous stress tensor τ_{ij} accounts for momentum transfer due to friction. For Newtonian fluids, i.e. if $\mu \neq f(\tau)$, Stokes' postulation (1845) states that the shear stresses are proportional to the velocity gradients, with the molecular viscosity as the proportionality factor:

$$\tau_{ij} = \mu \left(\frac{\partial v_i}{\partial x_j} + \frac{\partial v_j}{\partial x_i} \right) - \delta_{ij} \frac{2}{3} \mu \frac{\partial v_i}{\partial x_i}. \quad (4.12)$$

δ_{ij} is the Kronecker delta, and for incompressible flows ($\partial v_i / \partial x_i = 0$) the second term on the right hand side of Eq. 4.12 cancels out, such that Eq. 4.9 becomes

$$\rho \left(\frac{\partial v_j}{\partial t} + v_i \frac{\partial v_j}{\partial x_i} \right) = -\frac{\partial p}{\partial x_j} + \mu \left(\frac{\partial^2 v_j}{\partial x_i^2} \right) + \rho F_j, \quad j = 1, 2, 3. \quad (4.13)$$

Energy Conservation

In addition to mass and momentum conservation, energy conservation, i.e. the first law of thermodynamics, has to be solved to describe velocity, pressure and temperature distributions in non-isothermal flows. The general form of the energy equation for a control volume fixed in space reads

$$\frac{\partial E}{\partial t} = \dot{E}_{in} - \dot{E}_{out} + \dot{W}_g + \dot{W}_p + \dot{W}_\tau + \dot{Q}_{in} - \dot{Q}_{out}. \qquad (4.14)$$

That is, the rate of change in total energy E within the control volume is equal to the difference in energies transferred by entering and exiting mass flows, plus the work rates associated to gravitation, pressure and viscous forces, plus the net rate of heat transfer to the control volume.

The derivation of the each terms in Eq. 4.14 is shown in detail e.g. in ref. [14]. Neglecting the potential energy and utilizing the relation $h = u + p/\rho$, the thermal energy equation can be written by the formulation

$$\rho \, dx_1 \, dx_2 \, dx_3 \left(\frac{\partial h}{\partial t} + v_i \frac{\partial h}{\partial x_i} \right) = \dot{Q}_{in} - \dot{Q}_{out} + \dot{W}_{diss} \qquad (4.15)$$

after some rearrangements. $\dot{W}_{diss}$ indicates the work added to the control volume by dissipation. In most technical applications this term can be neglected. However, in high Mach number flows a significant temperature rise can be caused by dissipation. An example is the reentry of spacecrafts into the earth atmosphere.

Assuming that the heat transfer is solely caused by conduction, i.e. radiation is neglected, the exchange in x_1-direction becomes

$$\dot{Q}_{x_1,in} - \dot{Q}_{x_1,out} = -dx_2 \, dx_3 \frac{\partial \dot{q}_{x_1}}{\partial x_1} dx_1, \qquad (4.16)$$

where $\dot{q}$ indicates the heat flux in W/m². This heat flux can be expressed by Fourier's law as

$$\dot{q}_i = -k \frac{\partial T}{\partial x_i}, \qquad (4.17)$$

where k is the thermal conductivity which is often assumed to be approximately constant. Therefore, the thermal energy equation becomes

$$\rho \left(\frac{\partial h}{\partial t} + v_i \frac{\partial h}{\partial x_i} \right) = k \frac{\partial^2 T}{\partial x_i^2}, \qquad (4.18)$$

and if it is further assumed that the fluid can be described by the caloric equation of state for an ideal gas we obtain

$$\rho c_p \left(\frac{\partial T}{\partial t} + v_i \frac{\partial T}{\partial x_i} \right) = k \frac{\partial^2 T}{\partial x_i^2}. \qquad (4.19)$$

4.2 Numerical Methodology

Direct Numerical Simulation (DNS)

The Navier-Stokes equations are generally valid for both laminar and turbulent flows, and therefore the obvious and most accurate approach would be to solve the equations directly on a discretized grid that is fine enough to resolve the smallest length scales of the flow problem. This method is called direct numerical simulation and represents the simplest approach from a conceptual point of view. The smallest length scales are defined by the size of the smallest eddies, that are important for dissipation of turbulent kinetic energy and are characterized by the viscously determined Kolmogorov length scale. This length scale becomes smaller for increasing Reynolds numbers. An additional constraint for the grid spacing in two phase flows may be imposed by the size of the liquid droplets contained in the gas phase.

For typical conditions inside an engine cylinder the above criteria would require a grid point spacing of approx. 10 μm, resulting in a total of about 10^{12} grid points for a combustion chamber with a bore of 10 cm. However, a reasonable number of grid points that can be handled with today's computer systems is in the range of 10^6. Even with the expected rapid progress in computer technologies an increase of another six orders of magnitude in computer power will not be possible in the foreseeable future. Therefore, direct numerical simulation is not suitable for solving engineering problems, but it is constrained to fundamental research applications with relatively low Reynolds number flows and geometrically simple domains [3]. Nevertheless, DNS is useful in that it can provide detailed information about any variable of interest at a large number of grid points. These results may be regarded as equivalent of experimental data and can be used to produce statistical information or to create a numerical flow visualization. The wealth of information helps to get a better understanding of the complex flow phenomena and can be used to construct and validate other quantitative models, e.g. of LES or RANS type, that will allow other, similar flows to be computed.

Typical examples of DNS applications are the calculation of relatively low Reynolds number flows with a small number of species in a simple geometric domain, e.g. the two-dimensional calculation of an opposed jet hydrogen-air flame [7].

Large Eddy Simulation (LES)

Large eddy simulation is an approach in which only the large-scale eddies of the flow are resolved, in order to reduce the necessary number of grid points. Consequently, the behavior of the smaller eddies needs to be described by appropriate semi-empirical submodels. This procedure seems reasonable since the large eddies, that are of the order of the integral length scale, contain the major fraction of energy and are much more important in the transport of the conserved quantities than the smaller eddies. Moreover, the smaller eddies typically show a more isotropic behavior and are therefore easier to assess by modeling approaches than the large eddies directly resolved in LES. Figure 4.2 displays a qualitative comparison between direct numerical simulation and large eddy simulation. Whereas DNS resolves all spatial and temporal fluctuations of the flow, LES essentially represents a local average of the complete field. Obviously, DNS is the preferred method whenever it is feasible, because it is the more accurate method. LES is preferred in applications where the geometry is more complex and where the Reynolds number or the number of chemical species are too high for DNS.

The classification into large and small eddies is made by filtering the flow field by a specific cutoff length scale, referred to as Δ, that is appropriate for the problem. Since the small eddies are not resolved, the Navier-Stokes equations are expressed in terms of averaged quantities similar to the form obtained in the RANS equations described in the following subsection. Submodels are then needed to estimate the local mean fluctuations of the small-scale turbulence which is done by so-called subgrid-scale (SGS) Reynolds stress models. Several different types of such semi-empirical SGS models have been proposed in the past. An overview as well as references containing the details have been given in ref. [3].

It should be noted that a number of difficulties still exist in LES. To begin with, the selection of the cutoff length scale Δ is not straightforward. In many problems the Reynolds number and thus the characteristic length scales are not constant over the entire geometric domain. Moreover, the flow structure near walls is typically highly anisotropic. All this means that the appropriate Δ and the resulting grid resolution have to be carefully adjusted to a particular flow problem. A more thorough discussion of this issue can be found in refs. [3] and [10]. An additional problem with both DNS and LES is that there are no 'standard-models' available that are suitable for any purpose. Because of the significant requirements with respect to computer power and memory the programs are typically designed for a particular purpose, i.e. they are written for a specific geometry and contain special programming elements designed to obtain the highest performance on a particular machine.

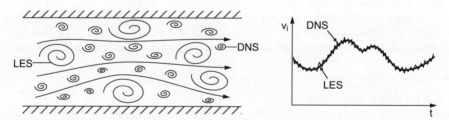

Fig. 4.2. Schematic representation of turbulent motion (left) and the time dependence of a velocity component at a specific spatial position (right) [3]

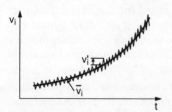

Fig. 4.3. Reynolds-averaging for an unsteady flow

While LES methods are less costly than DNS, they are nevertheless still very expensive in terms of computer time. Therefore, their application is currently limited to more fundamental research tasks as opposed to studies of general engineering problems. However, LES models have been applied to investigate specific problems in engines as well, e.g. turbulent diffusion combustion in diesel engines [11], and this trend is likely to be extended with more powerful computer technology becoming available in the future.

Reynolds-Averaged Navier-Stokes Equations (RANS)

Because of the problems encountered with DNS and LES when calculating complex flows with high turbulence levels, a third method of solving the Navier-Stokes equations is often applied in engineering problems. It is called the Reynolds-averaged method because it is based on ideas proposed by Osborne Reynolds over a century ago. In this approach the instantaneous values of the conserved quantities of the turbulent flow are split into an averaged value denoted by an overbar ($^-$) and a fluctuation about that value denoted by the superscript ('). A schematic illustration of this concept is shown in Fig. 4.3 for the velocity component v_i:

$$v_i(\vec{x},t) = \overline{v}_i(\vec{x},t) + v'_i(\vec{x},t) . \tag{4.20}$$

The procedure is similar to the one used in LES models, but in contrast to LES the RANS method represents an ensemble average over the entire range of turbulent eddies rather than local averages that still allow a prediction of large scale fluctuations. Hence, with the RANS approach it is not possible to predict specific eddies but only the mean effect of the entire turbulence spectrum. An additional consequence is that the RANS approach cannot be utilized in order to estimate cycle-by-cycle variations of the flow within a combustion chamber.

When the Reynolds approach, Eq. 4.20, is substituted into the conservation equations for mass, momentum and energy specified in Sect. 4.1, the following relations are obtained for the case of incompressible flow in the absence of external forces:

$$\frac{\partial \overline{v}_i}{\partial x_i} = 0 , \tag{4.21}$$

$$\rho \left[\frac{\partial \overline{v}_j}{\partial t} + \overline{v}_i \frac{\partial \overline{v}_j}{\partial x_i} \right] = -\frac{\partial \overline{p}}{\partial x_j} + \frac{\partial}{\partial x_i} \left(\overline{\tau}_{ij} - \rho \overline{v'_i v'_j} \right), \quad j = 1, 2, 3 \tag{4.22}$$

$$\rho c_p \left[\frac{\partial \overline{T}}{\partial t} + \overline{v}_i \frac{\partial \overline{T}}{\partial x_i} \right] = -\frac{\partial}{\partial x_i} \left(\overline{q}_i + \rho c_p \overline{v'_i T'} \right) . \tag{4.23}$$

$\overline{\tau}_{ij}$ and $\overline{q}_i$ are calculated according to Eqs. 4.12 and 4.17, respectively, however in terms of the averaged quantities instead of absolute quantities. Equations 4.21 to 4.23 have a similar form to the original set of conservation equations, Eqs. 4.8, 4.13 and 4.19, with the exception that two additional terms have been added due to the averaging process. These are the turbulent shear, or Reynolds stress tensor,

$$-\rho \overline{v'_i v'_j} \equiv \overline{\tau}_{ij,t}, \tag{4.24}$$

and the turbulent flux of a conserved scalar (in this case, the turbulent heat flux):

$$\rho c_p \overline{v'_i T'} \equiv \overline{\tilde{q}}_{i,t}. \tag{4.25}$$

The terms can be interpreted in a way, that the presence of turbulent eddies enhances the viscous stresses as well as the diffusion of a conserved scalar, i.e. heat conduction in the case of the energy equation, compared to a laminar flow with absolute properties equal to the ensemble averaged properties of the turbulent flow.

However, both the turbulent shear and the turbulent scalar flux cannot be evaluated based on first principles, which is commonly referred to as the closure problem, see e.g. ref [3]. Instead, some approximations are needed in order to model those quantities, usually in terms of mean quantities of the flow field. Most often the Boussinesq approximation (1877) is applied which is also referred to as eddy-viscosity model or, in case of a conserved scalar, eddy-diffusion model. It assumes that the Reynolds stress tensor $\overline{\tau}_{ij,t}$ may be modeled in analogy to the viscous stress tensor τ_{ij} in Eq. 4.12,

$$\overline{\tau}_{ij,t} = \mu_t \left(\frac{\partial \overline{v}_i}{\partial x_j} + \frac{\partial \overline{v}_j}{\partial x_i} \right) - \delta_{ij} \frac{2}{3} \rho k, \tag{4.26}$$

where μ_t is referred to as the turbulent viscosity and k is the turbulent kinetic energy which is further discussed in Sect. 4.3. It should be noted that while μ_t has the same dimensions as the molecular viscosity μ, it is not a fluid property but rather a property of the turbulence.

Accordingly, the turbulent scalar flux is modeled in analogy to Fourier's law, Eq. 4.17. For the energy equation the turbulent heat flux becomes

$$\overline{\tilde{q}}_{i,t} = -k_t \frac{\partial \overline{T}}{\partial x_i}, \tag{4.27}$$

where k_t is the turbulent conductivity which is, like the turbulent viscosity, not a fluid property but a turbulence property. In analogy to laminar flows, the ratio of turbulent viscosity to turbulent thermal diffusivity is defined as the turbulent Prandtl number

$$\Pr_t = \frac{\nu_t}{\alpha_t} = \frac{\mu_t c_p}{k_t}, \tag{4.28}$$

which is often assumed to be close to unity in undisturbed turbulent flows. Thus, it now becomes a task of approximating the turbulent viscosity of the flow field by so-called turbulence models. The turbulent conductivity can then be estimated based on Eq. 4.28.

4.3 Turbulence Models

It should be noted, that with the RANS approach the entire unsteady behavior of the turbulent flow field is included in the turbulent Reynolds stress and in the turbulent diffusivities that are determined as time-averaged quantities themselves. The complexity of turbulent flows makes it improbable, to say the least, that a single turbulence model is capable of predicting the non-linear behavior of the Navier-Stokes equations for all possible flow configurations and geometries. Hence, the turbulence models should be seen as approximations of the turbulent flow rather than general laws.

As noted above, the turbulent viscosity μ_t has the same dimension as the molecular viscosity μ and can therefore be expressed by the general formulation

$$\mu_t = C_\mu \cdot \rho \cdot l \cdot q, \tag{4.29}$$

where C_μ is a dimensionless constant and l and q are characteristic length scales and velocities, respectively, that need to be modeled appropriately. Several different modeling methods with different degrees of complexity have been suggested. Typically, they are distinguished by the number of partial differential equations that have to be solved. Algebraic models are referred to as zero-equation models, whereas one- and two-equation models require the solution of one and two partial differential equations for turbulent exchange properties, respectively.

Mixing-Length Model

One of the simplest and most ostensive turbulence models is the so-called mixing-length model which has been suggested by Prandtl in 1920. It is based on the idea that a turbulent eddy travels a mixing length l until it has completely mixed with its surroundings and lost its identity. In a purely algebraic approach, the velocity is assumed to be $q = l \cdot \partial v / \partial y$, and the turbulent viscosity becomes

$$\mu_t = \rho l^2 \left| \frac{\partial \overline{v}}{\partial y} \right|. \tag{4.30}$$

The mixing length l depends predominantly on the distance y from a wall, and several relations have been suggested for positions inside and outside the boundary layer [14]. However, while satisfying results may be achieved for simple flows, accurate prescription of l is hardly possible for highly three-dimensional flows. Thus, more detailed one- and two-equation turbulence models have been developed that rely on solving partial differential conservation equations for turbulence quantities in order to determine the turbulent viscosity.

k-ε Model

A widely applied turbulence model is the two-equation k-ε model that is based on partial differential equations for the turbulent kinetic energy k and its dissipation rate ε. In this model the characteristic velocity q is taken as proportional to the square root of the turbulent kinetic energy k, and Eq. 4.29 becomes

$$\mu_t = C_\mu \rho l \sqrt{k}. \tag{4.31}$$

k is defined as the kinetic energy of the turbulent fluctuations

$$k = \frac{1}{2}\overline{v'_i v'_i} = \frac{1}{2}\left(\overline{v'^2_1} + \overline{v'^2_2} + \overline{v'^2_3}\right), \tag{4.32}$$

and its conservation equation can be derived in analogy to the energy conservation equation in Sect. 4.1 [14]. After several mathematical rearrangements the form

$$\frac{\partial k}{\partial t} + \overline{v}_i \frac{\partial k}{\partial x_i} = \frac{\partial}{\partial x_i}\left(\frac{v_t}{C_k}\frac{\partial k}{\partial x_i}\right) + v_t \frac{\partial \overline{v}_j}{\partial x_i}\left(\frac{\partial \overline{v}_j}{\partial x_i} + \frac{\partial \overline{v}_i}{\partial x_j}\right) - \varepsilon \tag{4.33}$$

is obtained, where $v_t = \mu_t/\rho$. An additional equation is necessary to determine the length scale l of turbulence in Eq. 4.31. The choice is not obvious and a number of equations have been used for this purpose. The most popular one is based on the observation that the dissipation of turbulent kinetic energy is needed in the energy equation. In so-called equilibrium turbulent flows, that are characterized by near-balance production and dissipation rates of turbulent kinetic energy, the dissipation rate ε may be related to k and l by

$$\varepsilon = \frac{k^{3/2}}{l}. \tag{4.34}$$

This is based on the concept that there is an energy cascade from the largest scales to the smallest ones, and that the energy transferred to the smallest scale is dissipated.

The derivation of the conservation equation for the dissipation of turbulent kinetic energy from the Navier-Stokes equations is not straightforward, however it is possible and it can be shown that the following form results:

$$\frac{\partial \varepsilon}{\partial t} + \overline{v}_i \frac{\partial \varepsilon}{\partial x_i} = \frac{\partial}{\partial x_i}\left(\frac{v_t}{C_\varepsilon}\frac{\partial \varepsilon}{\partial x_i}\right) + C_1 \frac{\varepsilon}{k} v_t \frac{\partial \overline{v}_j}{\partial x_i}\left(\frac{\partial \overline{v}_j}{\partial x_i} + \frac{\partial \overline{v}_i}{\partial x_j}\right) - C_2 \frac{\varepsilon^2}{k}. \tag{4.35}$$

Combining Eqs. 4.31 and 4.34 yields the relation for the turbulent viscosity:

$$\mu_t = C_\mu \rho \frac{k^2}{\varepsilon}. \tag{4.36}$$

The five empirical constants C_i in the above equations are not universal though, and have to be adjusted to a specific problem. However, most often the following values are recommended:

$C_\mu = 0.09$, $\qquad C_1 = 1.44$, $\qquad C_2 = 1.92$, $\qquad C_k = 1.0$, $\qquad C_\varepsilon = 1.3$

Outlook on Further Turbulence Models

As pointed out earlier, the two-equation k-ε model is not general in nature but needs to be adjusted to a specific flow problem. While this can be done with reasonable success in two-dimensional flows, difficulties typically occur in flows that are three-dimensional and of transient nature. In order to overcome these deficiencies, a variety of other, more detailed turbulence models has been suggested.

Improvements of the linear k-ε model, based on the theory of renormalization groups (RNG), have been achieved by Yakhot and co-workers [16, 17] and have been successfully introduced to spray combustion modeling by Han and Reitz [4]. One of the features of the new model is that all constants of the standard k-ε model can be evaluated by the theory explicitly based on certain assumptions and mathematical development. Moreover, an additional term is included in the conservation equation for the dissipation rate ε that changes dynamically with the rate of strain of the turbulence. This results in more accurate predictions of flows with rapid distortion and anisotropic large scale eddies. An assessment of the model performance was made by simulating a direct injection diesel engine and it was found that considerable improvements compared to the standard k-ε model are obtained in spray and combustion calculations due to the higher strain rates associated with spray-generated turbulence [4]. For compression/expansion calculations and for high Reynolds number flows in general, the additional term in the ε-equation of the RNG model does not make significant contributions to the predictions.

Tanner et al. [13] included non-equilibrium turbulence considerations based on rapid distortion theory into the RNG k-ε model in order to better describe the transient behavior of turbulent flows in engine combustion chambers. The approach is motivated by the idea that the different turbulence scales within the flow respond with different time scales to changes in the macroscopic flow field and that the equilibrium assumption is not justified. The non-equilibrium turbulence model was tested by simulating compression and diffusion combustion in two direct injection diesel engines and it was shown that pressure traces and heat release rates could be calculated with less parameter adjustment than necessary with the equilibrium RNG k-ε model.

A number of so-called Reynolds stress models have been proposed based on a work by Launder et al. [6] in order to improve the predictive quality in three-dimensional flow problems. In anisotropic turbulent flows the eddy viscosity may no longer be treated as a scalar but in fact becomes a tensor quantity. Thus, each component of the Reynolds stress tensor $\overline{\tau}_{ij,t}$ is modeled by a separate algebraic or differential equation. This means that, because the tensor is symmetric, an additional six partial differential conservation equations have to be solved in the latter case. Consequently, the computational expenditure is increasing significantly compared to the k-ε model. It is obvious though, that the Reynolds stress models have a greater potential to predict three-dimensional, anisotropic turbulent flows as they may occur especially near walls. However, this potential has not been fully utilized yet and further research is still in progress. Application of Reynolds stress models to practical engine configurations, particularly to combustion modeling in engine flows, has been rare so far.

It should be noted, that there is not a single turbulence model known, that is generally applicable to all possible flow configurations and produces equally satisfying results for all problems. Instead, the choice of an appropriate turbulence model still very much depends on the specific nature of the investigated flow problem. For a more detailed review of turbulence models the reader is referred to Libby [8] or the referenced literature.

4.4 Boundary Layers and Convective Heat Transfer

An important topic in the modeling of engine flows is the convective heat transfer between cylinder gases and combustion chamber walls, as it directly affects thermal loads on essential engine components, engine efficiency as well as pollutant emissions. The convective heat transfer strongly depends on the flow velocity and turbulence level of the fluid phase and on the conditions within the velocity and thermal boundary layers that develop at the surface of a solid wall.

However, in high Reynolds number flows as they are encountered in in-cylinder flows of combustion engines the boundary layers are typically extremely thin. Therefore, a direct numerical simulation (DNS) that would require multiple grid cells to resolve the boundary layer is not feasible. Instead, the velocity and thermal boundary layers are often approximated by so-called wall functions [5]. They can be viewed as semi-empirical models that may be classified in between the DNS and the fully empirical models, that are based on Eq. 2.22, Fig. 4.4.

accuracy, complexity, comp. expenditure →

empirical models e.g. Woschni	semi-empirical models e.g. wall functions	direct numerical simulation (DNS)

← modeling degree

Fig. 4.4. Classification of wall heat transfer models

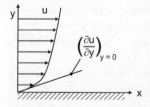

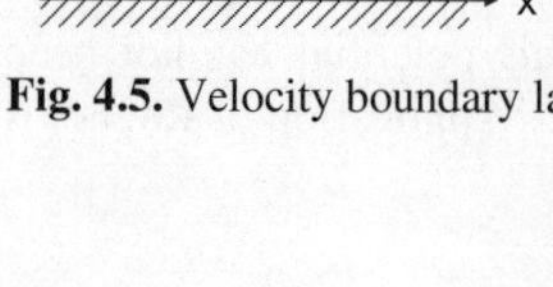

Fig. 4.5. Velocity boundary layer for turbulent flow above a solid wall

Velocity Boundary Layer

Assuming a two-dimensional steady turbulent flow of an incompressible fluid above a solid wall with a negligible pressure gradient parallel to the wall ($\partial p/\partial x=0$), the Reynolds-averaged mass and momentum equations within the boundary layer can be combined to

$$\frac{\partial}{\partial y}\left(\mu\frac{\partial \overline{u}}{\partial y} - \rho\overline{u'v'}\right) = 0, \qquad (4.37)$$

where u and v are the velocity components parallel and normal to the wall, respectively, Fig. 4.5 [2]. Integration yields

$$\mu\frac{d\overline{u}}{dy} - \rho\overline{u'v'} = \text{const} = \tau_w, \qquad (4.38)$$

where the integration constant becomes the wall shear stress τ_w. This is because the Reynolds stress $\rho\overline{u'v'}$ becomes zero at the wall ($y = 0$) and the effective shear stress, i.e. the sum of viscous and turbulent (Reynolds) stresses, is constant within the boundary layer:

$$\frac{\tau_w}{\rho} = \nu\frac{\partial \overline{u}}{\partial y} - \overline{u'v'}. \qquad (4.39)$$

By utilizing Eqs. 4.24 and 4.26 the above relation can also be written as

$$\frac{\tau_w}{\rho} = (\nu + \nu_t)\frac{\partial \overline{u}}{\partial y}. \qquad (4.40)$$

The left hand side of Eqs. 4.39 and 4.40 has the dimension of a velocity squared, and it is commonly referred to as the *shear velocity* u_τ:

$$u_\tau = \sqrt{\tau_w/\rho}. \qquad (4.41)$$

This quantity is utilized in order to bring the velocity and the distance from the wall to a dimensionless form,

$$u^+ = \frac{\overline{u}}{u_\tau}, \quad y^+ = \frac{u_\tau y}{\nu}. \qquad (4.42)$$

Equation 4.40 now becomes

$$\frac{\partial u^+}{\partial y^+} = \frac{1}{1+\nu_t/\nu}, \qquad (4.43)$$

and in order to integrate this relation the boundary layer is divided into two sublayers: a viscous sublayer directly at the wall and a fully turbulent layer on top of the viscous sublayer, Fig. 4.6. It is now assumed that within the viscous sublayer the molecular momentum transport is much greater than the turbulent one, i.e. ($\nu \gg \nu_t$), while in the fully turbulent sublayer the turbulent momentum transport is predominant, i.e. ($\nu \ll \nu_t$). Obviously, there must also be a transition regime with ($\nu \approx \nu_t$) between the two layers. However, this regime is usually neglected since it turns out that the two-layer model represents a very good approximation already.

For the viscous sublayer the right hand side of Eq. 4.43 reduces to unity and integration yields

$$\text{Viscous sublayer:} \qquad u^+ = y^+, \qquad (4.44)$$

i.e. the velocity increases linearly with the distance from the wall. In the fully turbulent region the right hand side of Eq. 4.43 reduces to ν/ν_t and utilizing Prandtl's mixing length model, Eq. 4.30, it can be shown that the logarithmic wall function,

$$\text{Fully turbulent layer:} \qquad u^+ = \frac{1}{\kappa} \ln y^+ + C, \qquad (4.45)$$

is obtained by integration [9]. κ is the so-called Karman constant ($\kappa = 0.4$) and C is an empirical constant related to the thickness of the viscous sublayer. It is typically set to $C = 5.5$. Figure 4.7 displays the velocity profile versus the wall distance on a logarithmic scale. It can be seen that the viscous layer is very thin ($y^+ \leq 5$) and that the fully turbulent region is reached for dimensionless wall distances greater than about 20. Beyond a dimensionless wall distance of about 200 the logarithmic wall function cannot be applied either.

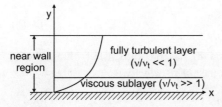

Fig. 4.6. Two-layer model for turbulent boundary layers

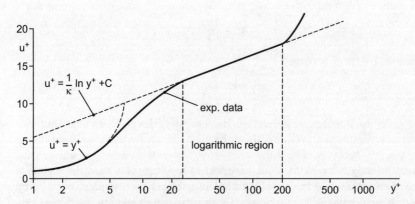

Fig. 4.7. Velocity profile in a turbulent boundary layer

With Eqs. 4.44 and 4.45 the dimensionless velocity is uniquely coupled to the dimensionless wall distance. However, in order to solve Eq. 4.42 for the absolute flow velocity $\bar{u}$, which is the unknown of primary interest, the shear velocity u_τ needs to be modeled. For this purpose it is often assumed that the flow is in local equilibrium, meaning the production and dissipation of turbulent kinetic energy are approximately equal. It this case it can be shown that u_τ depends on the turbulent kinetic energy k [9]:

$$u_\tau = C_\mu^{1/4} \sqrt{k} \ . \tag{4.46}$$

This formulation however, is not without controversy because it is based on the assumption of isotropic turbulence which is obviously critical in the immediate vicinity of a solid wall.

Thermal Boundary Layer

In general the thermal boundary layer behaves very similar to the velocity boundary layer, and consequently the heat flux is commonly expressed in analogy to Eq. 4.40 in terms of a molecular and a turbulent term,

$$\frac{\dot{q}_w}{\rho c_p} = (\alpha + \alpha_t) \frac{\partial \bar{T}}{\partial y} , \tag{4.47}$$

where α and α_t are the molecular and turbulent thermal diffusivities, respectively.

Again, the temperature can be brought into a dimensionless form as

$$T^+ = \frac{(\bar{T} - T_w) u_\tau}{\dot{q}_w / (\rho c_p)} , \tag{4.48}$$

and Eq. 4.47 becomes

$$\frac{\partial T^+}{\partial y^+} = \frac{1}{\dfrac{1}{\Pr} + \dfrac{\alpha_t}{\nu}} . \tag{4.49}$$

Applying the same boundary layer model with a viscous and a fully turbulent sub layer as displayed in Fig. 4.6, and neglecting the terms (α_t/ν) and (1/Pr) in the viscous and fully turbulent sublayers, respectively, integration yields

Viscous sublayer: $\quad\quad T^+ = \Pr y^+ , \tag{4.50}$

Fully turbulent layer: $\quad\quad T^+ = \dfrac{\Pr_t}{\kappa} \ln y^+ + C(\Pr) . \tag{4.51}$

Here $\Pr_t$ is the turbulent Prandtl number as defined in Eq. 4.28, and $C(\Pr)$ is

$$C(\Pr) = \Pr y_{visc}^+ - \frac{\Pr_t}{\kappa} \ln y_{visc}^+ , \tag{4.52}$$

where y^+_{visc} is the thickness of the thermal viscous sublayer which is approximately equal to 13.2. It should be noted that the thickness of the viscous sublayer may be different for the temperature profile than it is for the velocity profile because α_t may differ from ν_t, i.e. $Pr_t \neq 1$.

In CFD-codes Eq. 4.48 can now be solved for the unknown wall heat flux $\dot{q}_w$ if both the absolute and dimensionless temperatures $\overline{T}$ and T^+ are known at a certain distance from the wall. A requirement is however, that the gas temperature $\overline{T}$ is known at a distance from the wall that lies within the range of validity of the logarithmic wall function, i.e. the dimensionless wall distance y^+ needs to be less than approx. 200 as indicated in Fig. 4.7. This is a common source for uncertainties and grid dependencies in CFD-computations of combustion engines, since typical grid dimensions are often much greater than the boundary layer thickness. Therefore, the y^+ value corresponding to the first grid cell from the wall already exceeds the range of applicability of the logarithmic wall function.

4.5 Application to In-Cylinder Processes

The conservation equations for mass, momentum and energy, Eqs. 4.5, 4.9 and 4.19 are valid for single-component, single-phase flows. However, in engine combustion chambers there are chemical reactions taking place between multiple species. Furthermore, in direct injection engines two-phase flows with evaporating fuel droplets are encountered. As a consequence additional so-called transport- or source terms have to be included in the conservation equations that account for the energy release and change in species concentrations due to combustion as well as for the interactions between the liquid and gaseous phases. Moreover, it is no longer sufficient to solve only one equation for total mass conservation but separate conservation equations have to be applied to each species considered in the system. In general, any conserved scalar, e.g. a species concentration in a multi-component system, can be treated by a separate conservation equation in analogy to the energy equation in the foregoing sections. The turbulent conductivity in the energy equation of the RANS approach is then exchanged by a turbulent diffusivity of the respective scalar. Detailed formulations of the complete conservation equations for reactive multi-component two-phase flows have been presented by a number of authors, e.g. in refs. [1] and [12].

Whereas single-component single-phase flows can be modeled reasonably well with available CFD-codes, one of the challenges in today's modeling of combustion engines is to describe the influences of two-phase flow and combustion phenomena as accurately as possible. Specifically, source terms have to be included in the conservation equations for the various species concentrations in order to account for concentration changes due to spray evaporation and chemical reactions. The energy equation needs to be extended by terms describing heat transfer between the liquid and gaseous phases as well as heat released by chemical reactions. Finally, there may be momentum transfer between the two phases that has to be included in the Navier-Stokes equations for momentum conservation.

Taking all these phenomena into account, the set of conservation equations for the gas phase that needs to be solved in CFD-codes describing spray and combustion processes in engine combustion chambers can be summarized as follows. For every chemical species m in the system, mass conservation can be expressed by its species density ρ_m as

$$\frac{\partial \rho_m}{\partial t} + \frac{\partial (\rho_m v_i)}{\partial x_i} = \frac{\partial}{\partial x_i}\left(\rho D \frac{\partial (\rho_m/\rho)}{\partial x_i}\right) + \dot{\rho}_m^s + \dot{\rho}_m^c. \tag{4.53}$$

The first term on the right hand side of Eq. 4.53 is due to mass diffusion, whereas the second and third terms are the source terms due to spray effects and combustion, respectively.

In the presence of a liquid spray there can be momentum transfer between the two phases, such that the three gas phase momentum conservation equations (one for each spatial dimension j) become

$$\frac{\partial (\rho v_j)}{\partial t} + v_j \frac{\partial (\rho v_j)}{\partial x_i} = -\frac{\partial p}{\partial x_j} + \frac{\partial \tau_{ij}}{\partial x_i} + \rho F_j^s + \rho g_j. \quad j = 1, 2, 3 \tag{4.54}$$

Obviously, gravity acts only in the vertical direction such that ($g_1 = g_2 = 0$), and ρF_j^s indicates the momentum gain of the gas phase due to the spray.

Finally, the energy equation becomes

$$\rho c_p \left(\frac{\partial T}{\partial t} + v_i \frac{\partial T}{\partial x_i}\right) = k \frac{\partial^2 T}{\partial x_i^2} + \frac{\partial}{\partial x_i}\left(\rho D \sum_m h_m \frac{\partial (\rho_m/\rho)}{\partial x_i}\right) + \rho \varepsilon + \dot{Q}^s + \dot{Q}^c, \tag{4.55}$$

where the second term on the right hand side describes the enthalpy transfer associated with mass diffusion of species m in a multi-component system. The third term accounts for dissipation of turbulent kinetic energy and the fourth and fifth terms are the source terms due to spray and combustion.

A wide variety of models describing the transport processes caused by droplet-gas interactions and by combustion has been proposed in the literature. The models are partly based on different visions on physical and chemical subprocesses as well as on different numerical implementations. Chapters 5 and 6 will concentrate on discussing such submodels that are needed in order to determine the respective source terms.

References

[1] Amsden AA, O'Rourke PJ, Butler TD (1989) KIVA-II: A Computer Program for Chemically Reactive Flows with Sprays. Los Alamos National Laboratory, LA-11560-MS
[2] Baehr HD, Stephan K (1998) Heat and Mass Transfer. Springer, Berlin, Germany

[3] Ferziger JH, Peric M (2002) Computational Methods for Fluid Dynamics. 3rd edn, Springer, Berlin, Germany
[4] Han ZY, Reitz RD (1995) Turbulence Modeling of Internal Combustion Engines Using RNG k-ε Models. Combust Sci Tech, vol 106, pp 267–295
[5] Launder BE, Spalding DB (1974) The Numerical Computations of Turbulent Flows. Comp Meth Appl Mech Engng, vol 3, pp 269–289
[6] Launder BE, Reece G, Rodi W (1975) Progress in the Development of a Reynolds Stress Turbulence Closure. J Fluid Mech, vol 68
[7] Lee J, Frouzakis C, Boulouchos K (2000) Opposed-Jet Hydrogen/Air Flames: Transition from a Diffusion to an Edge Flame. Proc Combust Inst, vol 28, Edinburgh, UK
[8] Libby PA (1996) Introduction to Turbulence. Taylor and Francis, Washington, DC
[9] Merker GP (1987) Konvektive Wärmeübertragung. Springer, Berlin, Germany
[10] Piomelli U, Ferziger JH, Moin P, Kim J (1989) New Approximate Boundary Conditions for Large Eddy Simulations of Wall-Bounded Flows. Phys Fluids, A1, pp 1061–1068
[11] Rao S, Pomraning E, Rutland CJ (2001) A PDF Time-Scale Model for Diesel Combustion Using Large Eddy Simulation. Proc 11th Int Multidim Engine Modeling Users Group Meeting, Detroit, MI
[12] Reitz RD (1994) Computer Modeling of Sprays. Spray Technology Short Course, Pittsburgh, PA
[13] Tanner FX, Zhu GS, Reitz, RD (2000) Non-Equilibrium Turbulence Considerations for Combustion Processes in the Simulation of DI Diesel Engines. SAE Paper 2000-01-0586
[14] White FM (1991) Viscous Fluid Flow. 2nd edn, McGraw-Hill, New York, NY
[15] White FM (1999) Fluid Mechanics. 4th edn, McGraw-Hill, New York, NY
[16] Yakhot V, Orszag SA (1986) Renormalization Group Analysis of Turbulence. I. Basic Theory. J Sci Comp, vol 1, pp 3–51
[17] Yakhot V, Smith LM (1992) The Renormalization Group, the ε-Expansion and Derivation of Turbulence Models. J Sci Comp, vol 7, pp 35–61

5 Multidimensional Models of Spray Processes

5.1 General Considerations

5.1.1 Spray Processes in Combustion Engines

Spray processes play an important role in many technical systems and industrial applications. Examples are spray cooling, spray painting, crop spraying, humidification and spray combustion in furnaces, gas turbines, rockets, as well as diesel and gasoline engines, to name only a few. Typical drop sizes in sprays vary over several orders of magnitude for different applications. Figure 5.1 gives a qualitative classification of broad spray classes.

In internal combustion engines sprays are utilized in order to mix the liquid fuel with air and increase its surface area for rapid evaporation and combustion. For example, on a first order approximation the evaporation rate is proportional to the overall surface area of the liquid fuel, and thus, disintegrating a 2 mm drop into about eight million droplets of 10 µm increases the evaporation rate by a factor of 200. Moreover, in direct injection engines (both diesel and spark ignition), where the fuel is injected directly into the combustion chamber in order to form an ignitable mixture with air, the spray is one of the most effective measures to control the combustion process. The kinetic energy of the spray represents the main source for turbulence production within the combustion chamber, and therefore governs the microscale air-fuel mixing by turbulent diffusion as well as the flame speed of a premixed flame front. The spray significantly affects the ignition behavior, heat release and pollutant formation rates and thus the noise level, fuel consumption and exhaust emissions of an engine. Therefore, a thorough understanding of spray

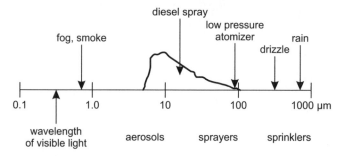

Fig. 5.1. Typical drop sizes for various classes of sprays. A representative size distribution is depicted for diesel sprays [79]

processes is vital for the design of modern combustion engines that are characterized by more and more flexible injection rate shapes.

However, at the same time spray phenomena are very complex because the liquid fuel droplets interact in multiple ways with the turbulent gas phase and with the flame itself. Moreover, engine combustion chambers wherein the sprays perform represent a hostile environment that is hard to access with appropriate measuring techniques without altering the geometric and thermal boundary conditions experienced in the production engine. This makes it difficult to assess the quality of spray models. Nevertheless – or maybe for this specific reason – numerical simulations represent a powerful tool that can provide valuable insight into spray processes and the complex interacting subprocesses involved. They allow access to any process or state variable at any position at any given point in time. Thus, if validated for a certain range of boundary conditions, spray models can effectively be utilized to interpret available experimental data, and also to execute precalculations for altered operating conditions.

The depth of analysis possible with spray models has increased significantly over the past decades, in part, due to dramatic advances in computer technology. This trend is likely to continue such that spray modeling with CFD codes will be routinely used in the development of new engines and combustion systems in the near future.

5.1.2 Spray Regimes

A typical two-phase flow originated from a pressure atomizer can be divided into different regimes as depicted in Fig. 5.2 [78]. Directly at the nozzle orifice an *intact core* of the liquid phase can be identified. It rapidly disintegrates into ligaments (*churning flow*) and further into droplets, but it still occupies a considerable fraction of the volume. Due to its density, which is significantly greater than the density of the gas phase, the contribution of the liquid phase to the total mass is even greater. This spray region is generally referred to as *thick* or *dense* spray.

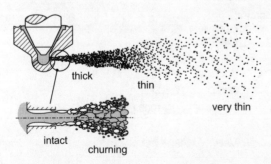

Fig. 5.2. Schematic illustration of different flow regimes

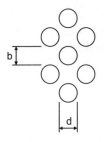

Fig. 5.3. Droplet arrangement with droplet spacing equal to droplet diameter ($b = d$)

Because of the conical spray shape and because of droplet evaporation, the average spacing between droplets expands further downstream of the nozzle, and the void fraction, i.e. the volume fraction occupied by the gas phase, increases and approaches unity. However, due to the liquid to gas density ratio, the mass fraction of the liquid phase may still be noticeable. This intermediate region of the spray is called *thin* spray. The *very thin* or *dilute* spray regime is finally characterized by both volume and mass fractions of the liquid phase that are negligible compared to the ones of the gas phase.

The behavior of various droplets within different regimes of a spray is quite different. Obviously, droplet-droplet interactions such as collision and coalescence can be significant close to the nozzle orifice. Moreover, when the droplet spacing is small the boundary layer around a droplet may be affected by an adjacent droplet. Consequently, it can no longer be assumed that there is an undisturbed gas phase around the droplet in order to calculate the exchange processes between liquid and gas. At the other extreme, in the dilute spray regime the droplet behavior can be calculated based on relations for an isolated droplet with good accuracy. There is still some mass, momentum and energy transfer between the droplets and the gas phase, but the influence that the droplets have on the gas phase is very small. Collisions between droplets are rare and typically neglected in the modeling.

In the intermediate thin spray regime the liquid phase still accounts for a noticeable mass fraction as noted above. Thus, there is considerable momentum transfer from the droplets to the gas phase, which in turn affects other droplets again. An example are the reduced drag forces on those droplets located in the wake of the spray tip that are decelerated by the gas less rapidly and may therefore overtake the droplets at the former spray tip that have been injected at an earlier timing.

While the qualitative difference in the behavior of the two-phase flow is distinct between the various spray regimes, the transitions between the regimes are continuous and their definitions are somewhat arbitrary. Typically, they are defined in terms of the void fraction defined as

$$\theta = 1 - \iiint f \frac{4}{3}\pi r^3 \, dr \, d\bar{v} \, dT_d \,, \tag{5.1}$$

where f is the probable number of droplets per unit volume in the spray. As will be further discussed in Sect. 5.2 f depends on time t, droplet radius r, droplet temperature T_d, the droplet position $\vec{x}$ (three coordinates) as well as the three droplet velocity components $\vec{v}$.

For a regular arrangement of spherical droplets with a spacing equal to their diameter as it is displayed in Fig. 5.3, it can be shown that the void fraction becomes $\theta \approx 0.92$ [60]. Consequently, it is often assumed that a spray behaves as a thick spray if the void fraction is less than about 0.9. O'Rourke [60] also considered an additional spray regime located between the intact core and the thick spray, termed *churning flow*, for void fractions less than 0.5. The reasoning is that for liquid volume fractions in excess of 50 percent, the liquid can no longer be assumed to be fully dispersed within a continuous gas phase and a different set of equations becomes necessary to describe the problem. However, there are still considerable uncertainties in the governing equations, such that this spray regime, which only applies to an extremely small volume close to the nozzle orifice, is often neglected in practical applications.

Most available CFD-codes utilized in combustion engine simulations are based on thin or dilute spray assumptions. These assumptions are justified when the computation begins someway downstream of the injection nozzle where the spray has already been diluted by the gas phase. But even if the injector is located within the computational domain, a thin spray may still be assumed when the computational grid cell is large compared to the size of the nozzle hole. In typical combustion engine applications, a grid cell as a dimension of about 1 mm whereas a modern passenger car diesel injector has a hole diameter of only about 0.15 mm. Approximations of thick spray effects can later be added by superimposing submodels to the conservation and exchange equations, see e.g. Sect. 5.6.

5.2 The Spray Equation

5.2.1 Equations and Exchange Terms

In typical diesel sprays the liquid fuel is atomized into a number of up to 10^8 droplets with average diameters in the ten-micrometer range. These numbers make it prohibitive to resolve each single droplet in numerical simulations. Instead, some kind of statistical averaging technique becomes necessary with additional submodels in order to describe the subscale processes.

Generally the problem can be defined by the so-called spray equation as formulated by Williams [108]. In this approach the probable number of drops per unit volume at time t, that are located between position $\vec{x}$ and $\vec{x} + d\vec{x}$ and characterized by a velocity between $\vec{v}$ and $\vec{v} + d\vec{v}$, a radius between r and $r + dr$ and a temperature between T_d and $T_d + dT_d$ is described with the probability density function f. Since both the droplet position $\vec{x}$ and its velocity $\vec{v}$ have three spatial coordinates, f has a total of nine independent variables:

$$\frac{\text{probable number of droplets}}{\text{unit volume}} = f(\vec{x},\vec{v},r,T_d,t)\,d\vec{v}\,dr\,dT_d. \tag{5.2}$$

It should be noted, that the above formulation is based on the assumption that the droplets are ideally spherical and that their size or mass is thus explicitly defined by their radius r. However, this assumption is valid only if the relative velocity between gas and droplet is small, which is not generally the case in typical engine sprays. Especially, in the vicinity of the injection nozzle considerable relative velocities between liquid and gas phases are encountered, such that aerodynamic forces cause droplet distortion and even droplet breakup. Therefore, in most CFD-codes applied for engine simulations with spray combustion, e.g. in the KIVA code [4, 5, 6], two additional independent variables are included in the distribution function: the droplet distortion parameter y and its temporal rate of change $\dot{y}$. Thus, f becomes a function of eleven independent variables, and Eq. 5.2 now reads

$$\frac{\text{probable number of droplets}}{\text{unit volume}} = f(\vec{x},\vec{v},r,T_d,y,\dot{y},t)\,d\vec{v}\,dr\,dT_d\,dy\,d\dot{y}. \tag{5.3}$$

The temporal and spatial evolution of the distribution function is described by a conservation equation which can be derived phenomenologically in analogy to the conservation equations of the gas phase [108]. It is commonly referred to as the spray equation and can be written as:

$$\frac{\partial f}{\partial t} = -\frac{\partial}{\partial x_i}(fv_i) - \frac{\partial}{\partial v_i}(fF_i) - \frac{\partial}{\partial r}(fR) - \frac{\partial}{\partial T_d}(f\dot{T}_d) - \frac{\partial}{\partial y}(f\dot{y}) - \frac{\partial}{\partial \dot{y}}(f\ddot{y})$$
$$+ \dot{f}_{coll} + \dot{f}_{bu}. \tag{5.4}$$

Equation 5.4 follows the Einstein convention (see Chap. 4). $\vec{F}$ denotes a force per unit mass, i.e. an acceleration. Thus the component F_i is the acceleration along the spatial coordinate x_i ($F_i = dv_i/dt$). R, $\dot{T}_d$ and $\ddot{y}$ are the time rates of change of droplet radius r, temperature T_d and oscillation velocity $\dot{y}$. The source terms $\dot{f}_{coll}$ and $\dot{f}_{bu}$ account for changes in the distribution function due to droplet collision and breakup, respectively.

By solving the spray equation, the so-called source or exchange terms can be obtained, that describe the interactions between the liquid and gas phases. In order to assure conservation of mass, momentum and energy of the total (two-phase) system, these terms need to be included in the gas phase conservation equations summarized in Chap. 4. Following Reitz [78], the source term accounting for mass evaporation of the liquid droplets becomes

$$\dot{\rho}^s = -\int f\rho_l 4\pi r^2 R\,d\vec{v}\,dr\,dT_d\,dy\,d\dot{y}. \tag{5.5}$$

The rate of momentum gain due to droplet drag, body forces and evaporation is

$$\rho_g \vec{F}^s = -\int f \rho_l \left(\frac{4}{3} \pi r^3 \vec{F}' + 4\pi r^2 R \vec{v} \right) d\vec{v}\, dr\, dT_d\, dy\, d\dot{y}, \tag{5.6}$$

the energy transfer between gas and droplets by evaporation, heat transfer into the droplet and work due to turbulent fluctuations is

$$\dot{Q}^s = -\int f \rho_l \left\{ 4\pi r^2 R \left[u_d + \frac{1}{2}(\vec{v}-\vec{u})^2 \right] + \frac{4}{3}\pi r^3 \left[c_{p,d} \dot{T}_d + \vec{F}'(\vec{v}-\vec{u}-\vec{u}') \right] \right\}$$
$$\cdot d\vec{v}\, dr\, dT_d\, dy\, d\dot{y}, \tag{5.7}$$

and the destruction of turbulent kinetic energy due to droplet dispersion is obtained by

$$\dot{W}^s = -\int f \rho_l \frac{4}{3}\pi r^3 \vec{F}' \cdot \vec{u}'\, d\vec{v}\, dr\, dT_d\, dy\, d\dot{y}. \tag{5.8}$$

In Eqs. 5.5 to 5.8 the superscript s indicates that the source terms are due to spray effects (as opposed to effects of chemical reactions that will be denoted by superscript c). $\vec{F}'$ is the difference between $\vec{F}$ and the gravitational acceleration $\vec{g}$, $(\vec{v}-\vec{u})$ is the relative velocity between droplets and gas phase, and $\vec{u}'$ is the turbulent fluctuation of the gas velocity. Note, that in Eq. 5.7 u_d denotes the specific internal energy of the droplet and is not to be mistaken for the gas velocity $\vec{u}$.

5.2.2 Numerical Implementation

There are generally two possible ways of solving the spray equation in addition to the gas phase conservation equations introduced in Chap. 4. The obvious method would be to directly solve the spray equation, Eq. 5.4, with a Eulerian finite difference or finite volume scheme similar to the numerical solution of the gas phase. This method has been applied in ref. [36] and has been termed the continuum droplet model (CDM) in the literature [41]. However, the CDM requires to discretize the droplet probability function f in all eleven independent dimensions, and thus imposes extremely high demands with respect to computer memory and power. For example, discretizing the problem on a coarse mesh with only ten grid points in each dimension results in a total of 10^{11} grid points. For this reason the CDM has been proven impractical for most technical applications.

An alternative and more practical approach is the so-called discrete droplet model (DDM) proposed by Dukowicz [28] and used in the KIVA CFD-code [6] and, in similar forms in most other CFD-codes applied for engine spray and combustion simulations, e.g. FIRE, FLUENT, STAR-CD, VECTIS, etc.. It features a Monte-Carlo based solution technique for the spray equation, that describes the spray droplets by stochastic particles which are usually referred to as parcels [19].

These parcels can be viewed as representative classes of identical, non-interacting droplets, and they are tracked through physical space in a Lagrangian manner. The collection of such parcels within the computational domain represents a discretized solution of the spray distribution function, and thus, as the number of spray parcels in the spray are increased the spray statistics are improved. It should be noted that while the spray parcels are usually viewed as groups of identical droplets, they are – in a strict mathematical sense – really statistical items describing the probability that the spray reacts in a certain way. The ostensive interpretation of droplet groups is probably due to the fact that due to available memory and CPU-power there are typically significantly fewer parcels considered than there are droplets in the spray (typically 10^3 to 10^4 parcels compared to about 10^8 droplets). However, the number of parcels necessary in order to obtain statistical significance is really independent of the number of droplets within the spray.

Account must be taken of the coupling between the Lagrangian liquid and the Eulerian gaseous phases in the DDM. While the KIVA code utilizes the non-iterative Dukowicz-method [28] where the drop motion equations (discussed in Sect. 5.3) are solved as functions of time, several other commercially available spray codes are based on the so-called particle-source-in-cell (PSI Cell) technique proposed by Crowe et al. [23]. This technique begins by solving the gas flow field neglecting the presence of any spray particles in the flow. The obtained gas phase results are then used in order to calculate trajectories of the droplets as well as mass, momentum and energy exchanges between the two phases. Thereafter the gas phase is recalculated, now including the source terms caused by the spray particles, and the whole procedure is repeated in an iterative manner until a certain convergence criterion is met. Consequently, the PSI cell method is particularly suited for calculating steady-state spray processes, but it is not so well suited for modeling droplet dispersion in turbulent flows since these processes are inherently unsteady [78].

Besides its many obvious advantages it must be mentioned that a significant difficulty in modeling sprays with the stochastic particle method exists, in that there is typically a strong influence of the numerical grid design on the simulation results. Therefore a wide experience is often necessary in order to design a numerical mesh that is appropriate for a given spray problem. This important topic will be discussed in more detail in Sect. 5.8.

Finally, it should be noted that with both approaches of solving the spray equation, i.e. with the continuum droplet model as well as with the discrete droplet model, the interface between a single droplet and the gas phase cannot be directly resolved because of computer limitations. Instead, an averaging of the flow process over a scale that is greater than the typical droplet diameter becomes necessary. Consequently, additional submodels are needed to describe the phase interactions at the droplet-gas interface. These submodels will be discussed in subsequent sections.

5.3 Droplet Kinematics

In the Lagrangian formulation of the discrete droplet model the position of a drop or actually the position of a parcel containing a group of identical drops is characterized by the vector $\vec{x}_p$. The movement of the drop during one computational time step dt is derived from

$$\frac{d}{dt}\vec{x}_p = \vec{v}, \qquad (5.9)$$

where the change in the drop velocity vector is determined from

$$\frac{d}{dt}\vec{v} = \vec{F}. \qquad (5.10)$$

The force $\vec{F}$ acting on the droplet is composed of body forces and the drag force caused by the relative velocity of the droplet to the surrounding gas phase. The latter force depends on the drop size and its drag coefficient (see Sect. 5.3.1) as well as on the mean gas velocity and its turbulent fluctuations (see Sect. 5.3.2).

The change in drop size over time is given by

$$\frac{d}{dt}r_p = \dot{R}, \qquad (5.11)$$

where the quantity $\dot{R}$ depends on vaporization of the droplet (see Sect. 5.7) and on droplet breakup and collisions (Sects. 5.5 and 5.6, respectively). The latter effects can lead to the change in the number of droplets in a specific size class and even to the appearance or disappearance of droplet classes from the computation. Thus, the number of parcels considered within a computation may change over time which can be expressed as

$$\frac{df}{dt} = \dot{f}_{coll} + \dot{f}_{bu}. \qquad (5.12)$$

5.3.1 Drop Drag and Deformation

The drag force acting upon a particle surrounded by gas of density ρ_g und velocity $\vec{u}$ can generally be expressed as

$$\rho_l V_p \vec{F}_D = \frac{1}{2}\rho_g C_D A_p \cdot |\vec{u}-\vec{v}| \cdot (\vec{u}-\vec{v}), \qquad (5.13)$$

where A_p is the frontal area of the particle, i.e. ($A_p = \pi r_p^2$) for a spherical droplet. The drag coefficient C_D is a mostly empirically determined parameter that depends on the geometrical shape of the particle as well as on the flow conditions and gas properties. For low relative velocities (Re ≤ 1) around a spherical particle separation effects of the gas flow around the particle are negligible and the drag force is mainly due to friction drag by viscous stress. For this regime the drag coefficient is formulated by Stoke's law as

$$C_D = \frac{24}{\text{Re}}, \tag{5.14}$$

where the Reynolds number is defined as

$$\text{Re} = \frac{2 r_p \rho_g \cdot |\vec{u} - \vec{v}|}{\mu_g}. \tag{5.15}$$

For greater Reynolds numbers, and thus for greater relative velocities as they are typically encountered in engine sprays, the gas flow separates from the particle surface and form drag becomes increasingly more important than viscous drag. The drag coefficient of the sphere is amplified which is usually expressed by the following relations:

$$C_D = \frac{24}{\text{Re}} \left(1 + \frac{1}{6} \text{Re}^{2/3} \right) \qquad \text{Re} \leq 1000 \tag{5.16}$$

$$C_D = 0.424 \qquad \text{Re} > 1000 \tag{5.17}$$

In order to account for thick spray effects that may increase the effective drag coefficient at positions close to the nozzle orifice, O'Rourke and Bracco [62] suggested to replace Eq. 5.16 by a similar relation that additionally includes the local void fraction θ:

$$C_D = \frac{24}{\text{Re}} \left(\theta^{-2.65} + \frac{1}{6} \text{Re}^{2/3} \theta^{-1.78} \right) \qquad \text{Re} \leq 1000 \tag{5.18}$$

Equation 5.18 was obtained from experiments on fluidized beds and other sources [60].

The above correlations for the drag coefficient are valid for ideally spherical (solid) particles. However, in engine sprays the liquid droplets are typically distorted from their ideal spherical shape prior to breakup. This will obviously have an effect on the drag coefficient which has been accounted for by Liu et al. [49], who applied the TAB model in order to determine the drop distortion parameter y. The TAB (Taylor-Analogy Breakup) model which will be discussed in more detail in Sect. 5.5 assumes a one-dimensional oscillation of the droplet in analogy to a spring-mass system. In this analogy the liquid viscosity acts as a damping element and the surface tension has the effect of a restoring force. The distortion parameter y is normalized by the droplet radius r and defined in accordance to Fig. 5.4.

The drag coefficient of the distorted droplet is now given as

$$c_D = c_{D,sphere} (1 + 2.632 \cdot y), \tag{5.19}$$

which is based on the consideration that for high Reynolds numbers the drag coefficient of a disc is approx. 3.6 times greater than that of a sphere.

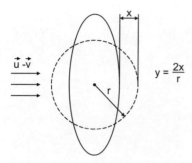

Fig. 5.4. Droplet distortion in the TAB model

5.3.2 Turbulent Dispersion / Diffusion

In turbulent sprays the liquid droplets are not only decelerated and deformed by the gas phase, but an additional dispersion or diffusion of the liquid phase can be observed that is caused by the turbulent eddies in the gas flow. On average, the random orientation of the turbulent velocity fluctuations leads to a quicker, more homogeneous dispersion of the liquid droplets than in a laminar gas flow. At the same time the momentum transfer between gas and liquid modulates the turbulence level within the gas phase.

This process has been reviewed in detail, e.g. by Faeth [31, 32]. The mechanism can be explained following the schematic diagram in Fig. 5.5. It shows a vortex structure (solid line), i.e. the track of an arbitrary gas molecule within a turbulent gas flow, as well as three possible droplet trajectories (dashed lines) that start out at the same position as the gas molecule. Typically, a particle is assumed to interact with an eddy for a time period taken as the smaller of either the eddy lifetime t_e or the transit time t_t required for the particle to pass through the eddy:

$$t_{int} = \min(t_e, t_t). \tag{5.20}$$

The characteristic eddy size is assumed to be the dissipation (integral) length scale

$$l_t = C_\mu^{3/4} \frac{k^{3/2}}{\varepsilon}, \tag{5.21}$$

where the constant C_μ is the same as in the k-ε turbulence model ($C_\mu = 0.09$, see Chap. 4). For isotropic turbulence the turbulent kinetic energy is related to the turbulence intensity by

$$k = \frac{1}{2}\left(u_x'^2 + u_y'^2 + u_z'^2\right) = \frac{3}{2}u'^2, \tag{5.22}$$

and thus, the eddy life time is expressed as

$$t_e = \frac{l}{u'} = \frac{C_\mu^{3/4}}{\sqrt{2/3}} \cdot \frac{k}{\varepsilon}. \tag{5.23}$$

The transit time can be estimated by linearizing the droplet momentum equations and is given by the expression

$$t_t = -\tau \ln\left[1 - \frac{l}{\tau|\vec{u}-\vec{v}|}\right]. \tag{5.24}$$

The particle relaxation time τ is defined by the particle acceleration,

$$\frac{d\vec{v}}{dt} \equiv \frac{\vec{u}-\vec{v}}{\tau}, \tag{5.25}$$

and can be derived from the particle's equation of motion, Eq. 5.13:

$$\rho_l V_p \frac{d\vec{v}}{dt} = \frac{1}{2}\rho_g C_D A_p (\vec{u}-\vec{v})^2. \tag{5.26}$$

Thus, the relaxation time becomes

$$\tau = \frac{8}{3}\frac{\rho_l r_p}{\rho_g C_D |\vec{u}-\vec{v}|}, \tag{5.27}$$

or, if Stoke's law is utilized for the drag coefficient, Eq. 5.14, we obtain

$$\tau = \frac{2}{9}\frac{\rho_l r_p^2}{\mu_g}. \tag{5.28}$$

When $l \geq \tau|\vec{u}-\vec{v}|$, Eq. 5.24 has no solution. This can be interpreted as the eddy having captured a particle so that the interaction time becomes equal to the eddy life time t_e.

Crowe et al. [24] have observed that small drops in a turbulent flow tend to follow the gas flow, whereas the larger drops with their smaller drag/inertia ratios leave the large-scale vortex structures. A time-scaling ratio was proposed which, assuming Stokes drag for the particles, gives the Stokes number:

$$\text{St} = \frac{\tau}{t_e}. \tag{5.29}$$

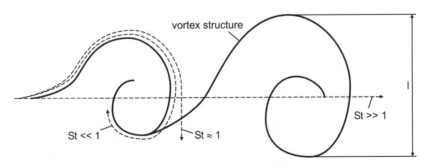

Fig. 5.5. Possible particle trajectories in a turbulent flow [78]

As depicted in Fig. 5.5, particles with Stokes numbers much less than unity, i.e. small particles, are trapped within the eddies and directly follow the vortex structure. Particles with very large Stokes numbers remain unaffected by the eddies, and for Stokes numbers in the order of unity a centrifuging effect may occur that leads to a particle dispersion which exceeds that of the gas phase.

The numerical implementation of the above phenomena in CFD-codes can be established by calculating the change in droplet motion for the interaction time period t_{int} as a function of the total gas velocity ($\vec{u} = \bar{\vec{u}} + \vec{u}'$) present at the beginning of the interaction. In order to yield a more realistic distribution of the turbulent dispersion effects, the fluctuating velocity component $\vec{u}'$ is typically sampled from a Gaussian distribution with a variance equal to the turbulence intensity $|\vec{u}'|$:

$$G(\vec{u}') = \frac{1}{\sqrt{2\pi} \cdot \sqrt{2k/3}} \cdot \exp\left(-\frac{|\vec{u}'|^2}{4k/3}\right). \tag{5.30}$$

Experiments by Modarress and co-workers [54, 55, 56] on turbulent round jets revealed that the spread rates of two-phase jets are smaller than those of single-phase jets. Moreover, it was shown that the turbulence level within two-phase jets depends on the liquid mass loading: it decreases for greater amounts of liquid mass dispersed in the turbulent jet. These results indicate that the droplet-turbulence interactions have a modulating influence on the gas phase turbulence. This is usually accounted for by adding the additional source term $\dot{W}^s$ in the k- and ε-conservation equations of the k-ε-turbulence model, Eqs. 4.33 and 4.35, respectively. $\dot{W}^s$ has been specified in Eq. 5.8, and for incompressible turbulence in the absence of gradients the conservation equations become

$$\rho \frac{dk}{dt} = \dot{W}^s, \tag{5.31}$$

$$\rho \frac{d\varepsilon}{dt} = C_s \frac{\varepsilon}{k} \dot{W}^s. \tag{5.32}$$

It can be shown that the turbulence length scale l_t given in Eq. 5.21 is unchanged by this turbulence modulation if the empirical constant in Eq. 5.32 is chosen as $C_s = 3/2$ [78].

5.4 Spray Atomization

In engine fuel injection systems the fuel typically leaves the injector nozzle in a more or less continuous liquid phase that can obviously not be reproduced with the Lagrangian discrete droplet approach. Therefore, additional submodels are necessary in order to describe the breakup processes that lead to the formation of droplets, before the DDM can be applied. This procedure seems reasonable since in high pressure injection systems the disintegration of the continuous liquid phase into small droplets starts very close to the nozzle orifice. Thus, the impact of the intact liquid core on the gas phase is extremely small compared to the influence that the dispersed liquid droplets have on the gas phase in the entire spray.

Two different types of liquid breakup into ligaments and droplets are typically distinguished, e.g. [98]. The first kind of breakup occurs at or in direct vicinity of the injection nozzle orifice, i.e. in the region that has been scaled up in the bottom part of Fig. 5.2. It is referred to as spray atomization or primary breakup and will be discussed in the present section. The primary breakup describes the breakup of the intact liquid phase into first ligaments and droplets. Later on, the relatively large initial droplets can be further distorted and subsequently broken up into smaller secondary droplets. This kind of breakup is termed secondary breakup and will be discussed in Sect. 5.5. Typically, the secondary breakup takes place a little further downstream of the nozzle, i.e. within the thick, thin, and very thin spray regimes indicated in the schematic illustration of Fig. 5.2.

5.4.1 Breakup Regimes

The primary breakup of liquid jets at the nozzle exit can be caused by a combination of three mechanisms: turbulence within the liquid phase, implosion of cavitation bubbles and aerodynamic forces acting on the liquid jet [8].

Due to the pressure drop across the injection nozzle the liquid fuel is accelerated within the small nozzle holes. Thereby a high level of turbulence is generated within the liquid phase that has a destabilizing effect on the jet once it exits the nozzle hole. Additionally, at sharp edges along the flow path inside the nozzle, e.g. at the inlet of the nozzle hole, the streamlines are contracted such that the effective cross-section the flow is reduced and its velocity is accelerated even more. According to Bernoulli's law this causes a reduction in the static pressure, and locally the static pressure may be decreased to a value as low as the vapor pressure of the fuel. This phenomenon is schematically shown in Fig. 5.6, where the theoretical (linear) pressure distribution inside the nozzle hole is compared to a more realistic distribution along a streamline. The effect is that cavitation bubbles are generated inside the injection nozzle. This can be seen in Fig. 5.7 which shows an exemplary photograph of a cavitating flow through an acrylic glass nozzle. The cavitation bubbles are swept out of the nozzle into the combustion chamber where they implode and contribute to the disintegration of the spray. The third mechanism is that the relative velocity between the liquid jet and the gas results in aerodynamic forces that act on the liquid surface. Therefore, surface disturbances develop and start to grow that lead to breakup as well.

Depending on injection parameters such as the relative velocity between liquid and gas, the liquid and gas densities and the liquid viscosity and surface tension, the relative contribution of each of the three above mechanisms to the spray breakup varies, and several different breakup modes can be identified. They are characterized mainly by different breakup lengths, i.e. the distance between the nozzle orifice and the breakup position, and by the sizes of the resulting droplets.

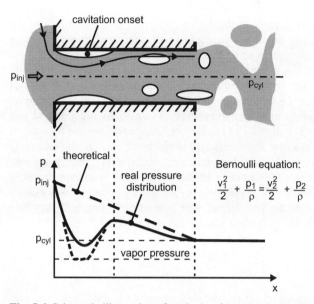

Fig. 5.6. Schematic illustration of cavitation formation inside the nozzle hole

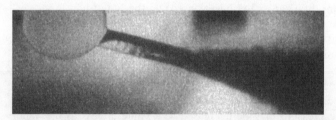

Fig. 5.7. Cavitation inside an acrylic glass diesel injection nozzle. The liquid phase is transparent, the gas phase is opaque. p_{inj} = MPa, p_{cyl} = MPa, d_{noz} = mm [15]

A widely agreed classification of breakup regimes has been proposed by Reitz and Bracco [81] in terms of the above injection parameters and fluid properties. For general applicability those quantities are expressed through the dimensionless Reynolds, Weber and Ohnesorge numbers, defined as

$$\text{Re} = \frac{\rho v_{inj} d_{noz}}{\mu}, \tag{5.33}$$

$$\text{We} = \frac{\rho v_{inj}^2 d_{noz}}{\sigma}, \tag{5.34}$$

$$Z = \frac{\mu}{\sqrt{\rho \sigma d_{noz}}}, \tag{5.35}$$

respectively.

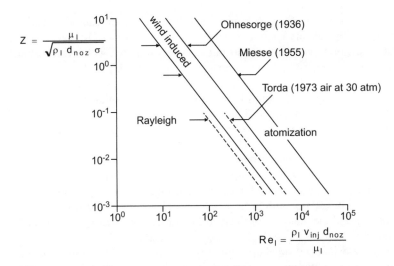

Fig. 5.8. Jet breakup regime boundaries by Miesse [53] and Ohnesorge [63]

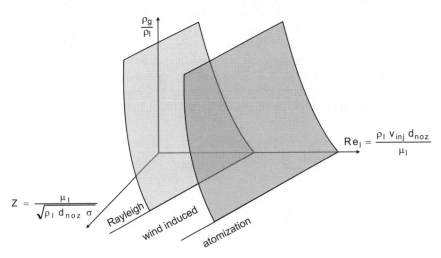

Fig. 5.9. Schematic chart of influence of gas density on breakup regime boundaries [75]

Four different spray regimes are typically distinguished, the Rayleigh, first wind induced, second wind induced and the atomization regime. Ohnesorge [63] and Miesse [53] suggested a definition by the liquid Reynolds and Ohnesorge numbers, i.e. the liquid phase properties are used in Eqs. 5.33 and 5.35. Figure 5.8 displays the results in which the first and second wind induced regimes have been combined. However, the definition in terms of the liquid properties implies that the effect of the gas phase on breakup is not taken into account in the classification, which is in contrast to the observation that the atomization can be enhanced by increasing the gas density (pressure) [102]. Therefore, Ranz [74] proposed a breakup classification in terms of the gas phase Weber number, which is based on the density of the gas phase ρ_g and the surface tension of the liquid σ_l. This approach however is incomplete as well, since now the influence of the liquid viscosity on breakup is neglected. In order to overcome these limitations Reitz [75] suggested to include the gaseous to liquid density ratio in the analysis, such that the two-dimensional plot in Fig. 5.8 now becomes a three-dimensional one as depicted in Fig. 5.9.

Figure 5.10 shows a schematic illustration of jet breakup in the characteristic breakup regimes. For relatively low injection velocities the Rayleigh breakup (a) is primarily governed by the inertia forces on the oscillating liquid and by its surface tension. The breakup length is far (many nozzle diameters) downstream of the nozzle orifice and the diameter of the resulting droplets is greater than the nozzle diameter.

In the first wind induced breakup regime (b) the inertia of the gas phase becomes more and more important. Surface disturbances are caused by the gas-liquid interactions that increase in amplitude and eventually lead to breakup. The average drop size decreases and is now in the range of the nozzle diameter. The breakup length is still a multiple of the nozzle diameter. For further increased injection velocities the second wind induced regime is reached. In Fig. 5.10 it has been combined with the first wind induced regime since the basic mechanisms leading to breakup are similar. The main difference is that as the relative velocity between liquid and gas increases the aerodynamic forces acting on the liquid surface are intensified and the wavelength of the disturbances becomes shorter. Therefore, the average droplet diameter is reduced in the second wind induced regime and the breakup length decreases compared to the first wind induced regime.

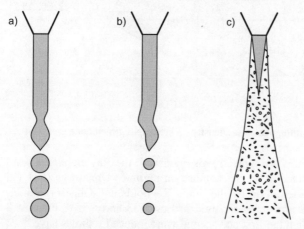

Fig. 5.10. Schematic depiction of breakup modes. a) Rayleigh breakup b) Wind induced breakup c) Atomization

Finally, for increased gas densities and large injection velocities the gas phase Weber number increases further and the atomization regime is reached (c). In this regime two different breakup lengths can be identified: the surface breakup begins directly at the nozzle orifice whereas an intact core may still be present several nozzle diameters downstream of the orifice. Furthermore, a conical shape of the overall spray is now observed. The mean droplet diameters are much smaller than the nozzle diameter.

The atomization regime is the most important for high pressure diesel injectors. However, a detailed experimental assessment of the primary spray breakup directly at the nozzle is extremely difficult because the spray is very dense and almost opaque at this position. Therefore, the detailed mechanisms that lead to primary spray breakup or atomization at the nozzle of modern diesel engine injection systems with injection pressures of up to 200 MPa are still not exactly understood. However, there is a general agreement that as the injection pressure increases, the effects of the inner nozzle flow such as the liquid phase turbulence and cavitation become more and more important.

5.4.2 Wave-Breakup Model

It has been discussed above that the development and aerodynamically driven growth of surface disturbances on the liquid phase is an important if not the dominant mechanism for breakup of jets. Vital contributions to deriving the theory behind this phenomenon and to establishing a coherent and continuous breakup model have been made by Reitz and co-workers, e.g. [75, 80, 81]. Since this so-called wave-breakup model (also referred to as Kelvin-Helmholtz breakup model) is widely applied in primary as well as in secondary breakup models it will be summarized here.

The analysis starts out from a cylindrical liquid jet of radius a that penetrates through a circular orifice into a stationary incompressible gas environment, Fig. 5.11. The liquid surface is subject to a number of infinitesimal perturbations with an initial amplitude of η_0 and a spectrum of wavelengths λ, typically expressed through the wave number $k = 2\pi/\lambda$. The initial disturbances may be caused by effects of the inner nozzle flow, e.g. by turbulence within the liquid phase. Their amplitudes will be increased exponentially by the liquid-gas interactions with a complex growth rate of $\omega = \omega_r + i\,\omega_i$:

$$\eta(t) = \mathrm{R}\left(\eta_0 \exp[ikx + \omega t]\right). \tag{5.36}$$

Assuming that the gas phase behaves as an inviscid fluid, i.e. there is free slip at the liquid-gas interface, and that the perturbations are much smaller than the jet radius ($\eta \ll a$) the so-called dispersion relation can be derived, that relates the growth rate ω to the wave number k [80]:

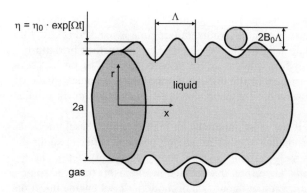

Fig. 5.11. Schematic growth of surface perturbations in the Wave-breakup model [78]

$$\omega^2 + 2v_l k^2 \omega \left[\frac{I_1'(ka)}{I_0(ka)} - \frac{2kl}{k^2+l^2} \frac{I_1(ka)}{I_0(ka)} \frac{I_1'(la)}{I_1(la)} \right] = \frac{\sigma k}{\rho_l a^2}(1-k^2 a^2)$$

$$\cdot \left(\frac{l^2 - k^2}{l^2 + k^2} \right) \frac{I_1(ka)}{I_0(ka)} + \frac{\rho_g}{\rho_l} \left(U - \frac{i\omega}{k} \right)^2 k^2 \left(\frac{l^2 - k^2}{l^2 + k^2} \right) \frac{I_1(ka) \cdot K_0(ka)}{I_0(ka) \cdot K_1(ka)}. \quad (5.37)$$

In the above equation I_n and K_n are the nth order modified Bessel functions of the first and second kind, respectively. The prime indicates differentiation, v_l is the kinematic viscosity of the liquid phase, U the gas velocity at the liquid surface, and $l^2 = k^2 + \omega/v_l$.

Even though the perturbations of different wave lengths will superpose each other in the real jet, it is assumed that only the fastest growing perturbation, indicated by growth rate Ω, that corresponds to the wave length Λ will ultimately lead to breakup. However, Eq. 5.37 is difficult to solve for a maximum value of ω since l is still a function of ω. To simplify the problem, Reitz [76] generated curve-fits of numerical solutions to Eq. 5.37 and obtained the following expressions for the maximum growth rate Ω and its corresponding wave length Λ:

$$\frac{\Lambda}{a} = 9.02 \frac{(1+0.45 Z^{0.5})(1+0.4 T^{0.7})}{(1+0.87 We_g^{1.67})^{0.6}}, \quad (5.38)$$

$$\Omega \left(\frac{\rho_l a^3}{\sigma} \right)^{0.5} = \frac{0.34 + 0.38 We_g^{1.5}}{(1+Z)(1+1.4 T^{0.6})}, \quad (5.39)$$

where $Z = \frac{We_l^{0.5}}{Re_l}$, $T = Z We_g^{0.5}$, $We_l = \frac{\rho_l U^2 a}{\sigma}$, $We_g = \frac{\rho_g U^2 a}{\sigma}$, $Re_l = \frac{Ua}{v_l}$.

The above relations have the effect that the growth rate increases and the corresponding wavelength decreases for increasing gas Weber numbers. This is in agreement with the experimental observation that for increasing injection velocities breakup is enhanced while the average diameters of the resulting droplets become smaller. The effect of liquid viscosity which appears in the Reynolds number Re and thus in the Ohnesorge number Z is that the wave growth rate is reduced.

In order to estimate the sizes of droplets formed by breakup, it is often assumed that there is a linear dependency between the resulting droplet radius r_d and the wave length Λ of the most unstable surface disturbance, e.g. [76, 99],

$$r_d = B\Lambda, \tag{5.40}$$

where B is of order unity. The length of the "intact" liquid core of the spray may be approximated by considerations of the mass removed from the jet by the atomization process,

$$L = \frac{c\,a}{f(\mathrm{T})}\sqrt{\frac{\rho_l}{\rho_g}}, \tag{5.41}$$

where T is defined as in Eqs. 5.38 and 5.39, and $f(\mathrm{T})$ asymptotically approaches $(3^{0.5}/6)$ for (T > 100), which is typically satisfied in diesel sprays. The constant c ranges from about 15 to 30 and accounts for various effects of the inner nozzle flow that are not resolved in detail.

The half-angle ($\alpha/2$) of the cone shaped spray observed in the atomization regime of high speed jets has been specified in ref. [81] based on the assumption that the droplet velocity component perpendicular to the spray direction $v_\perp$ is proportional to the wave growth rate of the most unstable wave:

$$\tan\left(\frac{\alpha}{2}\right) = \frac{v_\perp}{U} = \frac{\Omega\Lambda}{AU} = \frac{4\pi}{A}\sqrt{\rho_g/\rho_l}f(\mathrm{T}). \tag{5.42}$$

The expression $f(\mathrm{T})$ is the same as in Eq. 5.41, and the constant A accounts for the nozzle geometry. In ref. [78] it has been defined in terms of the length to diameter ratio of the nozzle hole as

$$A = 3.0 + \frac{l_{noz}/d_{noz}}{3.6}. \tag{5.43}$$

5.4.3 Blob-Injection Model

Reitz [76] applied the above Wave-breakup model to high speed diesel jets by assuming that during the injection duration there are continuously added large drops (so-called blobs) with a diameter comparable to the size of the nozzle hole to the gas phase. The frequency of the addition of new blobs is related to the fuel injection rate in a straightforward manner, assuming constant density of the liquid fuel and ideally spherical blobs. Immediately after injection the Kelvin-Helmholtz instabilities described by the wave-model start to grow on the blob surface, such that small secondary droplets are "sheared off" the blob surface as shown in Fig. 5.12.

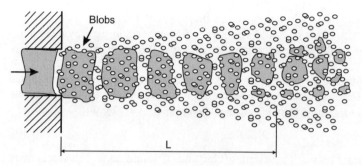

Fig. 5.12. Schematic illustration of the blob-injection model [76]

The calculation of the fastest growing wave length Λ and its corresponding growth rate Ω is executed in analogy to Eqs. 5.36 to 5.39. In ref. [76] the relation

$$r_d = B_0 \Lambda \tag{5.44}$$

was suggested to estimate the resulting droplet radii with a value of $B_0 = 0.61$ in the stripping breakup regime. For higher injection velocities and catastrophic breakup in the atomization regime, which is typical for diesel type injectors, Liu et al. [50] proposed the formulation

$$B_0 = 0.3 + 0.6 P, \tag{5.45}$$

where P is a random number within the interval between zero and one. By this method a distribution of droplet sizes is obtained which is more realistic for breakup of high speed jets. In both cases a bi-modal droplet size distribution is obtained for the complete spray, consisting of a number of larger droplets remaining from the original jet and a number of small droplets resulting from the above Kelvin-Helmholtz breakup.

Equations 5.44 and 5.45 may be used only if the resulting droplet diameter is less than the radius of the remaining parent drop, i.e. if $(B_0 \Lambda \leq a)$. Otherwise, the resulting radius of the newly formed droplet is estimated by

$$r_d = \min \begin{cases} (3\pi a^2 U / 2\Omega)^{1/3} \\ (3a^2 \Lambda / 4)^{1/3} \end{cases} \quad (B_0 L > a, \text{ one time only}), \tag{5.46}$$

which is based on the assumption that the jet disturbance has a frequency of $\Omega/2\pi$, i.e. one drop is formed each wave period, or that the drop size is determined from the volume of liquid contained under one surface wave.

Due to the breakup and generation of new small droplets, the size of the original blob is reduced. The temporal change in radius of this parent drop is given by

$$\frac{da}{dt} = -\frac{a - r_d}{\tau}, \tag{5.47}$$

where τ is the breakup time:

$$\tau = 3.726 B_1 \frac{a}{\Lambda\Omega}. \tag{5.48}$$

The constant B_1 has been introduced in order to account for effects of the inner nozzle flow on the breakup time that cannot be resolved directly. In ref. [76] a value of $B_1 = 20$ has been suggested whereas other references reported better results with values ranging from 1.73 [61] up to 30 [68]. This suggests that the inner nozzle flow has indeed an influence on primary spray breakup in addition to the liquid-gas interactions that is not yet included in the breakup analysis.

In order to reproduce the spray cone angle observed in diesel type sprays the child droplets separated from the initial blobs are equipped with a velocity component perpendicular to the main spray orientation. The maximum possible value of this component is obtained from the spray half angle specified in Eq. 5.42, and Reitz [76] suggests to choose an even distribution between zero and the maximum normal velocity for the various droplets in order to achieve a realistic droplet density within the spray.

In many recent applications of the blob injection method the above Wave- or Kelvin-Helmholtz breakup model has been combined with the so-called Rayleigh-Taylor breakup model in order to estimate the disintegration of the blobs into secondary droplets. The Rayleigh-Taylor model describes the instabilities that develop on a liquid-gas interface subject to strong normal accelerations pointed towards the gas phase. However, the disintegration of large drops into small droplets is considered a secondary breakup mechanism and therefore the Rayleigh-Taylor breakup will be discussed in Sect. 5.5.

It should be noted here, that the Kelvin-Helmholtz mechanism may be viewed as a secondary breakup mechanism as well, since it describes the breakup of large drops or blobs into smaller droplets. In fact, the model is used not only to estimate the disintegration of primary blobs but also to model the subsequent breakup of secondary droplets into even smaller droplets.

5.4.4 Turbulence and Cavitation Based Primary Breakup Model

In the above Wave-breakup model the influence of the inner nozzle flow on atomization of high speed jets cannot be predicted. The entire breakup analysis is based on aerodynamic interactions between the liquid and gas phases, and modified initial conditions that may be caused by different nozzle designs can only be included by adjusting empirical constants to experimentally obtained data. However, comprehensive studies on this subject show that effects of the inner nozzle flow such as liquid phase turbulence and cavitation do have an increasing influence on primary spray breakup for modern high pressure diesel injectors [9, 15, 29]. As an example, in Fig. 5.7 it can be observed that the spray angle on the upper side of the spray is greater than at the bottom side. This effect is likely to be caused by the cavitation inside the nozzle hole which is much more pronounced on the upper side of the hole because of the sharper edge at the nozzle hole inlet.

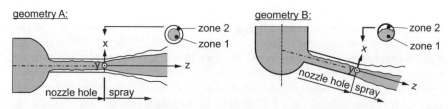

Fig. 5.13. Two-zone structure of the nozzle hole flow [16]

A variety of breakup models have been proposed in the literature that account for various effects of the inner nozzle flow on primary spray disintegration, e.g. [33, 43, 59, 97]. A detailed model that attempts to incorporate the most important findings of the above references has been presented by Baumgarten et al. [16]. This model will be summarized in the following.

The primary breakup model starts out from the observation that during the quasi-steady injection phase with full needle lift there is an almost stationary distribution of cavitation and liquid regions. Thus, the flow at the nozzle orifice is divided into two zones as indicated in Fig. 5.13. Zone 1 contains the liquid core of the jet and is characterized by a high momentum, whereas in zone 2 there is a mixture of liquid ligaments and cavitation bubbles with a significantly lower momentum than the one in zone 1. Due to a possible non-axis-symmetric nozzle hole design as it is shown for geometry B in Fig. 5.13, the liquid zone 1 does not have to be positioned at the center of the spray but may be displaced towards the nozzle hole side with less cavitation.

It should be noted that the described breakup model relies on the two zones as an initial condition. The inner nozzle flow that is the basis for the initialization of the zones at the nozzle orifice has to be either modeled with a separate suitable simulation code or it has to be determined by optical measurements. The initial conditions required for the primary breakup model are: the mass averaged values for the injection velocity v_{inj}, the turbulent kinetic energy k and its dissipation rate ε at the nozzle orifice, the exact position, shape and extension of zone 1 in the spray cross-section which does not have to be spherical (compare geometry B in Fig. 5.13), the mass flow rates of the two zones $\dot{m}_1$ and $\dot{m}_2$, and the void fraction θ within the cavitation zone 2.

A schematic illustration of the model is depicted in Fig. 5.14. A primary ligament containing both the intact liquid and the cavitation zone is injected into the combustion chamber. Due to the rise in static pressure the cavitation bubbles in zone 2 will implode such that energy is released and pressure waves are initiated that propagate both to the inner and outer surfaces of zone 2. It is now assumed that the fraction of energy that reaches the outer surface, i.e. the interface between zone 2 and cylinder gases, results in breakup of zone 2, whereas the remaining energy fraction reaching the interface between zones 1 and 2 increases the turbulence level in the liquid zone 1 and subsequently causes breakup of zone 1. The distribution of the total cavitation energy on the two zones is assumed to be proportional to the areas of the inner and outer interfaces of zone 2.

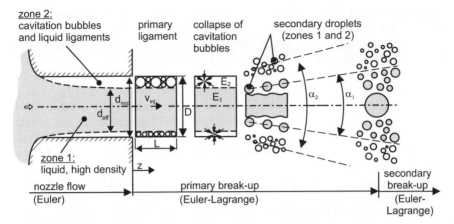

Fig. 5.14. Schematic illustration of the two-zone primary breakup model [16]

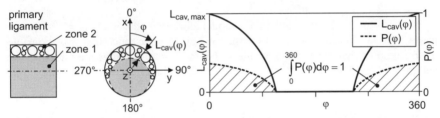

Fig. 5.15. Cross section of the spray in the two-zone model. The probability $P(\varphi)$ for a secondary parcel of zone 2 to be emitted at an angle φ depends on the radial thickness $L_{cav}(\varphi)$ of the cavitation zone [16]

In both zones 1 and 2, the breakup of the primary ligament into secondary droplets occurs once the collapse time of the cavitation bubbles has been exceeded. The breakup is due to the sum of turbulent kinetic energy and the energy induced by the collapse of the cavitation bubbles. The total amount of energy will be transferred into a combination of surface energy of secondary droplets and a velocity component that is perpendicular to the spray axis. The amount of surface energy controls the resulting droplet radius, and the radial velocity component is responsible for the visible spray angle. The orientation of this radial velocity component within the xy-plane, i.e. within the spray cross-section, is determined by sampling from a probability distribution that is proportional to the angular "thickness" of zone 2, indicated as L_{cav} in Fig. 5.15. Thus, the model is predictive in terms of the spray cone angle in a three-dimensional manner. This means that the spray does not have to be axis-symmetric, but that an increased spray angle at the nozzle hole side that is subject to stronger cavitation as it is shown in Fig. 5.7 can be estimated by the model.

At the breakup time the cavitation zone 2 disintegrates directly into a number of spherical secondary droplets of equal size, and the liquid zone 1 is broken up into a number of medium-sized spherical secondary droplets and one remaining cylindrical ligament that will be transformed into a sphere of greater diameter as well. All secondary droplets regardless of their origin (zone 1 or 2) will then be treated by secondary breakup models as they will be described in Sect. 5.5.

The detailed mathematical formulation of the primary breakup model is as follows. The primary ligament injected from the nozzle is characterized by an axial velocity v_{inj}, a diameter D that is equal to the nozzle diameter d_{noz} and a length L which is assumed to be equal to the effective flow diameter inside the nozzle, Fig. 5.14. Such an assumption is necessitated by the fact that in the Lagrangian parcel approach the primary ligaments have to be of finite dimensions, and here the length $L = d_{eff}$ of the cylindrical ligament is chosen in analogy to the blob-injection model where the initial blob diameter is also assumed to be equal to the effective nozzle diameter d_{eff}.

The history of turbulent kinetic energy k within each zone is estimated by solving the simplified conservation equations for k and its dissipation rate ε:

$$\frac{dk_i}{dt} = -\varepsilon_i, \quad \frac{d\varepsilon_i}{dt} = -C_2 \frac{\varepsilon_i^2}{k_i}, \quad i = 1, 2 \text{ (zonal index)}. \tag{5.49}$$

C_2 is a constant from the standard k-ε model and often chosen as $C_2 = 1.92$ (compare Chap. 4). In order to determine the total energy per zone i that is responsible for breakup of that zone, the absolute turbulent kinetic energies are added to the energies released by the implosion of cavitation bubbles:

$$E_i = E_{trb,i} + E_{cav,i} = m_i k_i + E_{cav,i}. \tag{5.50}$$

As mentioned above, the ratio of cavitation energy that is accounted for each zone is equal to the ratio of the outer surface areas of the respective zones:

$$\frac{E_{cav,1}}{E_{cav,2}} = \frac{A_1}{A_2}. \tag{5.51}$$

The total cavitation energy E_{cav} resulting from the collapse of the cavitation bubbles as well as the collapse time are estimated by utilizing the differential equation, that describes the bubble dynamics during breakup in a compressible environment [71]:

$$\left(1 - \frac{2\dot{R}}{a}\right) R\ddot{R} + \frac{3}{2}\left(1 - \frac{4\dot{R}}{3a}\right)\dot{R}^2 = \frac{1}{\rho_\infty}\left(p_v - \frac{2\sigma}{R} - \frac{4\mu}{R}\dot{R} - p_\infty\right). \tag{5.52}$$

R denotes the bubble radius and $\dot{R}$ and $\ddot{R}$ its first and second derivative, respectively. a is the averaged speed of sound in zone 2, μ the liquid viscosity, and p_v and p_∞ are the fuel vapor pressure and the back pressure of the gas environment, respectively. The kinetic energy of the fluid surrounding one bubble is estimated as

$$E_{kin,bubble} = 2\pi \rho_\infty \dot{R}^2 R^3. \tag{5.53}$$

Its maximum value which occurs shortly before the collapse of the bubble has to be multiplied by the number of bubbles in order to obtain the total cavitation energy E_{cav}.

For simplicity it is assumed that within one primary ligament all cavitation bubbles have the same size. However, each of the subsequently injected ligaments is characterized by a different initial bubble size which is obtained randomly from a Gaussian distribution with a mean and a standard deviation of both 10μm where the branches with values either less than 2μm or greater than $L_{cav,max}/2$ (see Fig. 5.15) are truncated.

The disintegration of the primary ligament into secondary droplets is calculated independently for the two zones with slightly different approaches. For the cavitation zone 2, it is assumed that its available energy E_2 is completely transformed into surface energy of the new droplets and into kinetic energy of these secondary droplets:

$$E_{surf,2} + E_{kin,2} = E_2, \qquad (5.54)$$

$$E_{surf,2} / E_{kin,2} = \kappa, \qquad (5.55)$$

$$E_{surf,2} = N_2 \sigma \pi d_{child,2}^2, \qquad (5.56)$$

$$E_{kin,2} = N_2 \frac{1}{2} m_{child,2} v_{rad,2}^2. \qquad (5.57)$$

In the above equations N_2 indicates the number of secondary (child) droplets resulting from zone 2, and $m_{child,2}$ is the mass of one these identical droplets. Obviously the constant κ which specifies the fraction of energy that is transformed into surface energy is a very important model parameter. In ref. [16] it is suggested to use a value of $\kappa = 0.003$ for a back pressure of $p_\infty = 5MPa$ and tests revealed that this value remains fairly constant for typical gas back pressures present in diesel engine applications.

The spray half angle becomes

$$\frac{\alpha_2}{2} = \operatorname{atan}\left(v_{rad} / v_{inj}\right), \qquad (5.58)$$

and the angle φ of a secondary droplet parcel within the xy-plane is determined by sampling from a probability distribution that is proportional to the radial "thickness" L_{cav} of zone 2, Fig. 5.15.

The breakup of the cylindrically shaped liquid zone 1 occurs at the same time as the breakup of zone 2, and it is calculated by assuming that the available energy E_1 causes turbulent fluctuations which induce a deformation force on the surface of zone 1. For isotropic turbulence, the turbulence intensity can be estimated as

$$u' = \sqrt{\frac{2}{3} \cdot \frac{E_1}{m_1}}, \qquad (5.59)$$

and following Knapp et al. [46], the deformation force is equal to the product of the dynamic pressure and the surface area,

$$F_{trb,1} = \frac{1}{2}\rho_1 u'^2 \pi d_{zone1} L, \qquad (5.60)$$

where d_{zone1} is approx. equal to d_{eff} (Fig. 5.14), and thus equal to L. Mass is split off zone 1 in form of secondary droplets until the surface tension force of the remaining cylindrical parent ligament becomes equal to the above deformation force:

$$F_{surf,1} = 2\sigma\left(d_{parent1} + L\right) \equiv F_{trb,1}. \qquad (5.61)$$

Combining Eqs. 5.59 to 5.61 the diameter of the remaining cylindrical parent ligament can be expressed as

$$d_{parent1} = \frac{\rho_1 E_1 \pi L^2}{6 m_1 \sigma} - L. \qquad (5.62)$$

After breakup the cylindrical parent drop is instantaneously transferred into a spherical drop of equal volume. From the remaining turbulent energy inside the drop the fraction necessary to form its surface is subtracted. The remaining part is converted into kinetic energy with a velocity component perpendicular to the spray axis. This is similar to the breakup of zone 2. However, in contrast to zone 2, that is much more dilute, not the total amount of kinetic energy is transformed into a radial motion of the parent droplet remaining from zone 1. Because the spray is typically very dense at the core where the primary breakup of zone 1 occurs it is assumed that a certain fraction of the kinetic energy is dissipated, e.g. by collisions with neighboring drops. Thus, an efficiency $\eta_{parent1}$ is introduced that specifies the fraction of available energy that is actually converted into a radial velocity component:

$$\eta_{parent1} = E_{kin,1,act} / E_{kin,1}. \qquad (5.63)$$

The value of $\eta_{parent1}$ is sampled from a uniform probability distribution between zero and one, where a value of 1.0 indicates the maximum possible spray angle and a value of 0.0 results in a droplet without a radial velocity component that stays directly on the spray axis. The orientation of the radial velocity component of the remaining parent droplet within the xy-plane is sampled from a uniform probability distribution between zero and 360°, since zone 1 is assumed to be almost axis-symmetric.

The mass of liquid fuel that is split off the parent ligament in order to form secondary droplets of zone 1 is treated in analogy to the breakup of zone 2, Eqs. 5.54 to 5.58. That is, a certain fraction of the available energy (defined by κ in Eq. 5.55) is again needed to come up for the surface energy and the remaining fraction is available for a radial momentum. However, because of the dense spray at the core of the spray an efficiency η_{child1} is again sampled from a uniform probability distribution between zero and one. η_{child1} is defined in analogy to $\eta_{parent1}$ in Eq. 5.63. The orientation of the radial velocity component within the xy-plane is also sampled from a uniform distribution between zero and 360° as it is done for the remaining parent droplet of zone 1. Thus, in contrast to the breakup of zone 2 that results in a three-dimensional droplet distribution, the breakup of zone 1 causes an axis-symmetric arrangement of droplets.

In ref. [16] the turbulence and cavitation based primary breakup model has been combined with the Kelvin-Helmholtz model for secondary breakup (compare Sect. 5.5) and applied to study a non-evaporating spray for the nozzle geometry B in Fig. 5.13. An injection pressure of 65 MPa was chosen, the air back pressure and temperature were set to 5 MPa and 298 K, respectively. In Fig. 5.16 a) the time averaged size distribution of droplets resulting from primary breakup is shown. It is clearly visible that the breakup model results in a bimodal distribution with small droplets stemming from breakup of the cavitation zone 2 and significantly larger droplets that are formed from the initially intact liquid zone 1. The peak at approx. 95 μm is caused by the remaining parent droplets of zone 1 whereas the even distribution between 55 and 95 μm results from the child droplets of zone 1 that are stripped of the parent droplet.

Figure 5.16 b) displays liquid mass concentrations along the x-axis indicated in Fig. 5.15 at several positions downstream of the nozzle orifice at a timing 3 ms after the start of injection. As expected the radial mass distribution is fairly symmetric at a position close to the nozzle. However, further downstream the peak in the mass distribution is shifted towards the upper side of the spray ($\varphi = 0°$) that is characterized by stronger cavitation within the nozzle. Thus, the model is capable of predicting non-axis symmetric spray angle and mass distributions that are in agreement with the optical results shown in Fig. 5.7. Finally, Fig. 5.17 displays the corresponding simulation results of the complete spray, as well as the xz-cut plane through the spray axis where the droplets resulting from zones 1 and 2 have been separated. Again, it can be seen that the droplets of the outer zone 2 are deflected in x-direction because of the non-symmetric cavitation effects.

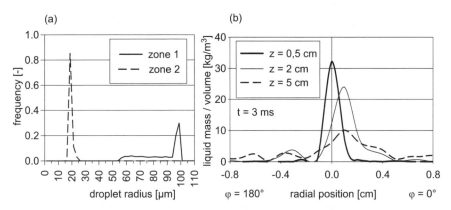

Fig. 5.16. Simulation results for p_{inj} = 65 MPa, p_{air} = 5 MPa, T_{air} = 298 K. a) time averaged droplet size distribution resulting from primary breakup b) Radial liquid mass distribution at various distances from the nozzle [16]

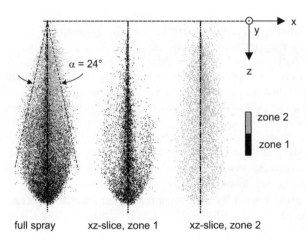

full spray   xz-slice, zone 1   xz-slice, zone 2

Fig. 5.17. Droplet distribution 3 ms after injection. Black droplets are formed from zone 1, gray droplets from zone 2 [16]

5.4.5 Sheet-Atomization Model for Hollow-Cone Sprays

In direct injection spark ignition (DISI) engines pressure swirl atomizers are often utilized in order to establish hollow cone sprays. These sprays are typically characterized by high atomization efficiencies, i.e. by small droplet diameters and effective fuel-air mixing that can be realized with only moderate injection pressures in the range of 5 to 10 MPa. A schematic illustration of such an injector is shown in Fig. 5.18. Due to tangentially arranged inflow ports the fuel is set into a rotational motion within the injector. The resulting centrifugal forces lead to the formation of a liquid film near the injector walls, surrounding an air core at the center of the injector. Outside the injector nozzle the tangential velocity component of the fuel is transformed into a mostly radial component such that a cone shaped sheet results. Due to mass conservation this sheet thins as it departs further from the nozzle and moreover, it is subject to aerodynamic instabilities that cause breakup into ligaments, Fig. 5.19. The ligaments then quickly breakup further into droplets. This process is driven by aerodynamically induced instabilities on the ligament surfaces as well, such that there is a general similarity to the diesel type breakup described by the wave-model above.

Meyer and Weihs [52] conducted a study on the effect of the inner to outer radius ratio of annular sheets, i.e. the relative sheet thickness compared to the curvature of the annulus, on the governing breakup mechanisms of such annular liquid sheets. They concluded that there is a critical sheet thickness, defined in terms of the surface tension, the gas density and the injection velocity as $t_{crit} = \sigma/(\rho_g U^2)$. For a thickness greater than t_{crit} the jet behaves like a solid cone diesel type jet, for a smaller thickness the annular jet may be treated as a thin planar (two-dimensional) sheet. The latter case typically applies to injectors utilized in DISI engines such that breakup can be treated in analogy to the findings of Squire [91], who showed that instability and breakup of planar sheets are caused by the growth of sinuous waves, as depicted in Fig. 5.20.

In many numerical studies on gasoline direct injection engines utilizing pressure swirl atomizers, e.g. [22, 25, 92], the primary spray breakup is modeled with the so-called LISA (linearized instability sheet atomization) model, that follows the above considerations and was presented in detail by Senecal et al. [89].

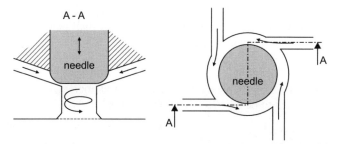

Fig. 5.18. Schematic illustration of a pressure swirl injector

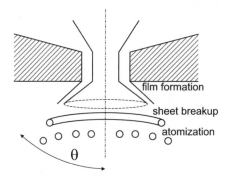

Fig. 5.19. Schematic sheet and spray formation with a pressure swirl injector [92]

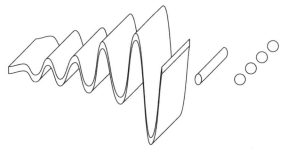

Fig. 5.20. Breakup mechanism of planar liquid sheets [78]

The mathematical formulation of the primary breakup, i.e. the disintegration of the liquid sheet into ligaments and the first generation of droplets, can be summarized as follows.

The total velocity of the fuel exiting the injector is related to the pressure drop across the injector exit by

$$U = c_D \sqrt{\frac{2\Delta p_{inj}}{\rho_l}}, \qquad (5.64)$$

and the axial velocity component can be determined by the cone half-angle θ of the spray, which is assumed to be known for a given injector:

$$u = U\cos(\theta). \qquad (5.65)$$

Based on similarity considerations between the swirl ports and nozzles, the discharge coefficient c_D is assumed to be 0.7. However, the expression

$$c_D = \max\left[0.7, \frac{4\dot{m}_{inj}}{\pi d_{noz}^2 \rho_l \cos(\theta)} \sqrt{\frac{\rho_l}{2\Delta p_{inj}}}\right] \qquad (5.66)$$

has to be obeyed, in order to make sure that the experimentally determined mass flow rate $\dot{m}_{inj}$ through the injector does not violate the continuity equation. In the general case, where the term on the right hand side of Eq. 5.66 is less than 0.7, it is assumed that an air-core exists in the center of the rotating flow as indicated in Fig. 5.19. The continuity equation relates the thickness t_f of the liquid film inside the injector to the measured mass flow rate:

$$\dot{m}_{inj} = \pi u t_f \left(d_{noz} - t_f\right). \qquad (5.67)$$

It is now assumed that at the nozzle orifice a two-dimensional, viscous, incompressible liquid sheet of thickness $2h$ moves with velocity U through a quiescent, inviscid, incompressible gas medium. A spectrum of infinitesimal disturbances is imposed on the sheet surface and the liquid-gas interaction causes the amplitudes of these disturbances to grow,

$$\eta(t) = \mathrm{R}\left(\eta_0 \exp[ikx + \omega t]\right), \qquad (5.68)$$

which is in direct analogy to the wave-breakup model for diesel sprays, Eq. 5.36. η_0 is the initial wave amplitude, $k = 2\pi/\lambda$ is the wave number, and $\omega = \omega_r + i\omega_i$ is the complex growth rate of the surface disturbances. The most unstable disturbance has the largest value of ω_r, denoted by Ω, and is assumed to be responsible for sheet breakup. Thus, it is desired to obtain a dispersion relation $\omega_r = \omega_r(k)$ from which the most unstable disturbance can be deduced.

In refs. [91] and [40] it has been shown that two solutions, or modes, exist which satisfy the liquid governing equations subject to the boundary conditions at the upper and lower interfaces of the sheet. For the first solution, called the sinuous mode, the waves at both surfaces are exactly in phase. On the other hand, for the varicose mode the waves are π radians out of phase. It has been suggested by Senecal et al. [89], that concentrating on the sinuous mode is sufficient for typical engine type applications. Moreover, it was concluded that a simplified form of the dispersion relation,

$$\omega_r = -2v_l k^2 + \sqrt{4v_l^2 k^4 + \frac{\rho_g}{\rho_l} U^2 k^2 - \frac{\sigma k^3}{\rho_l}}, \qquad (5.69)$$

can be used if three main assumptions are made: first of all, an order of magnitude analysis using typical values from the inviscid solutions shows that the terms of second order in viscosity can be neglected in comparison to all other terms. In addition, if a critical gas Weber number of $We_g=27/16$ (based on the relative velocity, the gas density and the sheet half-thickness) is exceeded, short waves will grow on the sheet surface, with a growth rate independent of the sheet thickness. Lastly, the gas to liquid density ratio has to be sufficiently small ($\rho_g/\rho_l \ll 1$). All of the above conditions are typically met by modern pressure-swirl atomizers in DISI engine type applications.

Once the disturbances on the sheet surface have reached a critical amplitude, ligaments are assumed to be formed. The breakup time τ_b for this process can be formulated based on an analogy with the breakup length of cylindrical liquid jets, e.g. [80],

$$\eta_b = \eta_0 \exp(\Omega \tau_b) \quad \Leftrightarrow \quad \tau_b = \frac{1}{\Omega} \ln\left(\frac{\eta_b}{\eta_0}\right), \qquad (5.70)$$

where η_b is the critical amplitude at breakup, η_0 is the amplitude of the initial disturbance and Ω is the maximum growth rate, that is obtained by numerically maximizing Eq. 5.69 as a function of the wave number k. The corresponding breakup length L can then be estimated by assuming a constant velocity for the liquid sheet:

$$L = U\tau_b = \frac{U}{\Omega} \ln\left(\frac{\eta_b}{\eta_0}\right). \qquad (5.71)$$

The quantity $\ln(\eta_b/\eta_0)$ is usually given a value of 12 as suggested in ref. [26]. The diameter of the ligaments formed at the point of breakup is obtained from a mass balance, assuming that that the ligaments are formed from tears in the sheet once per wavelength. The resulting diameter is given by

$$d_L = \sqrt{\frac{4}{\pi} \Lambda_s \cdot 2h} = \sqrt{\frac{16h}{K_s}}, \qquad (5.72)$$

where K_s is the wave number $K_s = 2\pi/\Lambda_s$ corresponding to the maximum growth rate Ω, that will lead to breakup of the sheet. Hence, the ligament diameter is a function of the sheet half-thickness h at the breakup position, which is related to its initial value h_0 at the nozzle orifice by

$$h = \frac{h_0 \left[d_{noz} - t_f \right]}{2L \sin(\theta) d_{noz} - t_f}, \quad (5.73)$$

and

$$h_0 \approx \frac{t_f}{2} \cos(\theta). \quad (5.74)$$

The further breakup of ligaments into droplets is calculated based on an analogy to Weber's result for growing waves on cylindrical, viscous liquid columns. The wave number K_L for the fastest growing wave on the ligament is [87]:

$$K_L d_L = \left[\frac{1}{2} + \frac{3\mu_l}{2\sqrt{\rho_l \sigma d_L}} \right]^{-1/2}. \quad (5.75)$$

If it is assumed that breakup occurs when the amplitude of the most unstable wave is equal to the radius of the ligament, one drop will be formed per wavelength. A mass balance then yields

$$d_{drop} = \left(\frac{3\pi d_L^2}{K_L} \right)^{1/3} \quad (5.76)$$

for the droplet diameter d_{drop}.

In ref. [92] the LISA sheet atomization model was combined with the TAB-model for secondary droplet breakup (compare Sect. 5.5) and applied to simulate a non-evaporating spray development in a pressure bomb. The injection parameters are summarized in Table 5.1, and in Figs. 5.21 and 5.22 the computational results are compared to spray photographs for three different timings after injection start and for two different back pressures.

Table 5.1. Injection parameters for sprays shown in Figs. 5.21 and 5.22

Injection parameter	Quantity
Spray half-cone angle θ [°]	54
Dispersion angle Δθ [°]	10
Fuel mass [mg]	44
Fuel temperature [K]	300
Air temperature [K]	300
Nozzle diameter [mm]	0.458
Injection pressure [MPa]	4.93
Injection duration [ms]	6.1
Liquid density [g/cm³]	0.76

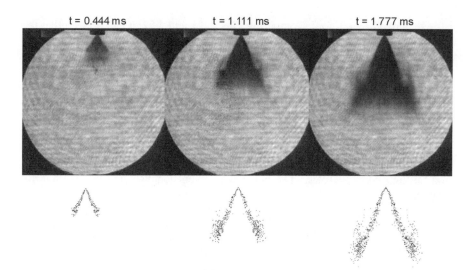

Fig. 5.21. Spray images. p_{air} = 101 kPa [92]

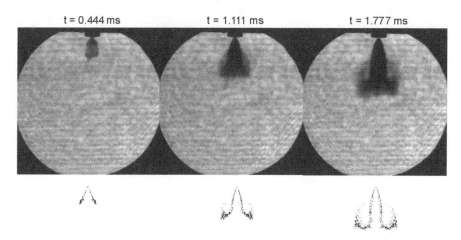

Fig. 5.22. Spray images. p_{air} = 366 kPa [92]

Both series of pictures suggest that the LISA spray model can predict the characteristic behavior of the spray very well. Just after the start of injection at $t = 0.444$ ms the spray has an almost perfect cone shape with the cone angle as specified in Table 5.1. During the progress of injection ($t = 1.111$ ms and $t = 1.777$ ms), the spray shape deviates from its initial perfect cone shape. The cone angle becomes narrower towards the spray front and a re-circulating vortex starts to form at the spray edges. This vortex is very well developed and is clearly visible at $t = 1.777$ ms. The influence of the increased backpressure on the spray is predicted very well, too. Comparing Figs. 5.21 and 5.22, it can be seen that the spray tip penetration decreases for the higher air pressure. Moreover, the cone angle becomes significantly narrower and the vortex at the spray edge is much more distinct for this case.

The latter effect, that the initially conical liquid sheet closes in on itself, may be even more distinct under certain boundary conditions, such that a so-called water bell can be formed, Fig. 5.23. The recombined liquid jet breaks up into a full-cone spray of large drops downstream of the bell, which is obviously not desirable in engine applications. Criteria for predicting the boundary between the two regimes can be formulated by using the results of studies on water bells generated by injection through poppet valves. The bell reattachment length y and the width x are given by

$$\frac{y}{F(\theta)} = \frac{x}{G(\theta)} = \frac{\rho_l U^2 \delta t_f}{2\sigma}, \qquad (5.77)$$

where

$$F(\theta) = 2\ln\left(\sec\theta + \sqrt{\sec^2\theta - 1}\right), \qquad (5.78)$$

$$G(\theta) = 1 - \cos\theta, \qquad (5.79)$$

and δ and t_f are the radius of the poppet valve (Fig. 5.23) and the sheet thickness at the nozzle, respectively.

The ratio of the reattachment length y to the breakup length L of the liquid sheet, that may be calculated by Eq. 5.71, serves as a predictor of the breakup regime. Reitz [78] suggests that a ratio of $y/L = 1$ is a reasonable criterion for estimating the location of the regime boundary. Thus, for good atomization in the sheet breakup regime small y/L-ratios are required. This is typically achieved with low surface tension, high injection pressure, and a large nozzle cone angle.

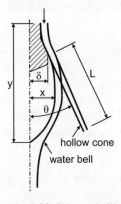

Fig. 5.23. Schematic illustration of sheet breakup and water bell from poppet nozzles [78]

5.5 Secondary Droplet Breakup

5.5.1 Drop Breakup Regimes

The secondary breakup of liquid fuel drops into even smaller droplets is primarily driven by aerodynamic forces employed on the drops by the surrounding gas phase. These forces cause a distortion of the initially spherical droplet that will eventually lead to breakup if the surface tension that counteracts the deformation is exceeded. Consequently, the dimensionless droplet Weber number, which relates the dynamic pressure to the surface tension and is defined as

$$\text{We} = \frac{\rho_g r_d v_{rel}^2}{\sigma}, \tag{5.80}$$

is a characteristic measure of the breakup behavior of liquid droplets. Depending on the relative velocity between droplet and gas phase, and thus depending on the Weber number, several different breakup mechanisms have been observed in experimental studies. Often the five different breakup regimes schematically shown in Fig. 5.24 are distinguished.

For very low Weber numbers near the critical value of about six the droplet executes an oscillation and may breakup into two new droplets of approximately equal size. If the Weber number is slightly increased the original drop will be deformed into a bag shape. After breakup a bimodal droplet size distribution will result with larger droplets originating from the rim and smaller ones originating from the trailing edge. For Weber numbers between approx. 10 and 25 an additional streamer-shaped interior may develop within the bag, leading to a class of droplets with a similar size to the ones resulting from the rim of the bag. Stripping breakup occurs for Weber numbers between 25 and 50. It is characterized by very small secondary droplets that are stripped or sheared off the surface of the bigger parent droplet. Finally, for large Weber numbers above about 50 the so-called catastrophic breakup takes place. It is dominated by surface instabilities that develop on a liquid-gas interface subject to strong accelerations in a direction normal to the interface. It should be noted though that there is some uncertainty about the limiting Weber numbers, especially towards the high-end of Weber numbers between the stripping and catastrophic regimes. For example, Arcoumanis et al. [8] distinguish two different kinds of stripping breakup, namely the sheet stripping and the wave crest stripping regimes, and extend their applicability to significantly greater Weber numbers such that catastrophic breakup does not occur until a value of We ≈ 500 is exceeded.

In high pressure diesel or gasoline sprays all of the above mechanisms may be present. Starting close to the injector orifice the relative velocity between droplets and gas and thus the Weber number is very large, such that catastrophic breakup is the dominant mechanism. However, further downstream of the nozzle droplets with significantly smaller Weber numbers may be present because of both lower relative velocities and smaller diameters resulting from previous breakup and evaporation.

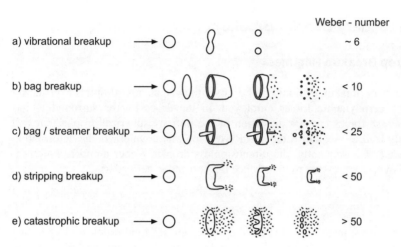

Fig. 5.24. Drop breakup regimes [107]

A variety of mathematical models for drop breakup have been proposed in the literature. Most of these models have been established in order to describe one particular of the above breakup mechanisms. Nevertheless, in engine spray simulations they are – for the sake of simplicity – often applied to the entire spectrum of breakup regimes. This is not entirely true though. In recent years it has become more and more standard to determine the governing breakup mechanism for a droplet class and then apply the more appropriate of at least two breakup models, e.g. a combination of the Kelvin-Helmholtz and the Rayleigh-Taylor model.

In the subsequent sections the secondary breakup models that are applied most often in engine spray simulations will be discussed.

5.5.2 The Reitz-Diwakar Model

In a first attempt to include secondary droplet breakup in CFD spray calculations Reitz and Diwakar [82] utilized the findings of Nicholls [58], who experimentally determined the criteria for bag and stripping breakup as:

$$\text{Bag breakup} \qquad \text{We} > 6 \qquad (5.81)$$

$$\text{Stripping breakup} \qquad \frac{\text{We}}{\text{Re}^{1/2}} > 0.5 \qquad (5.82)$$

The Weber number is defined as specified in Eq. 5.80, and the Reynolds number is

$$\text{Re} = \frac{2r_d v_{rel}}{v_g}. \qquad (5.83)$$

The corresponding lifetimes for unstable droplets are given as

$$t_{bag} = C_1 \sqrt{\frac{\rho_l r_d^3}{\sigma}}, \qquad (5.84)$$

$$t_{strip} = C_2 \frac{r_d}{v_{rel}} \sqrt{\frac{\rho_l}{\rho_g}}, \qquad (5.85)$$

where the constants C_1 and C_2 are of order unity.

Whenever one of the breakup criteria is satisfied for a droplet class for longer than the respective breakup time, it is assumed that the original droplet is disintegrated into a number of smaller droplets. All child droplets are of equal size. It is determined from equating the respective breakup criterion (Eq. 5.81 or Eq. 5.82) to its critical value and by solving it for the droplet diameter. Thus, it is assumed that the new child droplets are initially in a state that is just stable. The number of child droplets $N_{d,child}$ resulting from breakup of one class of parent droplets is determined by mass conservation principles:

$$N_{d,child} \cdot r_{d,child}^3 = N_{d,parent} \cdot r_{d,parent}^3. \qquad (5.86)$$

5.5.3 The Taylor-Analogy Breakup Model

Based on the Taylor analogy [101], that assumes that the droplet distortion can be described as a one-dimensional, forced, damped, harmonic oscillation similar to the one of a spring-mass system (see Fig. 5.24 a), O'Rourke and Amsden [61] proposed the so-called TAB (Taylor Analogy Breakup) model and implemented it into the KIVA-code. In this model the momentary droplet distortion is characterized by the dimensionless parameter $y = 2x/r$, where x describes the deviation of the droplet equator from its equilibrium position, see Fig. 5.4. Assuming that the liquid viscosity acts as a damping element and the surface tension as a restoring force, the equation of motion can be written as

$$\ddot{y} + \frac{5\mu_l}{\rho_l r^2} \dot{y} + \frac{8\sigma}{\rho_l r^3} y = \frac{2\rho_g v_{rel}^2}{3\rho_l r^2}. \qquad (5.87)$$

If it is further assumed that the relative velocity between droplet and gas remains constant, integration of Eq. 5.87 leads to the formulation of the time-dependent distortion amplitude,

$$y(t) = \frac{We}{12} + e^{-t/t_d} \cdot \left[\left(y_0 - \frac{We}{12} \right) \cos \omega t + \left(\frac{\dot{y}_0}{\omega} + \frac{y_0 - We/12}{\omega \, t_d} \right) \sin \omega t \right], \qquad (5.88)$$

where $t_d = \dfrac{2\rho_l r^2}{5\mu_l}$, $\omega^2 = \dfrac{8\sigma}{\rho_l r^3} - \dfrac{1}{t_d^2}$, and y_0 and $\dot{y}_0$ are the initial values of the distortion and its temporal change rate, respectively. Typically, y_0 and $\dot{y}_0$ are both taken as zero.

Breakup occurs if and only if the distortion parameter y exceeds unity, i.e. the equator deviation x becomes greater than half the droplet radius (Fig. 5.4). For an inviscid liquid it can be shown that this is the case when the Weber number exceeds a critical value of six, which is in agreement to the experimental findings described above. After implementing the breakup criterion $y > 1$ into Eq. 5.88, one can solve for the corresponding breakup time.

Two limiting cases can now be investigated: bag breakup for very low Weber numbers (We ≈ 6) and stripping breakup for very large Weber numbers. In the first case it is assumed that breakup occurs when $\omega t_{bu} = \pi$, and the breakup time becomes

$$t_{bu} = \pi \sqrt{\frac{\rho_l r^3}{8\sigma}}. \tag{5.89}$$

For high Weber numbers in the stripping breakup regime it is assumed that breakup ($y > 1$) occurs much earlier in the oscillation period, i.e. $\omega t_{bu} \ll \pi$. In this case the breakup time can be reduced to

$$t_{bu} = \sqrt{3}\frac{r}{v_{rel}}\sqrt{\frac{\rho_l}{\rho_g}}. \tag{5.90}$$

These results are identical to the findings of Nicholls [58], Eqs. 5.84 and 5.85, if the constants C_1 and C_2 are chosen as $\pi/(8)^{1/2}$ and $(3)^{1/2}$, respectively. Thus, even though the assumption of a one-dimensional droplet oscillation may not be exactly valid for an entire engine type spray, the TAB model predicts the same breakup durations for the limiting cases of low and high Weber numbers as the Reitz-Diwakar model.

The TAB model has also been used in order to determine the normal velocity component of child droplets after breakup and thus, the spray angle. At the time of droplet breakup the equator of the parent droplet moves with a velocity of $\dot{x} = \dot{y}r/2$ in a direction normal to the droplet path. This velocity is taken to be the normal velocity component of the child droplets, and the spray half angle can be estimated as

$$\tan\frac{\alpha}{2} = \frac{\dot{x}}{v_{rel}}. \tag{5.91}$$

The quantity of $\dot{y}$ at the time of breakup can be derived from Eq. 5.88, and for large Weber numbers that are typically encountered near the injector it reduces to

$$\dot{y} \approx \frac{We}{12}\omega^2 t_{bu}, \tag{5.92}$$

such that the spray angle finally becomes

$$\tan\frac{\alpha}{2} = C_v \cdot \frac{\sqrt{3}}{3} \cdot \sqrt{\frac{\rho_g}{\rho_l}}, \tag{5.93}$$

where C_v is a constant of order unity.

5.5 Secondary Droplet Breakup

The radius of the child droplets after breakup is calculated based on the concept that the sum of the surface energy and the energy bound in the distortion and oscillation of the parent drop is equal to the sum of surface energy and kinetic energy due to the normal velocity component of the child droplets, i.e.

$$4\pi r^2 \sigma + K \frac{\pi}{5} \rho_l r^3 (\dot{y}^2 + \omega^2 y^2) = 4\pi r^2 \sigma \frac{r}{r_{32}} + \frac{\pi}{6} r^5 \rho_l \dot{y}^2 . \tag{5.94}$$

K is a correction factor that accounts for the superposition of multiple oscillations in the real droplet and is suggested to be $K = 10/3$. Equation 5.94 implies that immediately after breakup the child drops are spherical and their distortion rate is zero. After several mathematical manipulations the above energy balance leads to the expression

$$\frac{r}{r_{32}} = \frac{7}{3} + \frac{\rho_l r^3}{8\sigma} \dot{y}^2 , \tag{5.95}$$

that relates the parent droplet radius r to the Sauter mean radius of the child droplets r_{32}. It is theoretically possible to assume that all child droplets have the same size, similar to the treatment in the Reitz-Diwakar model. However, more often a continuous size distribution is applied to the child droplets around the mean radius of r_{32} in order to obtain a more realistic spray of various drop sizes. Typically a chi-square or a Rosin-Rammler distribution is chosen for this purpose [61, 37].

In ref. [96] Tanner has modified the TAB model in order to obtain more realistic results for global spray parameters such as penetration, radial expansion and cross-sectional drop size distributions. In this so-called ETAB (enhanced TAB) model the major variation compared to the original TAB model is in the calculation of the size and number of child droplets after breakup. Here it is assumed that the rate of child droplet generation is proportional to the number of child droplets $N_{d,child}$:

$$\frac{dN_{d,child}(t)}{dt} = 3K_{bu} N_{d,child}(t) . \tag{5.96}$$

The proportionality constant ($3 K_{bu}$) depends on the breakup regime (the factor 3 has been introduced to simplify later expressions), and is specified as

$$K_{bu} = \begin{cases} k_1 \omega & \text{if} \quad We \leq We_t \\ k_2 \omega \sqrt{We} & \text{if} \quad We > We_t , \end{cases} \tag{5.97}$$

where a transition Weber number of $We_t = 80$ is suggested in order to distinguish between bag and stripping breakup. The values of k_1 and k_2 have been adjusted as $k_1 = k_2 = 2/9$ in order to match experimentally determined drop sizes. The number of child droplets can be written as

$$N_{d,child} = \frac{m_{d,parent}}{m_{d,child}} , \tag{5.98}$$

and differentiation yields

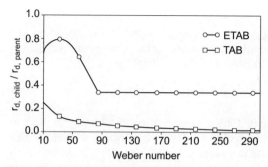

Fig. 5.25. Ratio of child to parent droplet radii predicted by TAB and ETAB models [96]

$$\frac{dN_{d,child}}{dt} = -\frac{m_{d,parent}}{m_{d,child}^2} \cdot \frac{dm_{d,child}}{dt}. \tag{5.99}$$

Implementing Eqs. 5.98 and 5.99 in Eq. 5.96 the expression

$$\frac{dm_{d,child}}{dt} = -3K_{bu}m_{d,child} \tag{5.100}$$

is obtained, and assuming a uniform distribution of child droplet sizes integration of Eq. 5.100 leads to the ratio of child to parent droplet radii:

$$\frac{r_{d,child}}{r_{d,parent}} = \exp(-K_{bu}t_{bu}). \tag{5.101}$$

The breakup time t_{bu} is obtained from the original TAB model, Eqs. 5.89 and 5.90.

The global effect of the ETAB model is that it generally predicts greater child droplet diameters than the original TAB model, Fig. 5.25. As a result, a more realistic droplet size distribution, especially in the thick spray regime close to the injection nozzle, is obtained.

5.5.4 The Kelvin-Helmholtz Breakup Model

Reitz [76] has shown that the wave-breakup theory describing the development of Kelvin-Helmholtz (KH) instabilities on a jet surface can be also be applied to model the secondary breakup of droplets, see Sects. 5.4.2 and 5.4.3. The complete formulation has already been given above, but it should be noted that for a specific adjustment of the breakup time constant B_1, the same results can be recovered than from the TAB model. The breakup time was given in Eq. 5.48 as

$$\tau = 3.726 B_1 \frac{r}{\Lambda\Omega}, \tag{5.102}$$

where the parent droplet radius r is equivalent to the jet radius a. Substituting Eqs. 5.38 and 5.39 for the wave length Λ and the wave growth rate Ω of the most unstable surface waves, respectively, into the above equation, the formulation

$$\tau = 0.82 B_1 \sqrt{\frac{\rho r^3}{\sigma}} \qquad (5.103)$$

is found for an inviscid liquid at the low Weber number limit (We = 6). This agrees to the result of the TAB model, Eq. 5.89, if the time constant is chosen as $B_1 = 1.35$.

For the other limiting case of very large Weber numbers, i.e. in the stripping breakup regime, Eq. 5.102 becomes

$$\tau = B_1 \frac{r}{v_{rel}} \sqrt{\frac{\rho_l}{\rho_g}}, \qquad (5.104)$$

which is the same result as in Eq. 5.90 if $B_1 = \sqrt{3}$. Moreover, the Kelvin-Helmholtz breakup model also predicts a normal velocity component of the secondary droplets after breakup that is determined from Eq. 5.42 and corresponds to Eq. 5.93 if the constants are chosen appropriately. However, as noted above there is considerable uncertainty about the value of B_1. In the literature values ranging from 1.73 up to 30 have been used, and in ref. [78] it was pointed out that B_1 may need to be adjusted to different initial disturbance levels of the droplet.

It should be noted though, that there is one significant difference in the numerical implementation of the TAB and KH breakup models. While in the TAB model one parcel of parent droplet is replaced by one parcel of identically sized child droplets after breakup, the KH breakup model results in a bimodal droplet size distribution with small droplets that are sheared off the surface of the parent droplets and larger droplets remaining from the original parent droplet. This effect is implemented into the numerical scheme of the CFD code by generating additional droplet parcels after breakup.

5.5.5 The Rayleigh-Taylor Breakup Model

The Rayleigh-Taylor (RT) breakup model is based on theoretical considerations of Taylor [100], who investigated the stability of liquid-gas interfaces when accelerated in a normal direction to the plane. Generally, it can be observed that the interface is stable when acceleration and density gradient point to the same direction, whereas Rayleigh-Taylor instabilities can develop if the fluid acceleration has an opposite direction to the density gradient. For a liquid droplet decelerated by drag forces in a gas phase this means, that instabilities may grow unstable at the trailing edge of the droplet, Fig. 5.26.

The acceleration (or deceleration) of a droplet is due to drag forces and follows from Eq. 5.13,

$$\left|\vec{F}\right| = \frac{3}{8} C_D \frac{\rho_g v_{rel}^2}{\rho_l r}, \tag{5.105}$$

where v_{rel} is the relative velocity between droplet and gas, and r is the droplet radius. Based on the assumption of linearized disturbance growth rates and negligible viscosity the frequency and wavelength of the fastest growing waves are

$$\Omega = \sqrt{\frac{2|\vec{F}|}{3}} \cdot \left[\frac{|\vec{F}|(\rho_l - \rho_g)}{3\sigma}\right]^{1/4} \tag{5.106}$$

and

$$\Lambda = 2\pi \sqrt{\frac{3\sigma}{|\vec{F}|(\rho_l - \rho_g)}}, \tag{5.107}$$

respectively [17]. In many applications of the RT-breakup model the gas density is neglected in the above equations because it is significantly smaller than that of the liquid. It is apparent from Eq. 5.106 that the acceleration is the prime factor causing a rapid growth of Rayleigh-Taylor instabilities, whereas the surface tension counteracts the breakup mechanism.

The breakup time is found as the reciprocal of the frequency of the fastest growing wave:

$$t_{bu} = \Omega^{-1}. \tag{5.108}$$

Furthermore, the size of the new child droplets is calculated in dependence of the RT-wavelength Λ, and breakup is only allowed when Λ is less than the diameter of the parent droplet [68]. In ref. [94] the number of new droplets is determined as the ratio of the maximum diameter of the deformed parent droplet to Λ, and the corresponding diameter of the child droplets is obtained from mass conservation principles.

Typically, the Rayleigh-Taylor breakup model is not applied as the only method to describe secondary droplet breakup, but it is rather used in combination with an additional breakup model, most often with the Kelvin-Helmholtz model describing stripping breakup. In that case the RT- and KH-models are implemented in a competing manner, i.e. the droplet breaks up by the mechanism that predicts a shorter breakup time. Close to the injector nozzle where the droplet velocities are highest, the RT-breakup is usually the governing mechanism, whereas the KH-breakup becomes more dominant further downstream.

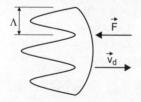

Fig. 5.26. Schematic illustration of Rayleigh-Taylor instabilities on a liquid droplet

However, one more constraint exists in most applications. In order to be able to reproduce experimentally obtained intact core or breakup lengths, e.g. given by the relation

$$L_{bu} = C \cdot \sqrt{\frac{\rho_l}{\rho_g}} \cdot d_{noz} ,\qquad(5.109)$$

the RT-breakup model that would predict extremely rapid breakup directly at the nozzle exit is switched off within this breakup length [20].

Since the RT-breakup model predicts the disintegration of a parent droplet into a number of equally sized child droplets, the combination of the RT- and KH-breakup models counteracts the formation of sprays with a distinct bimodal droplet size distribution as they will be predicted if the KH-model is applied as the only mechanism of secondary droplet breakup.

5.6 Droplet/Droplet and Spray/Wall Interactions

5.6.1 Droplet Collision and Coalescence

Droplet collisions may occur in almost all spray applications, but especially in high pressure sprays as utilized in combustion engines. These collisions have a strong influence on the mean droplet size and its spatial distribution and can therefore affect subprocesses of spray combustion such as mass, momentum and energy transfer between gas and droplets. The probability that two droplets collide obviously depends on their velocities and directions as well as on the local void fraction within the spray. Thus, the frequency of droplet collisions is greatest in the dense spray regions close to the nozzle and near the spray axis. It is also noticeable in the vicinity of combustion chamber walls where incoming and outgoing droplets may interact, or for droplets of significantly different velocity, e.g. because small droplets are decelerated more quickly by drag forces than larger ones.

Collision Regimes

The mechanisms behind droplet collisions are complex and many studies have been conducted on this topic. In general, it can be concluded that several types of droplet-droplet interaction are possible in collisions, e.g. [34, 72]: (i) the droplets may bounce apart almost elastically because of a dynamic pressure rise in the gas layer separating the droplets, (ii) they may coalesce permanently, (iii) they may coalesce but separate again shortly afterwards, or (iv) shattering may occur in which tiny droplets are expelled radially from the periphery of the interacting droplets. The third regime, temporary coalescence and subsequent separation, can be further subdivided into reflexive separation and stretching separation, depending on whether there is a "head-on" collision or an off-axis "grazing" collision, Fig. 5.27.

a) head-on collision (reflexive separation)

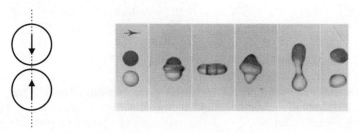

b) off-axis collision (stretching separation)

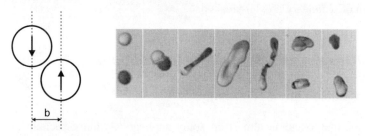

Fig. 5.27. Schematic diagrams and photographs of (a) head-on and (b) off-axis collision with separation of droplets [10]

The above collision regimes can be classified based on a collision Weber number and an impact parameter as qualitatively shown in Fig. 5.28. The collision Weber number is defined as

$$\text{We}_{col} = \frac{\rho_l d_2 v_{rel}^2}{\sigma}, \tag{5.110}$$

where the relative velocity v_{rel} between the droplets is determined as

$$v_{rel} = |\vec{v}_2 - \vec{v}_1| = \sqrt{v_1^2 + v_2^2 - 2v_1 v_2 \cos\alpha}. \tag{5.111}$$

The subscripts 1 and 2 refer to the larger and smaller of the two colliding droplets, respectively. According to Fig. 5.29 α is the angle between the paths of the droplets prior to breakup. The impact parameter x is defined as the dimensionless distance between the droplet centers measured perpendicular to the relative velocity vector:

$$x = \frac{b}{r_1 + r_2}. \tag{5.112}$$

Thus, $x = 0$ indicates a head-on collision, and $x = 1$ a "glancing" collision where the droplets just touch each other.

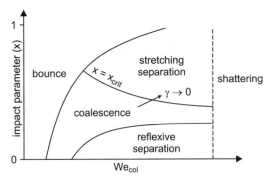

Fig. 5.28. Droplet collision regimes

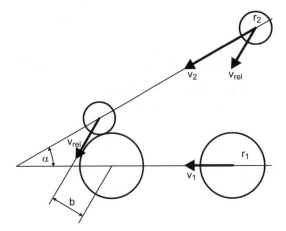

Fig. 5.29. Geometric collision parameters

Collision Modeling

While fairly detailed theories have been proposed to describe the various collision mechanisms included in Fig. 5.28, e.g. [11, 34, 70], up to now their application in numerical simulations has been mostly limited to fundamental studies. In most engine spray simulations the collision model by O'Rourke [60, 62] has been used as the standard collision model, probably because of two reasons. Firstly, the implementation of detailed collision models into CFD codes may be quite complex, and secondly, there is a general difficulty in even assessing the quality of a collision model by comparison to experimental spray data. The latter typically represents an integral result of both droplet breakup and droplet collision processes. And as noted above, there are still uncertainties encountered in the breakup modeling as well.

The O'Rourke collision model distinguishes only two spray regimes, coalescence and stretching separation. The physical criterion for separation is that the rotational energy of the temporarily coalesced drop pair has to exceed the surface energy required to re-form the original drops from the coalesced pair. The respective regime boundary has been indicated has x_{crit} in Fig. 5.28. It can be derived by equating the extra surface energy necessary for separation of the droplets to the rotational kinetic energy of the combined drop

$$\Delta\ k_{surf} = 4\pi\sigma\left(r_1^2 + r_2^2 - r_{eff}^2\right) \equiv k_{rot} = \frac{1}{2}\frac{L^2}{J}, \qquad (5.113)$$

where r_{eff} is the effective radius of the combined drop, L is the angular momentum about the center of mass, and J is the moment of inertia:

$$r_{eff} = \left(r_1^3 + r_2^3\right)^{1/3}, \qquad (5.114)$$

$$L = \frac{m_1 m_2}{m_1 + m_2} b v_{rel}, \qquad (5.115)$$

$$J = \frac{2}{5}(m_1 + m_2) r_{eff}^2. \qquad (5.116)$$

Substituting Eqs. 5.114 to 5.116 into Eq. 5.113 yields, after some mathematical rearrangements, the result for the critical impact parameter,

$$x_{crit}^2 = \frac{b_{crit}^2}{(r_1 + r_2)^2} = \frac{12 f(\gamma)}{5 \mathrm{We}_{col}}, \qquad (5.117)$$

where $(\gamma = r_2/r_1)$ is the ratio of the droplet radii ($0 < \gamma \leq 1$), and $f(\gamma)$ stands for the rather complex expression

$$f(\gamma) = \frac{(\gamma^3 + 1)^3}{\gamma^6 (\gamma + 1)^2} \cdot \left[\gamma^2 + 1 - (\gamma^3 + 1)^{2/3}\right]. \qquad (5.118)$$

Thus, x_{crit} is a function of the collision Weber number and of the droplet size ratio. In fact, Eq. 5.117 has the effect that for a constant Weber number the critical impact parameter increases as the radius ratio γ approaches zero. This can be explained ostensively by taking into account that for a constant Weber number (defined as a function of the smaller droplet) a decrease in γ means that the size of the larger drop increases. And the larger the size of large drop, the more easily it can absorb the energy of the small drop.

In case the two droplets permanently coalesce, i.e. if x is less than x_{crit}, the velocity $\vec{v}'$ of the combined droplet after collision is calculated as

$$\vec{v}' = \frac{m_1 \vec{v}_1 + m_2 \vec{v}_2}{m_1 + m_2}, \qquad (5.119)$$

and the new temperature becomes

$$T' = \frac{m_1 T_1 + m_2 T_2}{m_1 + m_2}. \qquad (5.120)$$

In the case of stretching separation, O'Rourke [60] derived the equations

$$\vec{v}'_1 = \frac{m_1\vec{v}_1 + m_2\vec{v}_2 + m_2\left(\vec{v}_1 - \vec{v}_2\right)}{m_1 + m_2} \cdot \frac{b - b_{crit}}{r_1 + r_2 - b_{crit}} \qquad (5.121)$$

and

$$\vec{v}'_2 = \frac{m_1\vec{v}_1 + m_2\vec{v}_2 + m_1\left(\vec{v}_1 - \vec{v}_2\right)}{m_1 + m_2} \cdot \frac{b - b_{crit}}{r_1 + r_2 - b_{crit}} \qquad (5.122)$$

for the velocities of droplets 1 and 2 after collision and subsequent separation, which implies some simplifying assumptions. Furthermore, the temperature of the initial droplets is assumed to be not affected by the collision.

Implementation of Collision Models in CFD Codes

In spray simulations with a large number of droplets it is not feasible to check for every drop whether its path crosses the path of any other drop during a time step such that collision occurs. Therefore, a statistical technique that yields a collision probability is necessary. Typically, this is achieved by assuming that the probable number of collisions k between N_1 droplets of parcel 1 and N_2 droplets of parcel 2, where parcels 1 and 2 are both located within the same computational grid cell, follows a Poisson distribution,

$$P(k) = \frac{\left(v_{12} \cdot \Delta t\right)^k}{k!} \exp\left(-v_{12} \cdot \Delta t\right), \qquad (5.123)$$

where Δt is the time increment of the computation and v_{12} is the time averaged collision frequency given as

$$v_{12} = \frac{N_2}{V_{cell}} \pi \left(r_1 + r_2\right)^2 \cdot \left|\vec{v}_1 - \vec{v}_2\right|. \qquad (5.124)$$

Again, subscript 2 indicates the smaller of the two droplets, and V_{cell} is the volume of the computational grid cell.

The numerical procedure is to sample the integrated distribution function of $P(k)$ from a uniform distribution between zero and one, and to solve it for the number of collisions k within the particular time step. In order to determine the impact parameter, the off-center distance b is determined from the relation

$$b^2 = q \cdot \left(r_1 + r_2\right)^2, \qquad (5.125)$$

where q is again sampled from a uniform distribution between zero and one.

5.6.2 Spray-Wall Impingement

Especially in modern passenger car diesel engines with compact combustion chambers and high pressure injection systems spray wall impingement is an inherent subprocess of mixture formation. However, its influence on combustion is inexplicit. On the one hand, it may intensify spray heating and vaporization because the droplet shattering at impact causes an increase in the overall spray surface area. On the other hand, a liquid wall film caused by fuel deposition is often subject to insufficient mixture preparation and thus contributes to an increase in unburned hydrocarbon and soot emissions. As a consequence, a detailed understanding of the subprocesses involved in spray wall impingement is necessary in order to asses the overall effects on combustion and engine performance.

Impingement Regimes

Depending on the impact conditions the outcome of droplet-wall interactions can be classified into several different regimes that are schematically shown in Fig. 5.30. The droplet may (i) stick to the wall in nearly spherical form, it may (ii) rebound because of either gas that is trapped between the drop and a liquid wall film or because of a fuel vapor layer that forms on a hot, dry wall, it may (iii) spread out to form a liquid wall film, it may breakup into a number of smaller droplets which can be either boiling induced (iv) or caused by the kinetic energy of the incoming droplet (v), or the drop may (vi) splash, in which a crown is formed, jets develop on the periphery of the crown and the jets become unstable and breakup into many fragments.

The impingement parameters governing the existence of the above regimes include fuel properties like density, surface tension and viscosity, the droplet size, incident angle and velocity, the wall temperature and roughness or, if present, the liquid wall layer thickness, as well as the characteristics of the gas boundary layer at the wall. Some of the above parameters can be combined to yield the following dimensionless parameters that are essential to the modeling of the impingement process.

The droplet Weber number is defined in terms of the incident droplet diameter and its velocity component normal to the wall. It represents a measure of the droplet kinetic energy to its surface energy:

$$\mathrm{We}_{in} = \frac{\rho_l d_{in} v_{in,n}^2}{\sigma}. \qquad (5.126)$$

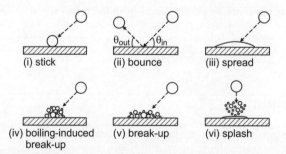

Fig. 5.30. Schematic illustration of different impact mechanisms [13]

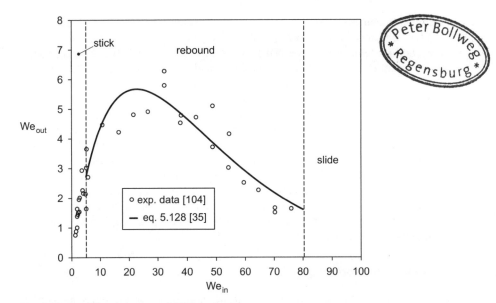

Fig. 5.31. Outgoing vs. approach Weber number

The droplet Laplace number relates the surface tension to the liquid viscosity:

$$\text{La} = \frac{\rho_l \sigma d_{in}}{\mu^2}. \tag{5.127}$$

Often this measure is also expressed by the Ohnesorge number which is related to the Laplace number by ($Z = \text{La}^{-1/2}$).

Wachters and Westerling [104] conducted an early experimental study about water droplets impinging on a heated wall and found out that the Weber number of the outgoing drop is strongly dependent on the Weber number of the incoming drop. Gonzalez et al. [35] utilized these experimental results and presented a curve fit for the outgoing Weber number,

$$\text{We}_{out} = 0.678\,\text{We}_{in}\exp(-0.044\,\text{We}_{in}), \qquad 5 < \text{We}_{in} < 80 \tag{5.128}$$

which is plotted in Fig. 5.31 in addition to the respective experimental data. It was proposed that for very low impact energies with ($\text{We}_{in} \leq 5$) the drop sticks on the wall. Within the range ($5 < \text{We}_{in} < 80$) there is droplet rebound. However, for incident Weber numbers greater than about 30 the outgoing Weber number is reduced again. This phenomenon is attributed to breakup of the incident droplet during impact which is associated with an increase in surface energy, and consequently a decrease in kinetic energy. For incident Weber numbers in excess of 80 it is assumed that the drops slide along the wall in the manner of a liquid jet.

Bai and Gosman [13] suggested a more detailed map of impingement regimes that includes the wall temperature in addition to the incident Weber number in order to determine the impact outcome. Furthermore, as indicated in Fig. 5.32 the additional regimes of spread, breakup and splash have been considered such that all impact mechanisms shown in Fig. 5.30 are accounted for. The characteristic wall temperatures T_b, T_N and T_{leid} that are indicated on the horizontal axis are the liquid boiling temperature, the Nakayama temperature at which a droplet reaches its maximum evaporation rate and the Leidenfrost temperature, respectively. The latter is characterized by a minimum evaporation rate because of an insulating stable vapor cushion that develops between the wall and the liquid.

Especially the splash regime that is observed for very high impact energies may be important to applications with a short distance between injector and wall, because these mechanisms contribute to a reduction in the mean droplet size and thus to an enhanced evaporation rate. On the other hand, in typical engine applications – especially in diesel engines where wall impingement is of greater importance because of the high injection pressures – the wall temperatures of the combustion chamber are often not significantly above the boiling temperature of the fuel. Thus, the two breakup regimes are often neglected in impingement models applied to engine sprays.

Impingement Modeling

Naber and Reitz [57] developed a model that is oriented on the three breakup regimes stick, reflect and slide as indicated in Fig. 5.31. As noted above, a drop is assumed to stick at the wall for very low incident Weber numbers. After impact it stays at the impingement location and continues to evaporate. In the so-called reflect regime applicable for intermediate incident Weber numbers the tangential velocity component of the outgoing droplet remains unchanged whereas the normal velocity component keeps its initial absolute value but changes its sign after impact. Note that this causes specular reflection and is in contrast to the experimental results in Fig. 5.31, where the outgoing Weber number is generally smaller than the incident Weber number. Consequently, there is no droplet breakup considered in this model. However, in a later study Eq. 5.128 was included in the reflect regime in order to account for energy dissipation during impingement instead of assuming specular reflection [35].

For the slide regime an empirical approach was chosen in analogy to a liquid jet impinging on an inclined wall (Fig. 5.33). The jet angle φ at which the impinging drop leaves the impact position in the plane of the wall was derived from mass and momentum conservation as

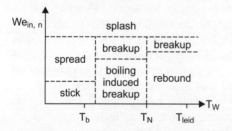

Fig. 5.32. Overview of impingement regimes for a dry wall[13]

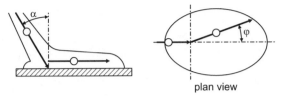

plan view

Fig. 5.33. Wall impingement in the slide regime [57]

$$\varphi = -\left(\frac{\pi}{\beta}\right)\cdot \ln\left[1 - P\left(1 - \exp(-\beta)\right)\right], \tag{5.129}$$

where β is determined from

$$\sin\alpha = \frac{\exp(\beta)+1}{\exp(\beta)-1} \cdot \frac{1}{1+(\pi/\beta)^2}, \tag{5.130}$$

and P is sampled from a uniform distribution between zero and one. The drop incident angle α is measured relative to the normal vector of the wall as indicated in Fig. 5.33. Again, breakup resulting from impingement is not considered in the model.

Bai et al. [13, 14] proposed a more detailed impingement model that distinguishes between dry and wetted walls. In the dry case the stick, spread and splash regimes are considered, where the former two have been combined to a so-called adhesion regime in which a liquid wall film is formed. This can be justified by the reasoning that in engine sprays there is seldom a single droplet impinging on a wall but rather a large number of droplets that will quickly form a wall film even if impact is in the stick regime. The transition criterion between adhesion and splash is given in terms of the Laplace number as

$$\text{Adhesion} \rightarrow \text{Splash} \qquad We_c = A \cdot La^{-0.18}, \tag{5.131}$$

where the parameter A depends on the surface roughness of the dry wall. For standard cases a value of $A = 2630$ has been suggested in ref. [14].

For a wall that has already been wetted by prior impingement four different regimes are distinguished: stick, rebound, spread and splash. The transition Weber numbers between these regimes are given as:

$$\text{Stick} \rightarrow \text{Rebound} \qquad We_c = 2 \tag{5.132}$$

$$\text{Rebound} \rightarrow \text{Spread} \qquad We_c = 20 \tag{5.133}$$

$$\text{Spread} \rightarrow \text{Splash} \qquad We_c = 1320 \cdot La^{-0.18} \tag{5.134}$$

In the stick and spread regimes the droplet mass is added to the already existing wall film. In the rebound regime the tangential and normal velocity components of the outgoing drop are calculated following the relations developed for solid particles bouncing on a solid wall [51]:

$$v_{out,t} = \frac{5}{7} v_{in,t},$$ (5.135)

$$v_{out,n} = -e \cdot v_{in,n}.$$ (5.136)

The quantity e is the restitution coefficient which is assumed to follow the relation

$$e = 0.993 - 1.76\theta + 1.56\theta^2 - 0.49\theta^3,$$ (5.137)

where θ is the incident angle with the wall, measured in radians ($\theta = \pi/2 - \alpha$).

As noted above, splashing will occur both on dry and on wetted walls if the impact energy is very high. In this case secondary droplets are formed that are smaller in size than the original model and do not need to be of uniform size themselves. In fact, it was suggested that the generation of two new droplet classes (parcels) for each original drop impinging on the wall is a reasonable compromise between accuracy and expenditure. Both groups of secondary droplets are of equal mass, i.e.

$$m_{sp,1} = m_{sp,2} = \frac{1}{2} m_{sp},$$ (5.138)

where the ratio of total mass splashed to incident droplet mass is determined from the empirical relation

$$\frac{m_{sp}}{m_{in}} = \begin{cases} 0.2 + 0.6p, & \text{for dry wall} \\ 0.2 + 0.9p, & \text{for wetted wall} \end{cases},$$ (5.139)

in that p is sampled from a uniform distribution within the interval (0,1). Hence, for a wetted wall the mass ratio can exceed unity, since splashing droplets may entrain liquid from the wall film.

Following observations from Stow and Stainer [93], the total number of droplets after impingement is assumed to be approx. equal to

$$N_{sp} = N_{sp,1} + N_{sp,2} \approx 5\left(\frac{We}{We_c} - 1\right),$$ (5.140)

where We_c is obtained from either Eq. 5.131 or Eq. 5.134 for dry or wetted walls, respectively. The droplet number of one class of droplets is then randomly chosen such that ($1 \leq N_{sp,1} \leq N_{sp}$), and the droplet number of the other class is obtained from solving the left part of Eq. 5.140 for $N_{sp,2}$. Once the droplet numbers of the two secondary droplet classes are known, the respective droplet sizes can be estimated from mass conservation requirements:

$$N_{sp,1} d_{sp,1}^3 = \frac{m_{sp}}{2m_{in}} d_{in}^3,$$ (5.141)

$$N_{sp,2} d_{sp,2}^3 = \frac{m_{sp}}{2m_{in}} d_{in}^3.$$ (5.142)

Energy conservation is applied in order to determine the velocity components of the secondary droplets. It requires that the sum of kinetic and surface energies of secondary droplets are equal to the splashing energy, which is assumed to be the difference between the kinetic energy of the incident drop and the critical kinetic energy below which no splashing occurs:

$$\frac{1}{4}m_{sp}\left(\left|\vec{v}_{sp,1}\right|^2+\left|\vec{v}_{sp,2}\right|^2\right)+\pi\sigma\left(N_{sp,1}d_{sp,1}^2+N_{sp,2}d_{sp,2}^2\right)=E_{k,sp}=E_{k,in}-E_{k,c}. \quad (5.143)$$

The latter can be evaluated from the critical Weber number as

$$E_{k,c}=\frac{We_c}{12}\pi\sigma d_{in}^2. \quad (5.144)$$

From experimental data on the size-velocity correlation of secondary droplets after splashing the relation

$$\frac{\left|\vec{v}_{sp,1}\right|}{\left|\vec{v}_{sp,2}\right|}\approx\ln\left(\frac{d_{sp,1}}{d_{in}}\right)/\ln\left(\frac{d_{sp,2}}{d_{in}}\right) \quad (5.145)$$

is deduced. It implies that the larger the size of the secondary droplets, the smaller the magnitude of their velocity.

Finally, application of the tangential momentum conservation law produces

$$\frac{m_{sp}}{2}\left|\vec{v}_{sp,1}\right|\cos\theta_{sp,1}+\frac{m_{sp}}{2}\left|\vec{v}_{sp,2}\right|\cos\theta_{sp,2}=c_f m_{in}\left|\vec{v}_{in}\right|\cos\theta_{in}. \quad (5.146)$$

Here c_f is the wall friction coefficient which is assumed to be in the range of 0.6 to 0.8. The quantities $\theta_{sp,1}$ and $\theta_{sp,2}$ are the ejection angles of the two secondary droplet classes, and θ_{in} is the respective angle of the incident droplet. $\theta_{sp,1}$ is randomly chosen within the assumed ejection cone (approx. $10°\le\theta_{sp,1}\le 160°$) and $\theta_{sp,2}$ is determined from Eq. 5.146. Obeying the fact that

$$\left|\vec{v}_j\right|^2=v_{j,n}^2+v_{j,t}^2, \qquad j=1,2 \quad (5.147)$$

the above set of equations can be solved for the normal and tangential velocity components of the two secondary droplet classes.

5.7 Fuel Evaporation

Fuel evaporation is a process of great importance especially in direct injection engines. It has a direct effect on the combustion rate since only vaporized fuel that has been mixed with air in a combustible ratio can chemically react with the oxygen contained in the fresh intake air. Thus, an engine's thermodynamic efficiency is affected by the evaporation rate. And moreover, the fuel evaporation can have a significant influence on the emission formation as well. Poor evaporation will typically cause increased soot and unburned hydrocarbon emissions, while very rapid evaporation, especially during the ignition delay in diesel engines, will cause an increase in nitrogen oxides because of rapid premixed combustion associated with high temperatures. Consequently, a thorough understanding of the dominating processes in fuel vaporization is a prerequisite in order to assess the overall quality of mixture formation and spray combustion.

In direct injection engines the major fraction of the fuel mass evaporates after the spray has broken up into small droplets. This is because the liquid fuel is typically below its boiling temperature when it exits the injection nozzle and its specific surface area is very small prior to atomization. Therefore, droplet evaporation is the most important part in evaporation modeling. However, under certain boundary conditions additional evaporation mechanisms such as wall film evaporation, e.g. after spray wall impingement, or flash boiling, e.g. for volatile fuels, may become important, too.

5.7.1 Droplet Evaporation

Droplet evaporation is governed by conductive, convective and radiative heat transfer from the hot gas to the colder droplet and by simultaneous diffusive and convective mass transfer of fuel vapor from the boundary layer at the drop surface into the gas environment. However, it is not feasible to directly resolve the flow field in and around the many droplets of a complete spray due to constraints with respect to computer power and memory. Therefore, in engine applications it is most often assumed that the droplets are ideally spherical and averaged flow conditions and transfer coefficients around the droplets are determined. Moreover, the radiative heat transfer between gas and drops is typically neglected as it is small compared to convection. Thus, the same set of equations that is utilized in phenomenological models and has been described above (see Eqs. 3.72 to 3.77) is also applied in standard CFD-calculations.

Nevertheless, several additional effects on droplet vaporization have been investigated in detail by various researchers. In ref. [83] it was shown that vaporization reduces both the heat transfer rate to the droplet and its drag coefficient. However, as reported in ref. [78] numerical studies performed with the CFD-code KIVA suggested that this effect is relatively small for diesel-type sprays.

Taking into account that the fuel drops are significantly distorted during the majority of their lifetime in diesel sprays, Gavaises and co-workers [8, 33] investigated the effect of non-spherical droplets on the fuel evaporation rate. They evaluated not only the drag coefficient as a function of this phenomenon but also accounted for a modification in the exchange area between drop and gas by assuming that the distorted drop becomes a spheroid that can be characterized by the maximum and minimum diameters calculated from the breakup model. Accordingly, the dimensionless Nusselt and Sherwood numbers were not estimated from Eqs. 3.76 and 3.77 but from more detailed correlations derived from heat transfer studies on spheroids. The reported results indicate that the consideration of droplet deformation effects yields an increased overall evaporation rate for complete sprays. Thus, there is an influence on the spatial equivalence ratio distribution, and ignition timings and locations are likely to be affected as well.

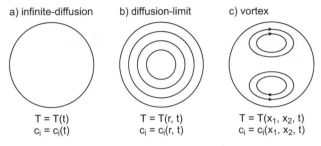

Fig. 5.34. Modeling approaches for the droplet interior

The above droplet evaporation models are all based on the lumped capacitance method, i.e. the temperature within the droplet is assumed to be spatially uniform and depends on time only. A measure whether this assumption is valid is given by the Biot number which relates the resistance to heat conduction within the droplet to the resistance to heat convection from the gas to the droplet, see e.g. [44]:

$$\mathrm{Bi} = \frac{R_{cond}}{R_{conv}} = \frac{hr_d}{k_l}. \tag{5.148}$$

For Biot numbers much less than unity the assumption of a uniform temperature within the droplet is justified. However, this may not always be the case in engine sprays as has been shown by several authors who investigated the evaporation process including the spatial and temporal evolutions of the droplet interior under diesel like conditions, e.g. [21, 30, 39]. A detailed review on this subject has been presented by Sirignano [90]. However, these sophisticated evaporation models that account for two- or even three-dimensional flows within the droplet interior are limited to studies of isolated droplets or simple arrays of droplets and cannot be used in complete spray simulations because of computer limitations. For this purpose three different degrees of simplification have been proposed and examined in the literature, Fig. 5.34.

Besides the simplest *infinite-diffusion* model (Eqs. 3.72–3.77), that is based on the lumped capacitance method and assumes a well-mixed droplet interior without spatial gradients at any time, the so-called *diffusion-limit* model has been proposed. It assumes a one-dimensional temperature and, in the case of a multi-component fuel, mass distribution as a function of the droplet radius. The heat and mass exchange processes inside the droplet are governed by conduction and diffusion, respectively, and are solved on a one-dimensional numerical grid of concentric shells. The most comprehensive of the simplified models is the *vortex model*. It considers an inviscid liquid flow region inside the droplet that is assumed to be given by Hill's vortex.

Aggarwal [3] compared the three above droplet models for varying boundary conditions and observed that the infinite-diffusion model yields results that are markedly different from the two more complex models. This is especially the case for multi-component fuel mixtures where the species concentrations of the more and less volatile compounds in the vapor phase are significantly overpredicted and underpredicted, respectively. The diffusion-limit and the vortex models were found to produce remarkably similar results under most conditions and thus, the use of the diffusion-limit model is recommended as it is easier to handle and computationally more efficient.

It should be noted though, that the simple droplet evaporation model described in Chap. 3, that assumes a spherical, well-mixed droplet and is based on a single-component fuel, can still be viewed as the standard in today's engine spray simulations. This is mainly due to its simplicity and because of computing time requirements. Nevertheless, the number of numerical studies including more comprehensive evaporation models is steadily increasing. The focus has especially shifted towards a more realistic description of real diesel and gasoline fuels that can hardly be characterized by a single-component model fuel since they are a mixture of several hundred different hydrocarbon components. These efforts will be discussed in the following section.

5.7.2 Multi-Component Fuels

Most early studies on multi-component fuel evaporation have concentrated on binary model fuels in that two representative hydrocarbon components are mixed in a discrete manner. For example, Jin and Borman [45] modeled a 50-50% pentane-octane mixture, and in ref. [3] a 50-50% mixture of n-hexane and n-decane was studied. In an attempt to describe conventional diesel fuel that also contains aromatic compounds more accurately, a two-component model fuel consisting of 70% n-decane and 30% α-methyl-naphthalene was investigated in ref. [42].

These discrete two-component models are based on phase equilibrium relations at the drop surface and, assuming ideal behavior of both the vapor and liquid phases, Raoult's law may be applied in order to specify the concentration of each fuel component in the vapor phase. Results obtained with such two-component fuel models indicate that there may indeed be a significant effect on the overall mixture formation and combustion process as the more volatile component generally evaporates more rapidly. Consequently, there is an inhomogeneous distribution of the components in the vapor phase such that the ignition delay and ignition location may be affected. As has been shown by Ayoub and Reitz [12], who investigated a 50-50% molar mixture of dodecane and hexadecane, this becomes especially important under cold-starting conditions. Figure 5.35 compares the fraction of the more volatile dodecane in the vapor phase for two different ambient temperatures as a function of time after start of injection. It can be seen that for the lower ambient temperature of 500 K the total vaporized fuel mass is comprised of about 90% dodecane during the time increment important for autoignition.

In the recent past it has become more and more popular to model more realistic fuels by means of continuous thermodynamics [109]. In this approach, that has been followed for both diesel and gasoline, e.g. [48, 67, 95, 110], the composition of the model fuel is described by a continuous distribution function that specifies the mole fraction in terms of the molecular weight.

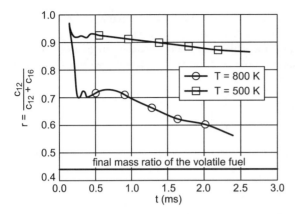

Fig. 5.35. Fraction of volatile mass in the vapor phase of a two-component fuel [12]

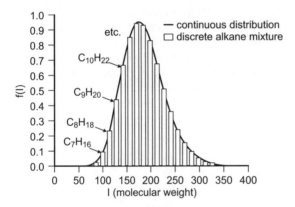

Fig. 5.36. Continuous and discrete representations of an alkane-mixture [67]

In Fig. 5.36 a typical distribution function, that has been chosen to model diesel fuels, is compared to an equivalent mixture of a discrete number of n-alkanes with corresponding molecular weights. In contrast to modeling and numerically solving the many differential equations required by a mixture of 15 or more discrete components, the continuous thermodynamics approach represents an elegant and more efficient method to describe a multi-component fuel. Here only three parameters are necessary in order to explicitly define the fuel state. These are the overall fuel mass, the mean molecular weight of the distribution function, and its second moment, i.e. a measure of its variance.

It should be noted, that the use of the continuous thermodynamics is only possible if the important fuel properties such as boiling and critical temperatures, density, surface tension, viscosity etc. can be explicitly expressed in terms of the molecular weight. This conditions is typically satisfied for a particular family of hydrocarbons, e.g. for n-alkanes, but not for components belonging to different categories. Therefore, it would not be possible to model a mixture of both alkanes and aromatic compounds with a single distribution function.

In order to approximate real fuels the following distribution of n-alkanes is chosen that specifies the relative mole fraction or frequency f within the liquid phase in terms of the molecular weight I:

$$f(I) = \frac{(I-\gamma)^{\alpha-1}}{\beta^{\alpha}\Gamma(\alpha)} \exp\left[-\left(\frac{I-\gamma}{\beta}\right)\right]. \tag{5.149}$$

$f(I)$ is normalized such that its integral becomes unity, and the Γ-function is defined as

$$\Gamma(\alpha) = \int_0^{\infty} e^{-t} t^{\alpha-1} dt. \tag{5.150}$$

α and β are shape parameters, that govern the mean θ and the variance σ of the distribution, whereas γ specifies its origin. In Table 5.2 the shape parameters as well as the resulting quantities of θ and σ are summarized for diesel fuel and gasoline. With an extremely narrow distribution it is also possible to describe a quasi-single-component fuel with the continuous thermodynamics method. The respective parameters of such a function for n-octane are included in Table 5.2 as well.

The change in the vapor phase composition can be determined by a separate conservation equation for each of the three conserved scalars, i.e. for the overall fuel mass fraction y_f, for its mean molecular weight θ, and for its second moment Ψ. Following Eq. 4.53, the respective relations can be written as

$$\frac{\partial}{\partial t}(y_f \rho) + \frac{\partial(y_f u_i \rho)}{\partial x_i} = \frac{\partial}{\partial x_i}\left(\rho \bar{D} \frac{\partial y_f}{\partial x_i}\right) + (y_f \dot{\rho})^s + (y_f \dot{\rho})^c, \tag{5.151}$$

$$\frac{\partial}{\partial t}(y_f \rho \theta) + \frac{\partial(y_f u_i \rho \theta)}{\partial x_i} = \frac{\partial}{\partial x_i}\left(\rho \tilde{D} \frac{\partial(y_f \theta)}{\partial x_i}\right) + (y_f \dot{\rho}\theta)^s + (y_f \dot{\rho}\theta)^c, \tag{5.152}$$

$$\frac{\partial}{\partial t}(y_f \rho \Psi) + \frac{\partial(y_f u_i \rho \Psi)}{\partial x_i} = \frac{\partial}{\partial x_i}\left(\rho \hat{D} \frac{\partial(y_f \Psi)}{\partial x_i}\right) + (y_f \dot{\rho}\Psi)^s + (y_f \dot{\rho}\Psi)^c. \tag{5.153}$$

Table 5.2. Distribution parameters for various fuels [48]

	n-Octane	Diesel	Gasoline
α	100	18.5	5.7
β	0.1	10	15
γ	104.2	0	0
θ	114.2	185	85.5
σ	1	43	35.8

In the above equations as throughout this text the Einstein notation has been utilized. The second and third terms on the right hand sides of Eqs. 5.151 through 5.153 are the source terms due to spray evaporation and combustion, respectively, and u is the velocity of the gas phase. The quantities $\overline{D}, \tilde{D}$ and $\hat{D}$ are the binary diffusion coefficients for the respective conserved scalars.

In order to determine the source terms originating from the spray, the evaporation of the liquid multi-component droplets has to be calculated. Most often this is done by assuming a well-mixed droplet interior. The molar flux over the droplet surface is defined as

$$\dot{n} = -\frac{1}{A_d}\frac{d}{dt}(c_l V_d), \qquad (5.154)$$

where V_d and A_d are the volume and surface area of the spherical droplet, respectively, and c_l indicates the liquid molar density in mol/cm^3. The change of droplet radius due to evaporation then becomes

$$\frac{dr}{dt} = \frac{r}{3c_l\theta_l}\left(c_l\frac{d\theta_l}{dt} - \frac{d\rho_l}{dt}\right) - \frac{\dot{n}}{c_l}, \qquad (5.155)$$

and the change of the mean and the second moment of the droplet composition are

$$\frac{d\theta_l}{dt} = \frac{3\dot{n}}{c_l r}\left[(\theta_l - Y_{f,s}\theta_s) + \frac{1}{B}(Y_{f,\infty}\theta_\infty - Y_{f,s}\theta_s)\right], \qquad (5.156)$$

$$\frac{d\Psi_l}{dt} = \frac{3\dot{n}}{c_l r}\left[(\Psi_l - Y_{f,s}\Psi_s) + \frac{1}{B}(Y_{f,\infty}\Psi_\infty - Y_{f,s}\Psi_s)\right], \qquad (5.157)$$

respectively. The fuel mole fraction is denoted by Y_f, and subscripts s and ∞ indicate the gas phase properties at the droplet surface and in the undisturbed surroundings, respectively. The Spalding transfer number B in the above equations is

$$B = \frac{Y_{f,s} - Y_{f,\infty}}{1 - Y_{f,s}}. \qquad (5.158)$$

The change in droplet temperature is governed by convective heat transfer and latent heat of evaporation. It can be estimated as

$$\frac{dT_l}{dt} = \frac{3}{c_l \tilde{c}_v r}(\dot{q}_{conv} - \dot{n}\tilde{h}_{fg}), \qquad (5.159)$$

where $\tilde{h}_{fg}$ is the molar heat of evaporation and $\tilde{c}_v$ is the molar specific heat of the liquid. The molar flux (Eq. 5.154) may be approximated by

$$\dot{n} = \frac{c_g \overline{D}}{2r}\ln(1+B)\left(2.0 + 0.6\,\mathrm{Re}^{1/2}\,\mathrm{Sc}^{1/3}\right), \qquad (5.160)$$

and the convective heat flux by

$$\dot{q}_{conv} = \frac{k(T_\infty - T_l)}{2r}\frac{\ln(1+B)}{B}\left(2.0 + 0.6\,\mathrm{Re}^{1/2}\,\mathrm{Sc}^{1/3}\right). \qquad (5.161)$$

In order to determine the gas phase properties at the drop surface (index s), phase equilibrium is assumed. Applying Raoult's law the fuel mole fraction in the vapor phase becomes

$$Y_f = \int_0^\infty f_i(I) \frac{p_i(I)}{p} dI, \qquad (5.162)$$

where $p_i(I)$ indicates the partial pressure of the virtual species i characterized by molecular weight I. Its value at the droplet surface is estimated by the Clausius-Clapeyron equation,

$$p_i = p_{ref} \exp\left[\frac{\tilde{s}_{fg}}{\tilde{R}}\left(1 - \frac{T_b}{T_l}\right)\right], \qquad (5.163)$$

where $p_{ref} = 101.325$ kPa, $\tilde{s}_{fg}$ is the entropy of evaporation, and T_b is the boiling temperature of the virtual component i. By assuming a linear variation of this boiling temperature as a function of the molecular weight I,

$$T_b(I) = a_b + b_b I, \qquad (5.164)$$

Eq. 5.162 can be simplified to the form

$$Y_{f,s} = \frac{p_{ref}}{p} \frac{\exp[A(1-\gamma B)]}{(1+AB\beta_l)^{\alpha_l}}. \qquad (5.165)$$

Here, A and B are defined as

$$A = \left(\frac{\tilde{s}_{fg}}{\tilde{R}}\right)\left(1 - \frac{a_b}{T_l}\right) \qquad (5.166)$$

and

$$B = \frac{b_b}{T_l - a_b}, \qquad (5.167)$$

respectively.

Typical results, that have been obtained with a continuous thermodynamics model by Pagel et al. [67], are shown below. In Fig. 5.37 the temporal evolution of the distribution function within an evaporating liquid droplet is plotted. It can be seen that during the course of evaporation the distribution becomes narrower and its mean molecular weight shifts towards greater values. This is due to the fact that the lighter components that are more volatile evaporate more quickly and only the heavier molecules remain in the liquid phase. The increase in the maximum of the distribution is simply caused by the fact that the area beneath the function has been normalized to unity.

In Fig. 5.38 the evaporation of a complete spray under diesel like conditions is compared for a single-component fuel (n-$C_{14}H_{30}$) and for the continuous thermodynamics approach. The plots show the overall fuel mass fraction in the vapor phase, and it becomes obvious that even though the mean properties of the multi-component fuel are very similar to the $C_{14}H_{30}$-properties, the overall mixture formation process is predicted quite differently. Specifically, with the multi-component fuel the spray tip penetration is distinctively shorter whereas close to the nozzle the fuel vapor distribution is broader and the gradients are not as steep as in the single-component case. These findings can be explained by the fact that a certain fraction of the multi-component fuel is more volatile than pure tetradecane. Thus, the initial evaporation rate in the vicinity of the injector is enhanced such that the average droplet diameter is decreased. Consequently, the specific aerodynamic forces are increased and the penetration length is reduced.

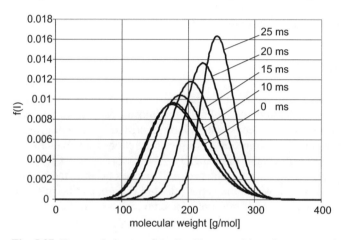

Fig. 5.37. Temporal change of the liquid composition of an evaporating droplet in nitrogen. $d_{d,0} = 100$ mm, $T_{d,0} = 300$, $T_{amb} = 973$ K, $p_{amb} = 0.1$ MPa, $v_{rel} = 0$ m/s [67]

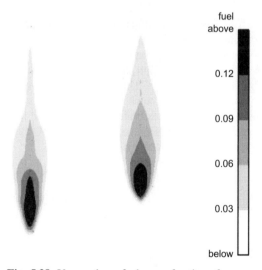

Fig. 5.38. Vapor phase fuel mass fractions for evaporating n-$C_{14}H_{30}$ (left) and continuous multi-component fuel (right) sprays in nitrogen after 3.5 ms. $m_{fuel} = 26.5$ mg, $T_{amb} = 800$ K, $p_{amb} = 5$ MPa, $p_{inj} = 40$ MPa [67]

It should be noted that up to now most evaporation models that are based on the continuous thermodynamics approach have utilized the infinite-diffusion (or well-mixed) model for the droplet interior. However, most recently a combination of the continuous thermodynamics with the diffusion-limit model is investigated as well, e.g. [66]. This will cause a significant increase in the computational effort, but as noted above, it has been shown that this more detailed treatment of the droplet interior has a considerable influence on the evaporation process of two-component fuels. Hence, it is likely that multi-component fuels will be described more realistically with the diffusion-limit model, too.

5.7.3 Flash Boiling

Flash boiling is a phenomenon that occurs when the fuel vapor pressure exceeds the ambient gas pressure at the time the fuel exits the injector nozzle. In this case the fuel evaporates almost instantaneously at the nozzle orifice. This will cause a tremendous increase in specific volume, and as a result the entire spray pattern and mixture formation process will change dramatically.

Flash boiling may have a significant influence in SI engines because the boiling curve of gasoline is relatively low, i.e. the vapor pressure is high, and because the ambient gas pressure is low. This is true both for conventional SI engines with fuel injection into the intake manifold and for homogeneously operated DISI engines with early fuel injection during the intake stroke while the cylinder pressure is still very low. In diesel engines, where the fuel boiling curve is much higher and the injection occurs shortly before TDC into compressed air, the occurrence of flash boiling is less likely.

As an example Fig. 5.39 shows four cases of gasoline injection into atmospheric air with fuel temperatures varying from 20°C to 120°C. It is obvious that the spray pattern changes dramatically for fuel temperatures greater than 100°C because of flash boiling. At this temperature the fuel vapor pressure is about 250 kPa which is in excess of the ambient gas pressure. The instantaneous evaporation causes a reduction in spray penetration and a much faster air-fuel mixing. The single jets originating from different nozzle holes that are observed for fuel temperatures below 80°C can no longer be distinguished, and for the 120°C case there is even the onset of a spray collapse visible, compare Fig.5.23.

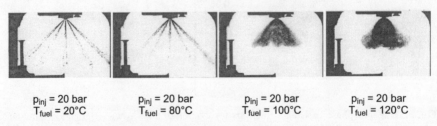

p_{inj} = 20 bar p_{inj} = 20 bar p_{inj} = 20 bar p_{inj} = 20 bar
T_{fuel} = 20°C T_{fuel} = 80°C T_{fuel} = 100°C T_{fuel} = 120°C

Fig. 5.39. Effect of flash boiling on spray formation [64]

While the fuel atomization and mixture formation process can generally be enhanced by flash boiling when the fuel is in a slightly superheated regime, a too high degree of superheat has to be avoided. This is because a so-called vapor-lock may decrease the nozzle mass flow rate when flash-boiling effects already start within the nozzle hole [77]. Several studies have been presented on the effect of flash boiling to engine applications, e.g. [7, 88], but the degree of superheat that is crucial for optimum engine performance is still extremely difficult to control. Hence, flash boiling is not yet utilized systematically in series production engines.

The modeling of flash boiling processes is very challenging for several reasons. Non-equilibrium effects play an important role since there is typically a time lag between the occurrence of the saturation vapor pressure and the onset of vaporization. Moreover, the number of inception points such as gas confinements in the liquid is difficult to determine, and a high spatial resolution is necessary in order to calculate the important mechanism of heat conduction within the liquid phase. Last but not least the multi-component nature of engine fuels complicates the problem because these fuels are not characterized by a distinct boiling temperature but rather by a continuous boiling curve.

For the above reasons most of the flash boiling models proposed in the literature have not concentrated on engine sprays but on different applications with simpler boundary conditions, e.g. [27, 84]. One example are water operated cooling systems that are important in nuclear reactor safety. Up to now no comprehensive flash boiling models specifically suited for engine spray applications have been included in commercially available CFD-codes. Studies on this subject have instead utilized conventional spray models and included the effects of flash boiling by altering the initial or boundary conditions of the spray computation. For example, VanDerWege et al. [103] obtained satisfying results by reducing the mean droplet diameter after primary breakup, by increasing the initial spray-cone angle and by fictitiously adding vapor to the spray core in an attempt to reproduce early vapor production in flash boiling conditions.

Nevertheless, it should be noted that flash boiling is a very important effect especially for future spray guided DISI engines with stratified charge operation. Thus, it will certainly be beneficial to develop comprehensive submodels for this process and include them in engine spray calculations.

5.8 Grid Dependencies

5.8.1 Problem Description

The term grid dependency describes the phenomenon that calculations executed under identical physical conditions often yield considerably different results when performed on different numerical grids. It is inherently linked with Eulerian-Lagrangian spray simulations, and probably everyone who has worked in this field has already experienced according problems. As an example, Fig. 5.40 shows a diesel type injection executed with the same code on both a coarse and a fine numerical grid under otherwise identical conditions. Almost all characteristic spray parameters such has spray tip penetration, mean droplet sizes, vapor phase concentrations, gas phase velocities etc. are influenced by the numerical grid. To make things worse the results obtained with the high resolution grid are not necessarily better than the ones obtained with the coarse grid, i.e. the results will not converge for successively refined grids. Moreover, grid dependent results are not only observed for different grid sizes but also for varying grid arrangements, e.g. for a change between polar and Cartesian grid cell arrangements, or even if the main spray axis is tilted relative to the grid structure.

Two main reasons can be identified for grid dependent simulation results: the lack of spatial resolution in typical engine simulations and the lack of statistical convergence in the Lagrangian treatment of the liquid phase. The first issue has been investigated among others by Abraham [1]. In addition to spray simulations he investigated transient gas jets in order to eliminate the statistical problems encountered with the stochastic particle technique of the liquid phase. An additional advantage is that for a single phase gas flow the grid resolution can simply by refined to any desired level without causing numerical instabilities that will typically arise in two-phase systems if the void fraction approaches zero.

Abraham's study shows that in regimes where the jet diffusivity is significantly greater than the turbulent diffusivity of the ambient gas, i.e. if the jet itself is the driving force for mixture formation, the orifice diameter is the relevant length scale that needs to be resolved by at least two grid cells in order to yield adequate results. If a coarser resolution is chosen with grid cells that are about five times as large as the nozzle orifice, which is imperative in typically spray simulations, the simulation results show an unphysical dependency on the ambient diffusivity that cannot be observed in experiments. In the other regime, where the ambient diffusivity is equal or greater than the jet diffusivity, it was shown that computed mixing rates are significantly underpredicted if a too coarse mesh resolution is chosen. And even with very fine grid resolutions it was found that the jet structure depends not only on the ambient diffusivity as one would expect, but also on the absolute values of the ambient turbulence length and time scales. These findings have yet to be assessed by experiments.

The issue of statistical convergence in the stochastic particle technique, that is utilized in literally all commercial CFD-codes for two-phase flows, has been discussed by Otto et al. [65]. In order to achieve such statistical convergence it is not even sufficient to increase the number of grid cells resolving the nozzle orifice, but in addition the ratio of stochastic particles (i.e. spray parcels) to grid cells needs to increase dramatically as well. This implicates an enormous number of particles and in typical engine spray simulations one is far from meeting such conditions. Furthermore, to have at least a chance of coming close to statistical convergence the various submodels included in the spray calculations need to be formulated in a numerically favorable manner. As an example, submodels that are based on interactions of two different particles, e.g. droplet collision models, are generally critical for statistical convergence, even if they are formulated in a physically correct way [85].

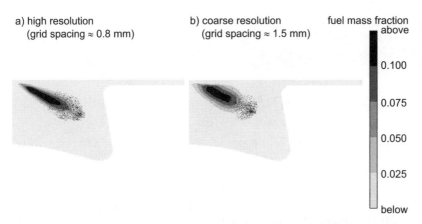

Fig. 5.40. Grid dependency in a high-pressure diesel spray simulation. a) high resolution (grid spacing ≈ 0.8 mm) b) coarse grid (grid spacing ≈ 1.5 mm)

5.8.2 Reduction of Grid Dependencies

General Measures

Because of the grid dependencies described above it is a common practice to tune model parameters in order to adjust a calculation to the specific numerical grid on which it is performed. However, often this adjustment to experimentally obtained data results in setting the parameters of submodels such as breakup or collision to values that are physically not meaningful. In other words, the submodels are trimmed to an unphysical behavior in order to overcome the deficiencies caused by inadequate grids. This procedure is obviously not acceptable, and instead more justifiable measures ought to be taken in order to achieve convergent spray simulations.

In ref. [65] a number of measures is suggested for this purpose. Those are, in an arbitrary order:

- Length scale limiter:
  The turbulence length scale needs to be limited to the jet diameter within the jet region. This is because in high pressure diesel-type jets the jet itself is the driving force for mixture formation and thus governs the relevant length scale.

- Spray adapted grid:
  As noted above, in spray simulations it is not feasible to directly resolve the nozzle orifice by several grid cells because of computer power and numerical stability concerns. However, the grid arrangement should be at least adjusted to the spray orientation, such that the spray penetrates more or less perpendicular into the cells, Fig. 5.41.

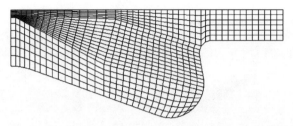

Fig. 5.41. Spray adapted grid for bowl-in-piston combustion chamber of a DI diesel engine

- Modified turbulence model:
  Due to insufficiencies of the standard k-ε-turbulence model, the spray angle is often overpredicted. This may be avoided by modifying the turbulence model, e.g. with the Pope-correction [69].

- Avoidance of too detailed subprocesses:
  As mentioned previously, submodels including two-particle correlations are generally critical for attaining statistical convergence. A typical example is the droplet collision model that has been shown to be an important source for grid dependencies.

The issue of the collision model's influence on numerical accuracy has been studied more closely by Schmidt and Senecal [86]. They utilize the No Time Counter (NTC) collision algorithm, which is otherwise popular in calculating intermolecular collisions in gas dynamics and has been extended to spray simulations in ref. [85]. In addition a separate, very fine collision mesh, which is automatically created and updated each time step, is superimposed on the regular numerical grid. It is cylindrical in shape and oriented along the spray axis and therefore better suited in order to describe the collision frequencies than the regular grid. The combination of the NTC algorithm and the collision mesh practically eliminates the grid dependency as a consequence of the collision algorithm.

However, the study still showed a grid dependency that is attributed to the coupling of the gas and liquid phases. This source of grid dependency could finally be reduced by replacing the mean gas velocity within the cell by a distance weighted average of the gas velocities at the eight vortices of the computational cell in order to estimate interphase exchange terms. Moreover, best results are obtained if the Cartesian components of the vortex gas velocities are transformed into polar components prior to the averaging process.

The ICAS-Concept

In addition to the above measures Wan and Peters [105, 106] proposed the so-called ICAS-model (Interactive Cross-sectional Averaged Spray model) for high pressure diesel injectors in an attempt to further reduce grid dependencies of numerical simulations. The model is based on the reasoning that the numerical uncertainties are generally most crucial in the dense spray regime close to the injector nozzle because particle interactions such as droplet collisions are most likely and because the statistical convergence is typically not given in this region. The latter reason is due to the fact that because of the high injection velocities the particle density in the first cells downstream of the injector is not sufficient.

To overcome these difficulties, in the ICAS approach the near field of the injector is not calculated with the conventional Eulerian/Lagrangian method (i.e. the discrete droplet model), but with a Eulerian description for both the gas and the liquid phases. As shown in Fig. 5.42, for this purpose a secondary numerical grid is superimposed on the regular grid for the first about 20-30 mm downstream of the nozzle orifice. On this secondary grid the liquid and gas phase conservation equations are now solved in a cross-sectionally integrated and averaged manner. Thus, the spray model becomes essentially one-dimensional within this domain, and numerically very efficient. This advantage along with the Eulerian formulation allows for a sufficiently high grid resolution while still maintaining acceptable calculation times. Numerical instabilities are avoided and statistical convergence is per definition not an issue in this Eulerian approach.

Transport equations are solved for the cross-sectionally averaged quantities of mass, momentum and energy for both the gas and liquid phases as well as for the liquid mean droplet diameter. As an example, the continuity equations for the gas and liquid phases read

$$\frac{\partial\left(\hat{\rho}(1-\hat{y}_l)b^2\right)}{\partial t}+\frac{\partial\left(\hat{\rho}\hat{u}_g(1-\hat{y}_l)b^2\right)}{\partial x}=\rho_g\beta b\hat{u}_g+\hat{\omega}_{vap}b^2, \quad (5.168)$$

$$\frac{\partial\left(\hat{\rho}\hat{y}_l b^2\right)}{\partial t}+\frac{\partial\left(\hat{\rho}\hat{u}_l\hat{y}_l b^2\right)}{\partial x}=-\hat{\omega}_{vap}b^2, \quad (5.169)$$

respectively. The superscript (^) denotes the cross-sectional average, $\hat{u}_g$ and $\hat{u}_l$ are the gas and liquid velocities, $\hat{\rho}$ the mixture density, $\hat{y}_l$ the liquid fuel mass fraction, $\hat{\omega}_{evap}$ the evaporation rate, and b is the spray radius. The so-called spreading coefficient β is a global quantity that models the turbulent exchange between the spray and the surroundings gas. It is related to the spray half angle $\alpha/2$, and in a stagnant environment it is equal to $\tan(\alpha/2)$ [106]. The axial spray coordinate, i.e. the distance from the nozzle orifice, is denoted by x.

In order to model the spatially resolved exchange terms between the gas and liquid phases within the spray region and in order to re-transform the results to the regular three-dimensional CFD-grid, the so-far one-dimensionally described spray needs to be expanded to three-dimensional form again. For this purpose a constant and predefined spray angle is assumed and furthermore, within the jet there are axis symmetric velocity and concentration distributions assumed. These distributions are typically expressed by a β-function in terms of the axial spray coordinate x and the dimensionless jet radius r/b, where $b = b(x)$:

$$f(x,r)=\left[\left(1+\frac{r}{b}\right)^2\left(1-\frac{r}{b}\right)^2\right]. \quad (5.169)$$

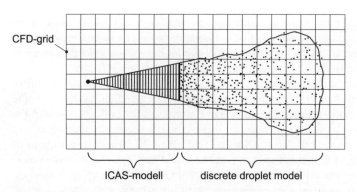

Fig. 5.42. Schematic illustration of the ICAS method

The one-dimensional description of the spray region has a disadvantage, in that the effect of multi-dimensional flow patterns such as swirl or tumble on the mixture formation can hardly be accounted for. Therefore, the ICAS method is applied only in the direct vicinity of the injector, where the jet itself is the dominant force for mixture formation rather than the ambient gas field. About 20 to 30 mm downstream of the injector the ICAS model is switched off and the calculation is continued with the conventional Eulerian/Lagrangian discrete droplet model, Fig. 5.42.

The ICAS model has been utilized in a number of studies on diesel sprays, e.g. [18, 38, 47]. In general very satisfying results have been reported that show a reduced grid dependency compared to the classic discrete droplet model. However, it should be noted that in order to utilize the full potential of the ICAS approach a spray adapted grid as shown in Fig. 5.41 is crucial. Otherwise, numerical diffusion within the coarse primary CFD-grid becomes an issue again and the overall results may not be better in terms of grid dependency than the ones obtained with a standard spray model. Designing a spray adapted grid obviously implies that an approximate spray structure has to be known a priori.

References

[1] Abraham J (1997) What is Adequate Resolution in the Numerical Computations of Transient Jets? SAE Paper 970051
[2] Abramzon B, Sirignano WA (1989) Droplet Vaporization Model for Spray Combustion Calculations. Int J Heat Mass Transfer, vol 32, no 9, pp 1605–1618
[3] Aggarwal SK (1987) Modeling of a Dilute Vaporizing Multicomponent Fuel Spray. Int J Heat Mass Transger, vol 30, no 9, pp 1949–1961
[4] Amsden AA (1993) KIVA-3: A KIVA Program with Block-Structured Mesh for Complex Geometries. LA-12503-MS, Los Alamos National Laboratories
[5] Amsden AA, Ramshaw JD, O'Rourke PJ, Butler TD (1985) KIVA: A Computer Program for Two- and Three-Dimensional Fluid Flows with Chemical Reactions and Fuel Sprays. LA-10245-MS, Los Alamos National Laboratories

[6] Amsden AA, O'Rourke PJ, Butler TD (1989) KIVA-II: A Computer Program for Chemically Reactive Flows with Sprays. LA-11560-MS, Los Alamos National Laboratories
[7] Aquino C, Plensdorf W, Lavoie G, Curtis E (1998) The Occurrence of Flash Boiling in a Port Injected Gasoline Engine. SAE Paper 982522
[8] Arcoumanis C, Gavaises M, French B (1997) Effect of Fuel Injection Process on the Structure of Diesel Sprays. SAE Paper 970799
[9] Arcoumanis C, Flora H, Gavaises M, Badami M (2000) Cavitation in Real-Size Multi-Hole Diesel Injector Nozzles. SAE Paper 2000-01-1249
[10] Ashgriz N (2002) http://www.mie.utoronto.ca/labas/mfl/collision-seminar_files/frame.htm
[11] Ashgriz N, Poo JY (1990) Coalescence and Separation in Binary Collisions of Liquid Drops. J Fluid Mechanics, vol 221, pp 183–204
[12] Ayoub NS, Reitz RD (1995) Multidimensional Computation of Multicomponent Spray Vaporization and Combustion. SAE Paper 950285
[13] Bai C, Gosman AD (1995) Development of Methodology for Spray Impingement Simulation. SAE Paper 950283
[14] Bai CX, Rusche H, Gosman AD (2000) Modeling of Gasoline Spray Impingement. Accepted for publication in Atomization and Sprays
[15] Baumgarten C, Shi Y, Busch R, Merker GP (2001) Numerical and Experimental Investigations of Cavitating Flow in High Pressure Diesel Nozzles, Proc 17th ILASS Europe Conf, pp 593–599, Zurich, Switzerland
[16] Baumgarten C, Stegemann J, Merker GP (2002) A New Model for Cavitation Induced Primary Breakup of Diesel Sprays. Proc 18th ILASS Europe Conf, pp 15–20, Zaragoza, Spain
[17] Bellman R, Pennington RH (1954) Effects of Surface Tension and Viscosity on Taylor Instability. Quarterly of Applied Mathematics, vol 12, pp 151–162
[18] v Berg E, Alajbegovic A, Tatschl R, Krüger C, Michels U (2001) Multiphase Modeling of Diesel Sprays with the Eulerian/Eulerian Approach. 17th ILASS Europe Conf, pp 443–448, Zurich, Switzerland
[19] Bracco FV (1985) Modeling of Engine Sprays. SAE Paper 850394
[20] Chan M, Das S, Reitz RD (1997) Modeling Multiple Injection and EGR Effects on Diesel Engine Emissions. SAE Paper 972864
[21] Chiang CH, Raju MS, Sirignano WA (1992) Numerical Analysis of Convecting, Vaporizing Fuel Droplet with Variable Properties. Int J Heat Mass Transfer, vol 35, no 5, pp 1307–1324
[22] Cousin J, Nuglisch HJ (2001) Modeling of Internal Flow in High Pressure Swirl Injectors. SAE Paper 2001-01-0963
[23] Crowe CT, Sharma MP, Stock DE (1977) The Particle-Source-in Cell Method for Gas Droplet Flow. ASME Journal of Fluids engineering, vol 99, pp 325–332
[24] Crowe CT, Chung JN, Troutt TR (1988) Particle Mixing in Free Shear Flows. Prog Energy Combust Sci, vol 14, pp 171–194
[25] Dognin C, Dupont A, Gastaldi P (2001) 3D Simulation: A Powerful Tool to Understand Mixture Preparation and Combustion in Direct Injection SI Engines. Proc 4th Cong Gasoline Direct Injection Engines, pp 80–109, Cologne, Germany
[26] Dombrowski N, Hooper PC (1962) The Effect of Ambient Density on Drop Formation in Sprays. Chem Eng Sci, vol 17, p 291

[27] Downar-Zapolski P, Bilicki Z, Bolle L, Franco J (1996) The Non-Equilibrium Relaxation Model for One-Dimensional Flashing Liquid Flow. Inj J Multiphase Flow, vol 22, no 3, pp 473–483
[28] Dukowicz JK (1980) A Particle-Fluid Numerical Model for Liquid Sprays. J Comp Physics, vol 35, pp 229–253
[29] Dumont N, Simonin O, Habchi C (2000) Cavitating Flow in Diesel Injectors and Atomization: a Bibliographical Review. 8th Int Conf on Liquid Atomization and Spray Systems (ICLASS), pp 314–322, Pasadena, CA
[30] Dwyer HA, Stapf P, Maly RR (2000) Unsteady Vaporization and Ignition of a Three-Dimensional Droplet Array. Combust Flame, vol 121, no 1-2, pp 181–194
[31] Faeth GM (1983) Evaporation and Combustion of Sprays. Prog Energy Combust Sci, vol 9, pp 1–76
[32] Faeth GM (1987) Mixing, Transport and Combustion in Sprays. Prog Energy Combust Sci, vol 13, pp 293–345
[33] Gavaises M (1997) Modelling of Diesel Fuel Injection Processes. Ph.D. Thesis, Imperial College, University of London
[34] Georjon TL, Reitz RD (1999) A Drop-Shattering Collision Model for Multidimensional Spray Computations. Atomization and Sprays, vol 9, no 3, pp 231–254
[35] Gonzalez MA, Borman GL, Reitz RD (1991) A Study of Diesel Cold Starting Using Both Cycle Analysis and Multidimensional Computations. SAE Paper 910108
[36] Gupta HC, Bracco FV (1978) Numerical Computations of Two-Dimensional Unsteady Sprays for Application to Engines. AIAA Journal, vol 16, no 10, pp 1053–1061
[37] Han Z, Parrish S, Farrell PV, Reitz RD (1997) Modeling Atomization Processes of Pressure-Swirl Hollow-Cone Fuel Sprays. Atomization and Sprays, vol 7, pp 663–684
[38] Hasse C, Peters N (2002) Eulerian Spray Modeling of Diesel Injection in a High-Pressure / High Temperature Chamber. Proc 12th Int Multidim Engine Modeling Users Group Meeting, Detroit, MI
[39] Haywood RJ, Nafziger R, Renksizbulut M (1989) A Detailed Examination of Gas an Liquid Phase Transient Processes in Convective Droplet Evaporation. ASME J. Heat Transfer, vol 111, pp 495–502
[40] Heagerty WW, Shea JF (1955) A Study of the Stability of Plane Fluid Sheets. J Appl Mech, vol 22, p 509
[41] Heywood JB (1988) Internal Combustion Engine Fundamentals. McGraw-Hill, New York, NY
[42] Hohmann, S, Klingsporn M, Renz U (1996) An Improved Model to Describe Spray Evaporation Under Diesel-Like Conditions. SAE Paper 960030
[43] Huh KY, Gosman AD (1991) A Phenomenological Model of Diesel Spray Atomization. Int Conf on Multiphase Flows, pp 515–518, Tsukuba, Japan
[44] Incropera FP, DeWitt DP (1996) Introduction to Heat Transfer. 3rd edn, Wiley, New York
[45] Jin JD, Borman GL (1985) A Model for Multicomponent Droplet Vaporization at High Ambient Pressures. SAE Paper 850264
[46] Knapp RT, Daily JW, Hammitt FG (1970) Cavitation. McGraw-Hill, New York, NY

[47] Krüger C, Otto F, Wirbeleit F, Willand J, Peters N (1999) Incorporation of the Interactive Cross-Sectional Average Methodology for diesel Spray Simulations into a 3D Code. Proc 9th Int Multidim Engine Modeling Users Group Meeting, Detroit, MI
[48] Lippert AM, Reitz RD (1997) Modeling of Multicomponent Fuels Using Continuous Distributions with Application to Droplet Evaporation and Sprays. SAE Paper 972882
[49] Liu AB, Mather D, Reitz RD (1993) Modeling the Effects of Drop Drag and Breakup on Fuel Sprays. SAE Paper 930072
[50] Liu Z, Obokata T, Reitz RD (1997) Modeling Drop Drag Effects on Fuel Spray Impingement in Direct Injection diesel Engines. SAE Paper 970879
[51] Matsumoto S, Saito S (1970) On the Mechanism of Suspension of Particles in Horizontal Conveying: Monte Carlo Simulation Based on the Irregular Bouncing Model. J Chem Eng, vol 3, pp 83–92
[52] Meyer J, Weihs D (1987) Capillary Instability of an Annular Liquid Jet. J Fluid Mech, vol 179, pp 531-545
[53] Miesse CC (1955) Correlation of Experimental Data on the Disintegration of Liquid Jets. Indust Engn Chem, vol 47, p 1690
[54] Modarress D, Wuerer J, Elghobashi S (1982) An Experimental Study of a Turbulent Round Two-Phase Jet. AIAA/ASME 3rd Joint Thermophysics, Fluid, Plasma and Heat Transfer Conf, St. Louis, MO, Paper AIAA-82-0964
[55] Modarress D, Tan H, Elghobashi S (1983) Two-Component LDA Measurements in a Two-Phase Turbulent Jet. AIAA 21st Aerospace Sciences Meeting, Reno, NV, Paper AIAA-83-0052
[56] Modarress D, Wuerer J, Elghobashi S (1984) An Experimental Study of a Turbulent Round Two-Phase Jet. Chem Engineering Communications, vol 28, pp 341–354
[57] Naber JD, Reitz RD (1988) Modeling Engine Spray/Wall Impingement. SAE Paper 880107
[58] Nicholls JA (1972) Stream and Droplet Breakup by Shock Waves, NASA-SP-194, pp 126–128
[59] Nishimura A, Assanis DN (2000) A Model for Primary Diesel Fuel Atomization Based on Cavitation Bubble Collapse Energy. 8th Int Conf on Liquid Atomization and Spray Systems, pp 1249–1256, Pasadena, CA
[60] O'Rourke PJ (1981) Collective Drop Effects on Vaporizing Liquid Sprays. Ph.D. Thesis, Princeton University
[61] O'Rourke PJ, Amsden AA (1987) The TAB Method for Numerical Calculation of Spray Droplet Breakup. SAE Paper 872089
[62] O'Rourke PJ, Bracco FV (1980) Modeling Drop Interactions in Thick Sprays and a Comparison with Experiments. Proc I Mech E, vol 9, pp 101–116
[63] Ohnesorge W (1936) Die Bildung von Tropfen an Düsen und Auflösung flüssiger Strahlen (Formation of Drops by Nozzles and Breakup of Liquid Jets). Z Angew Math Mech, vol 16, p 355
[64] Otto F (2001) Fluid Mechanical Simulation of Combustion Engine Processes. Class Notes, University of Hanover, Germany
[65] Otto F, Wirbeleit F, Willand J (2001) 3D-Simulation innermotorischer Prozesse. Proc 4th Dresdner Motorenkolloquium, pp 279–288, Dresden, Germany

[66] Pagel S, Merker GP (2003) Modellierung der Tropfenverdampfung eines Mehrkomponentenbrennstoffs mit detaillierter Betrachtung des Tropfeninneren. Accepted for publication, Chemie Ingenieur Technik
[67] Pagel S, Stiesch G, Merker GP (2002) Modeling the Evaporation of a Multicomponent Fuel. Proc 12th Int Heat Transfer Conf, pp 899–904, Grenoble, France
[68] Patterson MA, Reitz RD (1998) Modeling the Effects of Fuel Spray Characteristics on Diesel Engine Combustion and Emission. SAE Paper 980131
[69] Pope SB (1978) An Explanation for the Turbulent Round-Jet/Plane-Jet Anomaly. AIAA J, vol 16, pp 279–281
[70] Post SL, Abraham J (2002) Modeling the Outcome of Drop-Drop Collisions in Diesel Sprays. Int J Multiphase Flow, vol 28, no 6, pp 997–1019
[71] Prosperetti A, Lezzi A (1986) Bubble Dynamics in a Compressible Liquid, Part 1. First Order Theory. J Fluid Mech, vol 168, pp 457–478
[72] Qian J, Law C (1997) Regimes of Coalescence and Separation in Droplet Collision. J Fluid Mechanics, vol 331, pp 59–80
[73] Ramos JI [1989] Internal Combustion Engine Modeling. Hemisphere, New York, NY
[74] Ranz WE (1956) On Sprays and Spraying. Dept Engng Res, Bull 65, Penn State University
[75] Reitz RD (1978) Atomization and Other Breakup Regimes of a Liquid Jet. Ph.D. Thesis, Princeton University
[76] Reitz RD (1987) Modeling Atomization Processes in High-Pressure Vaporizing Sprays. Atomization and Spray Technology, vol 3, pp 309–337
[77] Reitz RD (1990) A Photographic Study of Flash-Boiling Atomization. Aerosol Science and Technology, no 12, pp 561–569
[78] Reitz RD (1994) Computer Modeling of Sprays. Spray Technology Short Course, Pittsburgh, PA
[79] Reitz RD (2000) Multiphase Flow and Heat Transfer. Class Notes, University of Wisconsin-Madison
[80] Reitz RD, Bracco FV (1982) Mechanism of Atomization of Liquid Jets. The Physics of Fluids, vol 25, p 1730–1742
[81] Reitz RD, Bracco FV (1986) Mechanisms of Breakup of Round Liquid Jets. The Encyclopedia of Fluid Mechanics, ed Cheremisinoff NP, vol 3, pp 233–249, Gulf Publishing, Houston, TX
[82] Reitz RD, Diwakar R (1986) Effect of Drop Breakup on Fuel Sprays. SAE Paper 860469
[83] Renksizbulut M, Yuen MC (1983) Experimental Study of Droplet Evaporation in a High Temperature Air Stream. ASME J Heat Transfer, vol 105, pp 384–388
[84] Saha P, Abuaf N, Wu BJC (1984) A Nonequilibrium Vapor Generation Model for Flashing Flows. Transactions ASME, vol 106, pp 198–203
[85] Schmidt DP, Rutland CJ (2000) A New Droplet Collision Algorithm. J Comp Physics, vol 164, pp 62–80
[86] Schmidt DP, Senecal PK (2002) Improving the Numerical Accuracy of Spray Simulations. SAE Paper 2002-01-1113
[87] Schmidt DP, Nouar I, Senecal PK, Rutland CJ, Martin JK, Reitz RD (1999) Pressure-Swirl Atomization in the Near Field. SAE Paper 1999-01-0496

[88] Schmitz I, Leipertz, A, Ipp W (2002) Flash Boiling Effects on the Development of Gasoline Direct-Injection Engine Sprays. SAE Paper 2002-01-2661
[89] Senecal PJ, Schmidt DP, Nouar I, Rutland CJ, Reitz RD, Corradini ML (1999) Modeling High-Speed Viscous Liquid Sheet Atomization. Int J Multiphase Flow, vol 25, no 6–7, pp 1073–1097
[90] Sirignano WA (1993) Fluid Dynamics of Sprays. ASME J Fluids Engineering, vol 115, pp 345–378
[91] Squire HB (1953) Investigation of the Instability of a Moving Liquid Film. Brit J Appl Phys, vol 4, p 167–169
[92] Stiesch G, Merker GP, Tan Z, Reitz RD (2001) Modeling the Effect of Split Injections on DISI Engine Performance. SAE Paper 2001-01-0965
[93] Stow CD, Stainer RD (1977) The Physical Products of a Splashing Water Drop. Journal of the Meteorological Society of Japan, vol 55, pp 518–531
[94] Su TF, Patterson MA, Reitz RD (1996) Experimental and Numerical Studies of High Pressure Multiple Injection Sprays. SAE Paper 960861
[95] Tamim J, Hallett WLH (1995) A Continuous Thermodynamics Model for Multicomponent Droplet Vaporization. Chem Eng Sci, vol 50, no 18, pp 2933–2942
[96] Tanner FX (1997) Liquid Jet Atomization and Droplet Breakup Modeling of Non-Evaporating Diesel Fuel Sprays. SAE Paper 970050
[97] Tatschl R, v Künsberg Sarre C, Alajbegovic A, Winklhofer E (2000) Diesel Spray Brekaup-Up Modeling Including Multidimensional Cavitating Nozzle Flow Effects. Proc 16th ILASS Europe Conf, pp I.9.1–I.9.9, Darmstadt, Germany
[98] Tatschl R, v Künsberg Sarre C, v Berg E (2002) IC-Engine Spray Modeling – Status and Outlook. Proc 12th Int Multidim Engine Modeling Users Group Meeting, Detroit, MI
[99] Taylor GI (1963) Generation of Ripples by Wind Blowing over a Viscous Fluid. in: Batchelor GK, The Scientific Papers of GI Talyor, vol 3, pp 244–254
[100] Taylor GI (1963) The Instability of Liquid Surfaces when Accelerated in a Direction Perpendicular to their Planes. in: Batchelor GK, The Scientific Papers of GI Taylor, vol 3, pp 532–536, University Press, Cambridge
[101] Taylor GI (1963) The Shape and Acceleration of a Drop in a High Speed Air Stream. in: Batchelor GK, The Scientific Papers of GI Taylor, vol 3, pp 457–464, University Press, Cambridge
[102] Torda TP (1973) Evaporation of Drops and the Breakup of Sprays. Astronautica Acta, vol 18, p 383
[103] VanDerWege BA, Lounsberry TH, Hochgreb S (2000) Numerical Modelihng of Fuel Sprays in DISI Engines Under Early-Injection Operating Conditions. SAE Paper 2000-01-0273
[104] Wachters LHJ, Westerling NAJ (1966) The Heat Transfer from a Hot Wall to Impinging Water Drops in the Spheroidal State. Chem Eng Sci, vol 21, pp 1047–1056
[105] Wan Y, Peters N (1997) Application of the Cross-Secctional Average Method to Calculations of the dense Spray Region in a Diesel Engine. SAE Paper 972866
[106] Wan Y, Peters N (1999) Scaling of Spray Penetration with Evaporation. Atomization and Sprays, vol 9, no 2, pp 111–132
[107] Wierzba A (1993) Deformation and Breakup of Liquid Drops in a Gas Stream at Nearly Critical Weber Numbers. Experiments in Fluids, vol 9, pp 59–64

[108] Williams FA (1985) Combustion Theory, 2nd edn, Benjamin/Cummings, Menlo Park, CA
[109] Willman B (1985) Continuous Thermodynamics of Fluid Mixtures. Ph.D. Thesis, Georgia Institute of Technology
[110] Zhu GS, Reitz RD, Xin J, Takabayashi T (2001) Characteristics of Vaporizing Continuous Multi-Component Fuel Sprays in a Port Fuel Injection Gasoline Engine. SAE Paper 2001-01-1231

6 Multidimensional Combustion Models

6.1 Combustion Fundamentals

6.1.1 Chemical Equilibrium

A chemical reaction between the reactant species A_a, A_b, etc., that forms the product species A_c, A_d, etc., is often written as

$$v_a A_a + v_b A_b + ... \rightarrow v_c A_c + v_d A_d + ... , \tag{6.1}$$

where the v_i are termed the stoichiometric coefficients of the reaction. Since every chemical reaction can generally proceed in both directions, the arrow in Eq. 6.1 may be replaced by an equal sign. The general form of the reaction equation now becomes

$$\sum_i v_i A_i = 0 , \tag{6.2}$$

where, by convention, the stoichiometric coefficients v_i are positive for the product species and negative for the reactant species.

Each reaction approaches its chemical equilibrium, that will be reached if there is sufficient time available. It can be interpreted as a situation in which both the forward and reverse reactions proceed with the same rate, such that the integral reaction rate becomes zero and the species concentrations remain constant. The equilibrium composition of a reacting mixture can be derived from the first and second laws of thermodynamics as follows.

For a closed compressible system with constant temperature and pressure as it is shown in Fig. 6.1 the first law of thermodynamics reads

$$dU = dQ + dW = dQ - pdV , \tag{6.3}$$

and the second law becomes

$$dS = \frac{dQ}{T} + dS_{irr} , \tag{6.4}$$

where dS_{irr} is the entropy production due to irreversibilities which is always greater than or equal to zero. Combination of Eqs. 6.3 and 6.4 yields

$$TdS - dU - pdV \geq 0 . \tag{6.5}$$

Introducing the Gibbs free energy,

$$G = H - TS = U + pV - TS, \tag{6.6}$$

differentiation and rearrangement leads to

$$dG - Vdp + SdT \leq 0. \tag{6.7}$$

Thus, for a closed control volume with constant temperature and pressure, the derivative of the Gibbs free energy is less or equal to zero, i.e. any change in composition by chemical reactions will reduce G. For chemical equilibrium the condition

$$dG|_{T,p} = 0 \tag{6.8}$$

has to be satisfied.

For multi-component single phase systems such as chemically reacting gases, the Gibbs free energy is a function of temperature, pressure and composition, i.e.

$$G = G(T, p, n_1, n_2, n_3, \ldots), \tag{6.9}$$

where the n_i indicate the number of moles of species i. The chemical potential μ_i of a particular component i is defined as the partial derivative of the Gibbs free energy with respect to the number of moles of that species:

$$\mu_i = \frac{\partial G}{\partial n_i}\bigg|_{T,p,n_j}, \quad j \neq i. \tag{6.10}$$

For an ideal gas mixture – which is a reasonable assumption for most combustion gases – it can be shown that the chemical potential is equal to the molar Gibbs function [36]

$$\mu_i = \tilde{g}_i(T, p_i) = \tilde{g}_i^\circ + \tilde{R}T \ln \frac{p_i}{p^\circ}, \tag{6.11}$$

where the superscript $^\circ$ indicates conditions at the reference pressure of 1 atm. The first term on the right hand side of Eq. 6.11 stands for

$$\tilde{g}_i^\circ = \tilde{h}_i(T) - T\tilde{s}_i^\circ(T), \tag{6.12}$$

and thus, it can be obtained from tabulated property data. Note that the molar enthalpy in the above equation consists of the enthalpy of formation of species i at reference conditions and a temperature dependent component:

$$\tilde{h}_i(T) = \tilde{h}_{f,i}^\circ + \Delta \tilde{h}_i(T). \tag{6.13}$$

Fig. 6.1. Closed reactor

With the above introduction of the chemical potential μ_i the equilibrium condition, Eq. 6.8, can be written as

$$dG\big|_{T,p} = \sum_i \mu_i dn_i = 0. \tag{6.14}$$

For a general chemical reaction of the form given in Eq. 6.2, the changes in the number of moles of the various species i are proportional to their respective stoichiometric coefficients. This can be expressed with the proportionality constant $d\varepsilon$ as

$$dn_i = v_i \cdot d\varepsilon. \tag{6.15}$$

Thus, the equilibrium condition can be simplified to the form

$$\sum_i \mu_i v_i = 0, \tag{6.16}$$

which contains all the information needed in order to solve for the equilibrium composition of a gas mixture. However, a solution to Eq. 6.16 is cumbersome as it can only be obtained by iteration.

To overcome this difficulty the concept of the equilibrium constant is adopted. By substituting Eq. 6.11 into Eq. 6.16 the relation

$$\sum_i v_i \tilde{g}_i^\circ + \tilde{R}T \ln \prod_i \left(\frac{p_i}{p^\circ}\right)^{v_i} = 0 \tag{6.17}$$

is obtained, in that the logarithmic term is referred to as the equilibrium constant,

$$K_p = \prod_i \left(\frac{p_i}{p^\circ}\right)^{v_i}, \tag{6.18}$$

that contains the information about the mixture composition in terms of the partial pressures p_i. As the first term on the left hand side of Eq. 6.17 depends on temperature only, it becomes obvious that the equilibrium constant solely depends on temperature as well. It can easily be calculated by thermodynamic property data and tabulated for any reaction of interest:

$$\ln K_p = \frac{-\sum_i v_i \tilde{g}_i^\circ}{\tilde{R}T} = \frac{-\Delta_R \tilde{g}^\circ}{\tilde{R}T}. \tag{6.19}$$

In Eq. 6.19 the numerator is commonly referred to as the free molar reaction enthalpy.

For a system subject to a single chemical reaction, e.g. ($CO + 0.5\ O_2 = CO_2$), the equilibrium composition for a specific temperature and pressure, i.e. the concentrations or partial pressures of the species CO, O_2 and CO_2, can be solved by Eq. 6.18. However, since there are three unknowns and so far only one equation in this example, two additional conditions have to be specified. These are the atom balances stating that the number of atoms of each element does not change in a chemical reaction, and the obvious requirement that the sum of all partial pressures has to equal the total system pressure.

In a more complex system consisting of multiple species undergoing multiple chemical reactions a situation is referred to as partial equilibrium, when several (but not necessarily all) reactions proceed so quickly, that the involved species can be assumed to be in chemical equilibrium with each other at any point in time. In this case the respective concentrations of the species in partial equilibrium can be evaluated in analogy to the procedure for a single reaction. However, since there are more unknowns, i.e. more species concentrations that need to be determined, one needs a greater number of equations as well. These equations can be obtained from additional chemical reactions between species that are in partial equilibrium with each other. For each additional reaction a new equilibrium constant can be determined from Eq. 6.19 and related to the partial pressures by Eq. 6.18.

Generally, the number of linearly independent reaction equations that are necessary in order to uniquely define a problem of partial equilibrium is given by

$$R = S - K, \tag{6.20}$$

where
- R: number of linearly independent chemical reactions,
- S: number of chemical species in partial equilibrium,
- K: number of indivisible elements (atoms) within the system.

As an example, the reactions between the chemical species CO, CO_2, H, H_2, H_2O, O, O_2 and OH are often assumed to be in partial equilibrium within the high temperature combustion products of hydrocarbon fuels. Since there are $S = 8$ species involved and the number of indivisible elements is $K = 3$ (the carbon, hydrogen and oxygen atoms), a total of $R = 5$ linearly independent reaction equations is necessary in order to describe the partial equilibrium. These are, for example

(i) $2H + M = H_2 + M$,
(ii) $2O + M = O_2 + M$,
(iii) $\frac{1}{2} H_2 + OH = H_2O$,
(iv) $\frac{1}{2} O_2 + H_2 = H_2O$,
(v) $\frac{1}{2} O_2 + CO = CO_2$,

where M is an arbitrary gas molecule that is not affected by the respective chemical reactions.

It should be noted that the above five reactions are not the only ones taking place on an elementary level in a real combustion system, but that additional reactions may take place as well. Nevertheless, the above set is sufficient to uniquely define the partial equilibrium composition. It would however be possible to replace one or several of the above reactions by different elementary reactions between the involved species as long the reaction system remains linearly independent.

6.1.2 Reaction Kinetics

On a microscopic scale, a chemical reaction such as given in Eq. 6.1 will generally proceed in both the forward and reverse directions, and the macroscopically visible direction of the reaction depends on their relative rates to each other. Thus, the chemical equilibrium discussed in the previous section is only a particular case, in that the forward and reverse reactions proceed at the same rate. The macroscopic reaction rate becomes zero, but on the microscopic scale there are still reactions taking place. While the macroscopic rate of any chemical reaction is always directed towards its chemical equilibrium, the equilibrium analysis itself does not provide any information about the reaction rates, i.e. about how long it will take to reach the equilibrium composition. These information can only be obtained from reaction kinetics.

For the chemical reaction given in Eq. 6.1 the temporal change of a species concentration, e.g. for $[A_c]$, can be specified by the empirical formulation

$$\frac{d[A_c]}{dt} = v_c \left(\underbrace{k_f [A_a]^{v_a} [A_b]^{v_b}}_{\text{forward}} - \underbrace{k_r [A_c]^{v_c} [A_d]^{v_d}}_{\text{reverse}} \right), \qquad (6.21)$$

where the first term on the right hand side indicates the rate of the forward reaction and the second term the rate of the reverse reaction. k_f and k_r are the rate coefficients for the forward and reverse reactions, respectively. They have to be obtained from experiments for each particular chemical reaction. These experiments are typically executed in shock tubes under well-defined boundary conditions. The reaction coefficients of most reactions are strongly temperature dependent, which is expressed by an Arrhenius-type equation of the form

$$k = A \cdot T^b \cdot \exp\left[-\frac{E_A}{\tilde{R}T} \right]. \qquad (6.22)$$

The constants A and b as well as the activation Energy E_A have been tabulated for many chemical reactions, especially for those taking place in the hydrocarbon oxidation, e.g. [65].

It is generally sufficient to specify only one rate coefficient, either the forward or the reverse reaction coefficient, by Eq. 6.22. The other coefficient can be derived from the knowledge of the corresponding equilibrium constant. This becomes clear if we consider that in the case of chemical equilibrium the rate of change of a species concentration reduces to zero because both the forward and the reverse reactions proceed at the same rate. Substituting this condition into Eq. 6.21, the relation

$$\frac{k_f}{k_r} = \frac{[A_c]^{v_c} [A_d]^{v_d}}{[A_a]^{v_a} [A_b]^{v_b}} \equiv K_c \qquad (6.23)$$

is obtained, where K_c is the equilibrium constant defined in terms of species concentrations. It is explicitly related to the equilibrium constant K_p defined in terms of partial pressures as introduced in Eqs. 6.18 and 6.19:

$$K_c = K_p \left(\frac{p^o}{\tilde{R}T} \right)^{\sum_i v_i}. \qquad (6.24)$$

Because both the reaction rate coefficients as well as the equilibrium constants depend on temperature only, but not on the actual species concentrations $[A_i]$ in the mixture, the equation

$$\frac{k_f}{k_r} = K_p \left(\frac{p^o}{\tilde{R}T} \right)^{\sum_i \nu_i} \tag{6.25}$$

is not only valid for the equilibrium state but rather universally applicable. Hence, if one of either k_f or k_r is known the other rate constant can be calculated from Eq. 6.25 in a straightforward manner.

6.1.3 Reaction Mechanisms for Hydrocarbon Flames

The combustion of even simple hydrocarbons such as methane (CH_4) is subject to very complex reaction mechanisms that involve numerous different species and even more elementary reactions between those species. For example Frenklach et al. [12] proposed a methane combustion scheme that includes 149 elementary reactions between 33 chemical species. For longer hydrocarbon molecules as they are typical for internal combustion engine fuels, the respective reaction mechanisms become even more complicated. For n-heptane, which is often used as a single-component substitute for diesel fuel because of its similar ignition qualities, detailed reaction mechanisms include several hundred or even a thousand elementary reactions between more than a hundred different species, e.g. [3, 8].

While modern computing power is sufficient to solve such complex reaction mechanisms in spatially homogeneous reactors and there are even commercial codes available to pursue this task, e.g. [35], application of these complex mechanisms to three-dimensional turbulent flows is not yet a standard because of excessive CPU times. Consequently, it is desirable to come up with reduced mechanisms with much smaller numbers of species and reactions that are still able to reflect the most important combustion characteristics to a desired level of detail. In the simplest case a global single-step mechanism of the form (Fuel + Oxidizer → Products) can be specified. For simple hydrocarbons the general form reads

$$C_xH_y + \left(x + \frac{y}{4}\right) O_2 \rightarrow x\, CO_2 + \frac{y}{2}\, H_2O. \tag{6.26}$$

As an example, for octane it becomes

$$C_8H_{18} + 12.5\, O_2 \rightarrow 8\, CO_2 + 9\, H_2O. \tag{6.27}$$

Westbrook and Dryer [69] approximated the global reaction rates of a variety of hydrocarbons by the Arrhenius equation

$$\frac{d[C_xH_y]}{dt} = -A \cdot \exp\left(-\frac{E_A}{RT}\right) \cdot [C_xH_y]^m \cdot [O_2]^n, \tag{6.28}$$

Table 6.1. Single-step reaction rate parameters for use with Eq. 6.28 [69]

Fuel	A (mol, cm, s)	Activation temp. E_A/R (K)	m (-)	n (-)
CH_4	$8.3 \cdot 10^5$	15,098	-0.3	1.3
C_2H_6	$1.1 \cdot 10^{12}$	15,098	0.1	1.65
C_3H_8	$8.6 \cdot 10^{11}$	15,098	0.1	1.65
C_4H_{10}	$7.4 \cdot 10^{11}$	15,098	0.15	1.6
C_5H_{12}	$6.4 \cdot 10^{11}$	15,098	0.25	1.5
C_6H_{14}	$5.7 \cdot 10^{11}$	15,098	0.25	1.5
C_7H_{16}	$5.1 \cdot 10^{11}$	15,098	0.25	1.5
C_8H_{18}	$4.6 \cdot 10^{11}$	15,098	0.25	1.5
C_9H_{20}	$4.2 \cdot 10^{11}$	15,098	0.25	1.5
$C_{10}H_{22}$	$3.8 \cdot 10^{11}$	15,098	0.25	1.5
C_2H_4	$2.0 \cdot 10^{12}$	15,098	0.1	1.65
C_3H_6	$4.2 \cdot 10^{11}$	15,098	-0.1	1.85
C_2H_2	$6.5 \cdot 10^{12}$	15,098	0.5	1.25
CH_3OH	$3.2 \cdot 10^{12}$	15,098	0.25	1.5
C_2H_5OH	$1.5 \cdot 10^{12}$	15,098	0.15	1.6
C_6H_6	$2.0 \cdot 10^{11}$	15,098	-0.1	1.85
C_7H_8	$1.6 \cdot 10^{11}$	15,098	-0.1	1.85

where the parameters A, m and n as well as the activation temperature E_A/R that are summarized in Table 6.1 have been chosen to provide best agreement between experimental and predicted flame speeds and flammability limits.

These global single-step mechanisms allow rough estimations about integral reaction and heat release rates, but they cannot provide more detailed insight into the formation and oxidation of any intermediate species which can become important for the formation of exhaust emissions. For such information, quasi-global multi-step mechanisms are necessary that describe the oxidation of hydrocarbon fuels by a set of at least two or more global and sometimes also elementary reactions. These multi-step mechanisms have to include the rate controlling reaction steps as well as the characteristic intermediate species that become important for pollutant formation reactions.

The procedure of systematically identifying the rate controlling steps of a complex elementary reaction mechanism and reducing it to a simpler and computationally more efficient quasi-global multi-step mechanism has been described in detail in the respective literature [43, 65]. It is based on the observation that the various elementary reactions contained in complex mechanisms are proceeding with partly very different reaction rates. They can differ by as much as several orders of magnitude. As a result the sensitivity of the overall reaction rate, i.e. the complete oxidation of a hydrocarbon fuel into carbon dioxide and water, to the rates of the involved elementary reactions is greatly different. This allows to neglect the less important elementary reactions and concentrate on the ones with greater sensitivities, that are rate controlling for the global reaction. Moreover, an analysis of the main reaction paths can be executed by determining how much of a certain species is formed by a specific reaction. As an example Figs. 6.2 and 6.3 show the main reactions paths for a methane combustion under stoichiometric and rich boundary conditions, respectively. It is clearly visible that the main reaction path can depend strongly on boundary conditions such as the equivalence ratio. It should be noted that this is a common limitation to reduced reaction schemes. They are often not generally valid but limited to certain flame types (premixed or diffusion), equivalence ratios or temperature and pressure conditions.

The oxidation of higher paraffins can coarsely by divided into three consecutive steps. First the fuel molecule is attacked by O and H atoms and forms primarily olefins and hydrogen. The hydrogen oxidizes to water, subject to available oxygen. In the second step the unsaturated olefins further oxidize to CO and H_2. Essentially all of the H_2 is converted to water. Finally, the carbon monoxide burns out to CO_2. It is only in this step that the major fraction of heat associated with the entire combustion is released.

Once the main reaction paths and the rate controlling steps are identified, further simplification can be obtained by assuming that several species are in partial equilibrium with each other because reactions between these species are extremely fast. This allows to directly solve for their concentrations as explained in Sect. 6.1.1, without the need of numerical integration of the differential equations describing the chemical kinetics. Additionally, other species may be assumed to be quasi-steady [65]. This status describes the fact that the concentration of a certain species is almost constant over time when its formation reaction is very slow compared to its decomposition reactions. Thus, all species that is newly formed will immediately be decomposed by subsequent reactions such that the overall concentration change is close to zero. A prominent example for quasi-steadiness is the nitrogen atom in the formation of thermal nitrogen oxides, see Chap. 7.

A wide variety of reduced reaction mechanisms has been proposed in the literature for various hydrocarbon fuels. As noted above, n-heptane is of great interest for internal combustion engine applications as it shows similar ignition behavior to actual diesel fuel. And even for this model fuel the variety of suggested reduced mechanisms is wide, ranging from relatively simple two-step mechanisms, that describe the oxidation of C_7H_{16} to CO_2 and H_2O via CO, to mechanisms that still have a great level of detail and include 10 or more intermediate species and tens or even hundreds of both quasi-global and elementary reactions.

As an example Bollig et al. [5] proposed a reduced reaction mechanism for non-premixed n-heptane combustion including seven quasi-global reaction steps between the intermediate species C_3H_6, C_2H_4, C_2H_2 and CO, which become important for subsequent pollutant formation reactions:

$$
\begin{align}
\text{(i)} \quad & C_7H_{16} = C_3H_6 + 2\,C_2H_4 + H_2 \\
\text{(ii)} \quad & 2\,C_3H_6 = 3\,C_2H_4 \\
\text{(iii)} \quad & C_2H_4 = C_2H_2 + H_2 \\
\text{(iv)} \quad & C_2H_2 + 2\,H_2O + 2\,H = 4\,H_2 + 2\,CO \\
\text{(v)} \quad & CO + H_2O = CO_2 + H_2 \\
\text{(vi)} \quad & 2\,H + M = H_2 + M \\
\text{(vii)} \quad & O_2 + 3\,H_2 = 2\,H + 2\,H_2O
\end{align}
$$

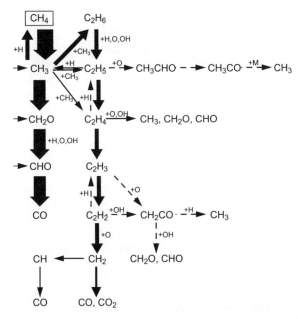

Fig. 6.2. Main reaction paths in premixed stoichiometric methane-air flames at $p = 100$ kPa, $T = 298$ K. [64]

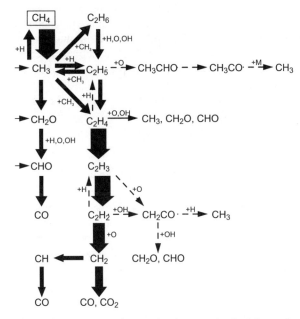

Fig. 6.3. Main reaction paths in premixed rich methane-air flames at $p = 100$ kPa, $T = 298$ K. [64]

However, it is not straightforward to specify the reaction rates of the above seven quasi-global reactions from known kinetic data of elementary reactions. This would be a prerequisite for integrating the time dependent concentrations of the involved intermediate species. To achieve this task, steady-state assumptions have to be introduced for further 22 species, namely for O, OH, HO_2, CH, CHO, CH_2OH, 3-CH_2, 1-CH_2, CH_3, CH_4, CH_2O, C_2H_6, CH_3O, CH_3OH, HCCO, C_2H_3, C_2H_5, C_3H_4, C_3H_5, n-C_3H_7, p-C_4H_9 and 2-C_7H_{15}. These steady-state assumptions lead to a rather complex system of non-linear equations, which needs to be solved numerically. However, as reported in [5], additional simplifying assumptions are necessary in order to secure numerical convergence. As a consequence, the solution of such multi-step reaction mechanisms is still relatively costly and not yet standard in engine modeling. Even in CFD-calculations of internal combustion engines one- and two-step schemes are still widely applied to model the combustion process. Nevertheless, a number of numerical studies on turbulent combustion in IC engines have been executed based on detailed chemistry, e.g. [46], and often improved results compared to simpler reaction mechanisms have been reported.

6.1.4 Combustion Regimes and Flame Types

Several different flame regimes can be identified depending on the mixture state and the interactions between chemistry, turbulence and molecular diffusion. In the most general sense one can distinguish between *premixed* and *non-premixed* flames, depending on whether fuel and oxidizer are already mixed homogeneously prior to combustion or whether mixing and combustion take place simultaneously. The latter case is also referred to as *diffusion* combustion, which is however somewhat misleading as diffusion plays an important role in premixed combustion as well. Depending on the flow conditions, both flame types can further be divided into laminar and turbulent flames. Some prominent examples for the above flame types are:

laminar premixed flame:	Bunsen flame
turbulent premixed flame:	SI engine
laminar non-premixed flame:	candle
turbulent non-premixed flame:	aircraft turbine

In engine combustion chambers there is generally a high level of turbulence such that the effect of this turbulence on the flame has to be considered. The departure of the flame front from a laminar plane to an increasingly three-dimensional structure can be displayed in a Borghi diagram, Fig. 6.4 [6, 70]. Different regimes can be identified in this diagram, based on the dimensionless turbulent Reynolds, Karlovitz and Damköhler numbers.

The turbulent Reynolds number,

$$\mathrm{Re}_t = \frac{u' l_t}{v}, \qquad (6.29)$$

relates the inertia to viscous forces, and is thus a measure of how much the larger eddies of integral length scale ($l_I = C_\mu k^{3/2}/\varepsilon$) are damped by molecular viscosity.

The Karlovitz number,

$$\text{Ka} = \frac{t_F}{t_K}, \quad (6.30)$$

is the ratio of the time scale of the laminar flame t_F to the Kolmogorov time scale t_K. These time scales are defined as

$$t_F = \delta_l / s_l, \quad (6.31)$$

$$t_K = \sqrt{\nu/\varepsilon}, \quad (6.32)$$

where δ_l is the laminar flame thickness, s_l the laminar flame speed, ν the kinematic viscosity and ε the dissipation rate of turbulent energy. On the Kolmogorov scale, the time for an eddy of size $l_K = (\nu^3/\varepsilon)^{1/4}$ to rotate is equal to the time to diffuse across that eddy. Thus, the Kolmogorov length scale describes the smallest eddies in turbulent flows, and for dimensions smaller than l_K, the flow is considered laminar [41].

The Damköhler number describes the ratio of the macroscopic turbulent time scale to the flame time,

$$\text{Da} = \frac{t_0}{t_F}, \quad (6.33)$$

where

$$t_0 = l_I / u'. \quad (6.34)$$

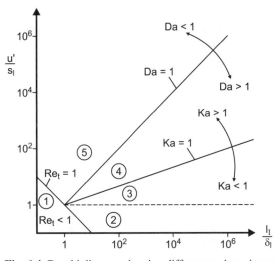

Fig. 6.4. Borghi diagram showing different regimes in turbulent premixed combustion

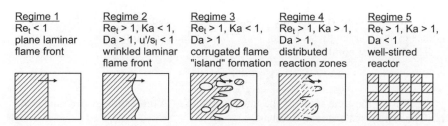

Fig. 6.5. Schematic illustration of different flame types in turbulent premixed combustion

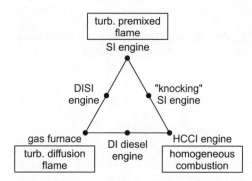

Fig. 6.6. Flame types in combustion engines [37]

The different flame types that correspond to the combustion regimes in the Borghi diagram are schematically shown in Fig. 6.5. In regime 1 ($Re_t < 1$) a plane laminar flame front can be observed. In regime 2, where the turbulent Reynolds number is greater than unity but where the turbulence intensity is less than the laminar flame speed, the flame is still laminar, but its surface becomes more and more wrinkled by large scale eddies. In regime 3, u' exceeds s_l such that "islands" can be ripped out of the flame front and a corrugated flame results. However, in contrast to regime 4 the Karlovitz number is still less than unity. This means, that the flame stretch is sufficiently weak such that the flame front is not yet torn into pieces. The latter phenomenon will result in the formation of multiple distributed reaction zones (regime 4). Finally, regime 5 is characterized by Damköhler numbers less than unity. Thus, chemistry is slow compared to the turbulent flow processes and the system can be treated as a well stirred reactor. The presence of a distinct flame front can no longer be identified in this regime.

In engine combustion chambers the gas flow is generally turbulent. Three out of the above flame categories and combinations between those categories are of special interest. These are turbulent premixed flames, turbulent non-premixed or diffusion flames as well as homogeneous combustion. Figure 6.6 indicates engine related applications where the respective combustion types can be observed. As it will be discussed further below, the different flame types of various engine concepts may require different modeling approaches to appropriately describe the respective combustion phenomena.

6.2 Ignition Processes

6.2.1 Ignition Fundamentals

Ignition processes may be classified into two categories: (i) thermal explosions and (ii) chemical or chain explosions.

A thermal explosion can be explained following Semenov's analysis [55]: the temperature change within a system is described by the difference between heat production by chemical reactions (index P) and heat loss to the surroundings (index L):

$$\rho V c_p \frac{\partial T}{\partial t} = \dot{Q}_P - \dot{Q}_L. \tag{6.35}$$

For simplicity, the heat production term is expressed by an Arrhenius equation for a single-step reaction from fuel to products,

$$\dot{Q}_P \sim [\text{fuel}]^{v_F} \cdot A \cdot \exp\left(-\frac{E_A}{RT}\right), \tag{6.36}$$

and the heat loss is estimated based on Newton's law for convective heat exchange,

$$\dot{Q}_L = h A_w (T - T_w), \tag{6.37}$$

where h is the convective heat transfer coefficient and A_w denotes the wall area.

Hence, the heat production term increases exponentially with temperature whereas the heat loss term is a linear function of the system temperature. This is schematically shown in Fig. 6.7, where two different heat production curves $\dot{Q}_{P1}$ and $\dot{Q}_{P2}$ have been compared to the linear energy loss function $\dot{Q}_L$. For the lower production function $\dot{Q}_{P1}$, there are two stationary intersection points, S_1 and S_2, with the heat loss function. System temperatures below T_{S1} will lead to an increase in temperature since the production term exceeds the loss term. Contrarily, for temperatures $T_{S1} < T < T_{S2}$, the heat production is less than the heat loss such that the system temperature will decrease towards T_{S1}. Therefore, T_{S1} referred to as a stable, stationary point.

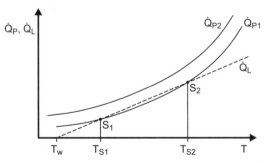

Fig. 6.7. Temperature dependencies of heat production $\dot{Q}_P$ and heat losses $\dot{Q}_L$

At temperature T_{S2}, there is a second stationary point with heat production and losses of equal size. However, in contrast to S_1, this point is unstable as for only slightly increased temperatures the system will be heated further and further. The surplus heat production will increase the more the temperature rises, and a thermal explosion results.

Figure 6.7 also indicates that there may exist curves such as $\dot{Q}_{P2}$, that do not intersect with the heat loss function at all. In this case there are no stationary points, and heat production always exceeds cooling. Therefore, the system will explode for any initial temperature. An example is an adiabatic system in that the heat loss is zero. Consequently, any exothermic reaction will lead to explosion.

In contrast to a thermal explosion, a chemical or chain explosion is not directly a function of the system temperature but rather of the reaction path. This is because a reaction mechanism and the associated heat release can be accelerated when the number of radicals in the system is increased by chain branching reactions. As an example, the most important steps with respect to ignition in the hydrogen-oxygen-system can be written as:

(i) $\qquad H_2 + O_2 = 2\,\dot{O}H$

(ii) $\qquad \dot{O}H + H_2 = H_2O + \dot{H}$

(iii) $\qquad \dot{H} + O_2 = \dot{O}H + \dot{O}$

(iv) $\qquad \dot{O} + H_2 = \dot{O}H + \dot{H}$

(v) $\qquad \dot{H} = \frac{1}{2} H_2$

(vi) $\qquad \dot{H} + O_2 + M = HO_2 + M$

The reaction (i) is termed the *chain initiation* because radicals are formed from stable molecules. Reaction (ii) is a *chain propagation* where the number of radicals is conserved, and reactions (iii) and (iv) are the *chain branching* steps that increase the number of reactive species in the system and cause the chain explosion. Finally, *chain termination* may occur either at a wall (v), or within the gas phase by reaction (vi).

The ignition of hydrocarbon fuels can be viewed as a chain branching process in analogy to the H_2-O_2-system described above. However, the reaction mechanisms for ignition of hydrocarbons are much more complicated in that significantly greater numbers of species and reactions are involved. Nevertheless, the same characteristic behavior is observed. As for other chain explosions, ignition only takes place after a certain induction time (ignition delay) has been passed. This is in contrast to thermal explosions that start instantaneously. During the induction time of a chemical explosion the chain initiation proceeds relatively slowly, because the involved reactants are stable molecules. Thus, the system temperature rises only marginally. Only after a certain threshold in the radical concentration is surpassed, additional radicals can be formed by chain propagation and chain branching, leading to explosion. The induction time for chain reactions is strongly temperature dependent and can thus be significantly reduced by external addition of energy. A prominent example is the spark energy in SI engines, that rises the local temperature and greatly increases the radical concentration within the volume between the spark plug electrodes.

The flammability or explosion limits indicate the mixing (equivalence) ratios, for which a fuel-air mixture is inflammable, i.e. for which either autoignition is possible or for which a self sustained flame can proceed after externally induced ignition. The explosion limits are a function of both temperature and pressure and can be displayed in a p-T explosion diagram, Fig. 6.8. For a given temperature and a very low pressure (below the first explosion limit) the mixture is not ignitable since all radicals that are formed are quickly diffused to the system walls where they recombine to stable molecules. Diffusion is inversely proportional to the system pressure and therefore very fast in this regime. As the pressure is increased (T is kept constant), diffusion rates are reduced such that the formation of new radicals exceeds their recombination at the *first explosion limit*. Chain branching reactions can proceed and chain explosion is observed.

By further increasing the system pressure, the *second explosion limit* may be passed and the mixture becomes stable again. This is because chain terminating steps in the gas phase (e.g. reaction (vi) in the above H_2-O_2-system) become dominant compared to the chain branching reactions (reaction (iii)). The *third explosion limit* indicates the thermal explosion limit. It is governed by the competition of heat production by chemical reactions and heat losses to the system walls, and has been discussed above, see Fig. 6.7. The heat production per unit volume increases with increasing pressure, such that at high pressures a transition to explosion is observed.

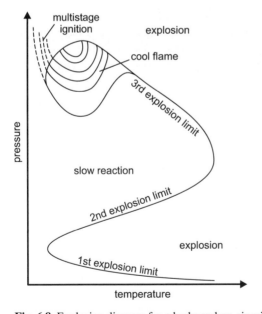

Fig. 6.8. Explosion diagram for a hydrocarbon-air mixture[65]

For hydrocarbon fuels additional and more complex ignition regions can be observed, especially under low temperature and high pressure conditions. In the *multistage ignition* regime ignition takes place only after the emission of short light pulses [65]. In the *cool flame regime*, that is typically observed for gas temperatures between 600 and 800 K, combustion proceeds slowly with only a small temperature rise of about 100 K. In this regime the production of hydroperoxides plays an important role via the reactions

(i) $\dot{R} + O_2 = \dot{R}O_2$,
(ii) $\dot{R}O_2 + RH \rightarrow RO_2H + \dot{R}$,

where RH represents the fuel and $\dot{R}$ is a hydrocarbon radical. The hydroperoxides react rapidly and produce additional radicals:

(iii) $RO_2H \rightarrow RO + \dot{O}H$.

However, for increasing temperatures above about 600 K the reverse reaction of (i) becomes faster. As a result the hydrocarbon radical now reacts preferentially to olefins and hydrogenperoxide:

(iv) $\dot{R} + O_2 \rightarrow olefin + HO_2$,
(v) $HO_2 + RH \rightarrow H_2O_2 + \dot{R}$.

The hydrogenperoxide can further be decomposed into two hydroxyl radicals which is referred to as a degenerated chain branching reaction:

(vi) $H_2O_2 \rightarrow 2\ \dot{O}H$.

However, this reaction proceeds at considerable rates only for temperatures above 800 K. Therefore, a so-called negative temperature coefficient can be observed in hydrocarbon ignition, that is characterized by an increased ignition delay for increased temperatures in the cool flame regime, Fig. 6.9.

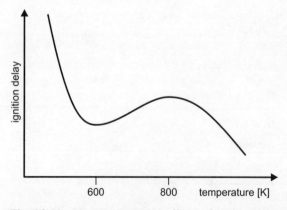

Fig. 6.9. Negative temperature coefficient for hydrocarbon ignition

6.2.2 Autoignition Modeling

In combustion engines the autoignition process takes place on a time scale that is relatively long compared to the relevant fluid dynamical time scales. Therefore, the effects of convective and diffusive species transport during the ignition delay cannot be neglected in the modeling. This is typically taken into account by solving mass conservation equations for certain radical indicator species. These conservation equations include convection, molecular and turbulent diffusion as well as source terms representing the chemical reactions. The latter are commonly expressed by Arrhenius type reaction rates. In order to determine the ignition timing, it is assumed that a certain threshold value for the radical indicator species concentration has to be reached. In diesel engines, the ignition model is then typically switched off and the calculation is continued with a combustion model that concentrates on turbulent mixture formation rather than on detailed reaction kinetics. Such models will be described in more detail in Sect. 6.4.

Single-Step Mechanism

The simplest method to describe the chemical production of a virtual radical species is to assume a single-step reaction expressed by an Arrhenius equation. This procedure is identical to the one utilized in most phenomenological combustion models (see Sect. 3.2.3). Within the framework of CFD-codes the production rate of the radical species becomes the reciprocal value of the formal ignition delay. The latter is again expressed as

$$\tau_{id} = C_{id} \frac{1}{\phi \cdot p^2} \exp(E_{id}/T), \qquad (6.38)$$

where the constant E_{id} can be viewed as an activation temperature. For standard calculations of DI diesel engines with ignition mainly taking place in the high temperature regime, this simple approach can yield satisfactory results as has been pointed out in [38]. However, it should be noted, that the model constants C_{id} and E_{id} are not of fundamental nature and have to be adjusted when the temperature, pressure or equivalence ratio ranges are changed, within that ignition occurs. Specifically, the "negative temperature coefficient" of the ignition delay of hydrocarbon fuels, i.e. the increase of the ignition delay for an increasing temperature, that is observed in the cool flame region because of degenerated chain branching reactions, cannot be predicted with a simple single-step model.

The Shell-Model

Probably the most widely used autoignition model in CFD calculations is the so-called Shell-model. It was originally developed by Halstead et al. [14] in order to predict knock in spark ignition engines, and was later adjusted and applied to model diesel ignition as well, e.g. [21, 54]. The name Shell-model does not describe the nature of the model but simply stems from the affiliation of the authors of the original study.

The Shell ignition model includes eight reaction steps between five species. However, it cannot be viewed as a formally reduced reaction mechanism with global reactions between actual species as discussed in Sect. 6.1.3. It rather represents a virtual mechanism between generic species, that attempts to reflect the actual ignition behavior of hydrocarbon air mixtures including multistage ignition and cool flames.

The eight reaction steps are specified as:

(i) $\quad$ RH + O$_2$ $\xrightarrow{k_q}$ 2 R*

(ii) $\quad$ R* $\xrightarrow{k_p}$ R* + P + Heat

(iii) $\quad$ R* $\xrightarrow{f_1 k_p}$ R* + B

(iv) $\quad$ R* $\xrightarrow{f_4 k_p}$ R* + Q

(v) $\quad$ R* + Q $\xrightarrow{f_2 k_p}$ R* + B

(vi) $\quad$ B $\xrightarrow{k_b}$ 2 R*

(vii) $\quad$ R* $\xrightarrow{f_3 k_p}$ termination

(viii) $\quad$ 2 R* $\xrightarrow{k_t}$ termination

where RH indicates the fuel, R* is the generalized radical, Q is an unstable intermediate agent, B is the branching agent, and P denotes oxidized products such as CO, CO$_2$ and H$_2$O. Thus, reaction (i) represents the chain initiation, (ii) to (v) are chain propagation reactions, (vi) is the chain branching step, and (vii) and (viii) represent linear and quadratic terminations, respectively.

The involved species concentrations can be solved by numerically integrating differential equations for their respective change rates. Following Eq. 6.21, the change rates of the intermediate species become

$$\frac{d[\text{R}^*]}{dt} = 2k_q[\text{RH}][\text{O}_2] + 2k_b[\text{B}] - f_3 k_p[\text{R}^*] - k_t[\text{R}^*]^2, \tag{6.39}$$

$$\frac{d[\text{B}]}{dt} = f_1 k_p[\text{R}^*] + f_2 k_p[\text{R}^*][\text{Q}] - k_b[\text{B}], \tag{6.40}$$

$$\frac{d[\text{Q}]}{dt} = f_4 k_p[\text{R}^*] - f_2 k_p[\text{R}^*][\text{Q}]. \tag{6.41}$$

The change rates of oxygen and the fuel can be written as

$$\frac{d[\text{O}_2]}{dt} = -p k_p[\text{R}^*], \tag{6.42}$$

$$\frac{d[\text{RH}]}{dt} = \frac{[\text{O}_2] - [\text{O}_2]_{(t=0)}}{p \cdot m} + [\text{RH}]_{(t=0)}, \tag{6.43}$$

where m is related to the number of hydrogen atoms in the original fuel molecule C$_n$H$_{2m}$, and p is obtained from

$$p = \frac{n(2-\gamma) + m}{2m}. \tag{6.44}$$

γ is the CO/CO$_2$ ratio, which is often assumed to be approximately equal to 0.67. The rate coefficients in the above mechanism are again of Arrhenius type,

$$f_1 = A_{f1} \exp(-E_{f1}/RT) \, [O_2]^{x1} \, [RH]^{y1} \,, \tag{6.45}$$

$$f_2 = A_{f2} \exp(-E_{f2}/RT) \,, \tag{6.46}$$

$$f_3 = A_{f3} \exp(-E_{f3}/RT) \, [O_2]^{x3} \, [RH]^{y3} \,, \tag{6.47}$$

$$f_4 = A_{f4} \exp(-E_{f4}/RT) \, [O_2]^{x4} \, [RH]^{y4} \,, \tag{6.48}$$

$$k_i = A_i \exp(-E_i/RT) \,, \tag{6.49}$$

where index i stands for ($i = 1, 2, 3, q, b, t$), and

$$k_p = \left[\frac{1}{k_1[O_2]} + \frac{1}{k_2} + \frac{1}{k_3[RH]} \right]^{-1}. \tag{6.50}$$

There are a total of 26 parameters in the above equations that need to be adjusted in order to represent the ignition behavior of a certain fuel. In Table 6.2 these parameters are summarized for three hydrocarbon fuels of different octane ratings.

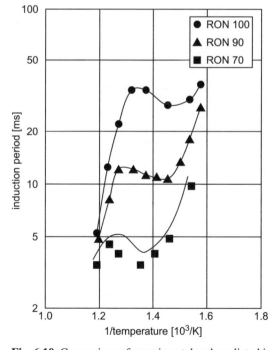

Fig. 6.10. Comparison of experimental and predicted ignition delays [14]

Table 6.2. Parameters of the Shell autoignition model [14, 21]

Parameter	70 RON	90 RON	100 RON
A_{f1}	$1.6\ 10^{-6}$	$7.3\ 10^{-4}$	$7.3\ 10^{-4}$
A_{f2}	$1.8\ 10^{2}$	$1.8\ 10^{2}$	$1.8\ 10^{2}$
A_{f3}	0.75	1.47	2.205
A_{f4}	$1.21\ 10^{6}$	$1.88\ 10^{4}$	$1.7\ 10^{4}$
A_{1}	$1.0\ 10^{12}$	$1.0\ 10^{12}$	$1.0\ 10^{12}$
A_{2}	$1.0\ 10^{11}$	$1.0\ 10^{11}$	$1.0\ 10^{11}$
A_{3}	$1.0\ 10^{13}$	$1.0\ 10^{13}$	$1.0\ 10^{13}$
A_{q}	$6.96\ 10^{11}$	$1.2\ 10^{12}$	$3.96\ 10^{13}$
A_{b}	$3.35\ 10^{18}$	$4.4\ 10^{17}$	$6.512\ 10^{15}$
A_{t}	$2.5\ 10^{12}$	$3.0\ 10^{12}$	$3.51\ 10^{12}$
E_{f1}	$-1.5\ 10^{4}$	$-1.5\ 10^{4}$	$-1.5\ 10^{4}$
E_{f2}	$-7.0\ 10^{3}$	$-7.0\ 10^{3}$	$-7.0\ 10^{3}$
E_{f3}	$1.0\ 10^{4}$	$1.0\ 10^{4}$	$1.0\ 10^{4}$
E_{f4}	$3.0\ 10^{4}$	$3.0\ 10^{4}$	$3.0\ 10^{4}$
E_{1}	0.0	0.0	0.0
E_{2}	$1.5\ 10^{4}$	$1.5\ 10^{4}$	$1.5\ 10^{4}$
E_{3}	$8.5\ 10^{2}$	$8.5\ 10^{2}$	$8.5\ 10^{2}$
E_{q}	$3.5\ 10^{4}$	$3.5\ 10^{4}$	$4.0\ 10^{4}$
E_{b}	$4.7\ 10^{4}$	$4.5\ 10^{4}$	$4.0\ 10^{4}$
E_{t}	0.0	0.0	0.0
x1	1.0	1.0	1.0
x3	0.0	0.0	0.0
x4	-1.3	-1.0	-1.0
y1	-0.5	0.0	0.0
y3	0.0	0.0	0.0
y4	1.0	0.35	0.35

Pre-exponential factors in [cm, mol, s]; activation energies in [cal/mol]

Various studies that incorporated the Shell model have shown that under engine like conditions it is capable of predicting the negative temperature coefficient commonly observed in autoignition phenomena. Moreover, the dependence of the ignition delay on pressure, temperature and mixture stoichiometry is described with reasonable accuracy. An overview over such studies has been given in [50]. As an example, Fig. 6.10 compares experimentally determined and computed ignition delays for three different reference fuels as a function of temperature [14]. Note, that in contrast to Fig. 6.9 the x-axis displays the reciprocal temperature. Thus, the graph appears to be side-inverted.

While the Shell model can be viewed as the standard for modeling diesel engine autoignition and SI engine knock phenomena, it is typically not suitable in order to precisely describe autoignition in modern homogeneous charge compression ignition (HCCI) engine concepts, where the induction period spans over a very long time interval because of low temperatures during the intake and much of the compression stroke. In order to reasonably predict ignition timings and heat release rates in the cool flame regime under such challenging boundary conditions, it is crucial to model the hydrocarbon chemistry as detailed as possible. Hence, more complex reaction mechanisms than the one utilized in the Shell model become necessary. Just a few examples for an increasing number of numerical studies on HCCI engines that are based on detailed chemistry are refs. [24, 26, 30, 72]. It should be noted though, that practically all such studies are based on single-component model fuels like n-heptane. For realistic fuels, that are mixtures of many different hydrocarbon components, an adequate description of the HCCI combustion is even more difficult.

6.2.3 Spark-Ignition Modeling

Phenomenology

In spark ignition engines combustion is initiated by energy addition through an electrical spark into a premixed, essentially stoichiometric air-fuel mixture. The purpose of the spark is twofold: it has to heat the mixture at the spark plug above the third explosion limit (see Fig. 6.8) in order to initiate a thermal explosion and moreover, it has to supply a high concentration of radicals to the reaction zone and thus promote a chain explosion. Both measures contribute to establishing a self-sustained flame, which can produce sufficient thermal energy in order to propagate from the spark location into the combustion chamber without quenching.

The many subprocesses involved in spark ignition are extremely complicated as has been discussed in detail in [17]. The most characteristic phenomena are that during the *breakdown* phase a slim, highly ionized plasma channel is formed between the spark plug electrodes with temperatures of about 60,000 K [28]. This takes place on a nanosecond time scale, and energy transfer efficiencies to the gas are close to 100 %. Thereafter the discharge continues in the *arc phase*, where the plasma expands largely due to heat conduction and diffusion such that the shape of the ignition kernel becomes spherical. Moreover, chemical energy is released more or less instantaneously within the ignition kernel because of the high temperatures encountered. This process takes place on a micro- to millisecond time scale. The energy transfer efficiency to the gas is significantly reduced to about 50 %, primarily because of heat losses to the spark plug electrodes. Due to these heat losses and also because of expansion, the temperature inside the ignition kernel decreases rapidly to a range of approx. 4000 to 10000 K. Thermal dissociation of the gas molecules is now dominant over ionization. Finally, a *glow discharge phase* may follow, during that, depending on the details of the ignition system, the energy storage device, e.g. the ignition coil, will dump its energy into the discharge circuit. The heat losses to the electrodes are most significant in this phase, and the energy transfer efficiency is only about 30 %. The temperature within the ignition kernel is reduced to approx. 3000 K. The flame front that starts to develop from the ignition kernel will initially propagate with about laminar flame speed. Only as it becomes larger in size, its surface is more and more affected by turbulence such that the flame speed increases. This will be discussed in more detail below.

While the above spark ignition process has been numerically investigated by several authors in great detail, e.g. [29, 61, 62], those studies have been mostly limited to fundamental investigations. In the modeling of an entire engine combustion process, it is hardly feasible to resolve the necessary length and time scales for all the involved subprocesses. Instead, it is typically sufficient to describe the initial growth of the ignition kernel by a phenomenological approach. Once this kernel has reached a critical size, that is related to the turbulence level in the spark plug region, the simulation can be continued with a separate combustion model for fully developed turbulent flames, see Sect. 6.3.

Thermodynamic Ignition Kernel Analysis

The probably least comprehensive approach to model spark ignition is to simply add energy to the computational grid cell where the spark gap is located. This has to be continued until the gas temperature is sufficiently high, i.e. until the third explosion limit is exceeded, such that a flame front can develop that produces enough heat by chemical reactions to sustain itself. However, since this approach does not include any details that can nevertheless be of importance for the ignition process, e.g. thermal expansion of the ignition kernel, flame/turbulence interactions, heat losses to the electrodes, etc., unrealistic inputs for the total spark energy as well as for its temporal history may be required in order to describe the induction period with reasonable accuracy.

Herweg and Maly [16] performed a more detailed analysis on the early flame kernel development based on an energy balance of an assumed spherical kernel. Treating the kernel as an open system, its enthalpy change is given by the change in internal energy and the expansion work,

$$\frac{dH_k}{dt} = m_k \frac{dh_k}{dt} + h_k \frac{dm_k}{dt} = \frac{dU_k}{dt} + p\frac{dV_k}{dt} + V_k \frac{dp}{dt}, \qquad (6.51)$$

where subscript k indicates mass-averaged kernel properties.

Since the time increment of flame kernel formation is typically small, and since there is only a small amount of heat released by combustion in this process, it is reasonable to assume that the pressure increase due to combustion may be neglected during this period. Hence, the increase in the mean kernel temperature is related to the enthalpy increase by the equation

$$\frac{dh_k}{dt} = c_{p,k} \frac{dT_k}{dt}. \qquad (6.52)$$

The mass burning rate and thus the increase in the flame kernel mass becomes the product of the unburned mixture density ρ_u, the flame kernel surface are A_k, and an effective kernel growth speed s_{eff}. The latter is the sum of the turbulent flame speed s_t and a so-called plasma velocity s_{pl}, which results from the addition of the spark energy. A derivation of these two components will be given further below.

$$\frac{dm_k}{dt} = \rho_u A_k s_{eff} = \rho_u A_k \left(s_t + s_{pl}\right) \qquad (6.53)$$

The change in the kernel internal energy is given by the first law (Fig. 6.11) as

$$\frac{dU_k}{dt} = \frac{dW_{sp}}{dt} + \frac{dQ_{ch}}{dt} - \frac{dQ_L}{dt} - p\frac{dV_k}{dt},\qquad(6.54)$$

where W_{sp} is the electrical energy transferred to the spark, Q_L is the heat loss to the electrodes, and Q_{ch} is the heat of reaction, which is expressed as the product of the mass burning rate and the enthalpy of combustion products h_b:

$$\frac{dQ_{ch}}{dt} = \frac{dm_k}{dt} \cdot h_b.\qquad(6.55)$$

By combining Eqs. 6.51 to 6.55 and performing several mathematical rearrangements, the equation

$$\frac{dT_k}{dt} = \frac{1}{m_k c_{p,k}} \left[\frac{dW_{sp}}{dt} + (h_b - h_k)\rho_u A_k S_{eff} - \frac{dQ_L}{dt} + V_k \frac{dp}{dt} \right]\qquad(6.56)$$

is obtained. Further using ($m_k = \rho_k V_k$), the mass burning rate can be written as

$$\frac{dm_k}{dt} = \rho_k \frac{dV_k}{dt} + V_k \frac{d\rho_k}{dt} = \rho_u A_k S_{eff}.\qquad(6.57)$$

Solving this relation for the change in kernel volume and utilizing the ideal gas law,

$$\frac{1}{\rho_k}\frac{d\rho_k}{dt} = \frac{1}{p}\frac{dp}{dt} - \frac{1}{T_k}\frac{dT_k}{dt},\qquad(6.58)$$

the following equation is obtained:

$$\frac{dV_k}{dt} = \frac{\rho_u}{\rho_k} A_k S_{eff} + V_k \left[\frac{1}{T_k}\frac{dT_k}{dt} - \frac{1}{p}\frac{dp}{dt} \right].\qquad(6.59)$$

Finally, the change in kernel radius results from normalizing Eq. 6.59 by the kernel surface area A_k:

$$\frac{dr_k}{dt} = \frac{1}{A_k}\frac{dV_k}{dt} = \frac{\rho_u}{\rho_k} S_{eff} + \frac{V_k}{A_k}\left[\frac{1}{T_k}\frac{dT_k}{dt} - \frac{1}{p}\frac{dp}{dt} \right].\qquad(6.60)$$

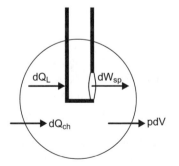

Fig. 6.11. Thermodynamic system of an ignition kernel

The set of differential Eqs. 6.56, 6.59 and 6.60 can now be integrated numerically for the kernel properties T_k, V_k and r_k. The only unknowns remaining are the spark energy W_{sp}, which is typically supplied as a time dependent input from experimental data, and the heat loss to the spark plug electrodes Q_L. This quantity is obtained from

$$\frac{dQ_L}{dt} = h A_{sp} \left(T_k - T_{sp} \right), \qquad (6.61)$$

where T_{sp} is the temperature of the electrodes, A_k the contact area between the kernel and the electrodes, and h is an overall transfer coefficient that is often assumed to be approx. 2000 W/m^{2}K.

Pischinger and Heywood [45] utilized a very similar approach to describe the thermodynamics of the ignition kernel. However, instead of assuming a spherical kernel, they allowed for a rotationally symmetric ellipsoid. The reasoning for this measure stems from Schlieren photographs showing that the ignition kernel can stay attached to the spark gap even though the developing flame is quickly convected away from its origin by the turbulent gas flow. One principal axis of the ellipsoid is determined from the distance of the flame center to its origin, which is given as an input from the respective experimental data. The other two axes are determined from the kernel volume which is calculated in analogy to Eq. 6.59. The benefit of this approach is that the contact area between the flame kernel and the spark electrodes can be calculated more accurately than for an assumed spherical kernel.

The model was applied in order to simulate ignition and early flame development for various engine cycles that were subject to significant cyclic-by-cycle variations. Provided that the shape and orientation of the ellipsoid flame kernel as well as the amount of discharge energy were given as inputs, the predicted flame development was generally in good agreement with experimental data. These results suggest that the cycle-by-cycle variations observed in spark ignition engines are indeed caused to a significant extent by varying heat losses to the spark plug. They can differ in each engine cycle because the turbulent flow is inherently irregular, and the flame kernel is not deflected into the same directions in every cycle.

Ignition Kernel Growth and Flame Speed

While the effective growth speed s_{eff} of the ignition kernel in Eq. 6.53 was assumed to be equal to the laminar flame speed in [45], a more comprehensive approach was followed in ref. [16]. It describes the effective kernel growth rate as the sum of the turbulent flame speed and a spark energy enhanced expansion rate:

$$s_{eff} = s_t + s_{pl}. \qquad (6.62)$$

The turbulent flame speed s_t depends on the laminar flame speed s_l and on the turbulence level within the combustion chamber. As schematically shown in Fig. 6.12, the flame front is wrinkled by turbulent eddies, such that its effective surface A_l is increased. Locally, the flame front still propagates with laminar flame speed s_l, however, due to the area increase by wrinkling the volume swept by the flame front per unit time and thus the mass burning rate is increased. The effect is that the apparent flame speed, termed the turbulent flame speed s_t, is enhanced as well. The turbulent wrinkling typically occurs on a length scale that cannot be resolved by numerical grids utilized in engine CFD simulations. Instead, for modeling purposes the smaller area of a plane flame front, termed turbulent flame area A_t, is utilized and multiplied with the turbulent flame speed that exceeds the laminar flame speed by the same factor by that the wrinkled (laminar) flame surface area A_l exceeds the plane flame front area A_t. Hence, the volumetric burning rate remains unaffected:

$$\dot{V}_b = A_l s_l = A_t s_t \,. \tag{6.63}$$

The laminar flame speed s_l is commonly calculated based on the relations of Keck and co-workers [33, 34, 52] that have already been discussed in Chap. 3 (Eqs. 3.139 to 3.142, Table 3.4). The increase in flame speed due to turbulence can in principle be described in the same way as it is done in the phenomenological models, i.e. with either Eq. 3.143 or Eq. 3.145. However, it is also possible to choose a more detailed formulation for the turbulent to laminar flame speed ratio, because more details of the turbulent flow field are available in the framework of a CFD-code. In fact, Herweg and Maly [16] proposed the relation

$$\frac{s_t}{s_l} = \underbrace{I_0 + I_0^{1/2}}_{\text{i: strain}} \cdot \underbrace{\left(\frac{\left[\bar{u}^2 + u'^2 \right]^{1/2}}{\left[\bar{u}^2 + u'^2 \right]^{1/2} + s_l} \right)^{1/2}}_{\text{ii: effective turbulence factor}} \cdot \underbrace{\left(1 - \exp\left(-\frac{r_k}{l_I} \right) \right)^{1/2}}_{\text{iii: size dependent integral length scale}}$$

$$\cdot \underbrace{\left(1 - \exp\left(-\frac{\left[\bar{u}^2 + u'^2 \right]^{1/2} + s_l}{l_I} \cdot t \right) \right)^{1/2}}_{\text{iv: time dependent integral time scale}} \cdot \underbrace{\left(\frac{u'}{s_l} \right)^{5/6}}_{\substack{\text{v: fully developed} \\ \text{turbulent flame}}} . \tag{6.64}$$

Fig. 6.12. Flame speed enhancement due to turbulent wrinkling of the flame front

It describes a continuous transition from a very small flame kernel to a fully developed turbulent flame. Initially, the flame may propagate with a velocity that is even slower than laminar flame speed because of strain effects evoked by flame curvature, that enter the equation through term (i). The second term accounts for turbulent wrinkling of the reaction sheet, where $\bar{u}$ denotes the mean (Reynolds-averaged) flow velocity and u' is the turbulence intensity. The terms (iii) and (iv) are taking into account that the effects of the turbulent length and time scales on the burning velocity become more pronounced only, if the respective scales of the flame kernel are at least of similar magnitude. For example, a flame kernel that is smaller in size than the turbulence integral length scale will not be wrinkled by a turbulent eddy but rather convected as a whole. Only as the flame kernel grows to a size larger than the eddies, its surface area can be increased by wrinkling such that the flame speed is enhanced. A similar behavior can be observed for the time scales of the turbulence and the flame kernel as well, which is accounted for by term (iv). Term (v) finally describes the behavior of a fully developed, freely expanding turbulent flame.

The strain rate I_0 can be approximated as

$$I_0 = 1 - \left(\frac{\delta_l}{15\ l_I}\right)^{1/2} \left(\frac{u'}{s_l}\right)^{3/2} - 2\frac{\delta_l}{r_k}\frac{\rho_u}{\rho_k}, \qquad (6.65)$$

where δ_l is the laminar flame thickness, l_I the turbulence integral length scale, and r_k the radius of the developing flame kernel. Typical values for these and other important flame related parameters in stoichiometrically operated SI engines have been collected by Heywood [18] from various sources in the literature. They are summarized in Table 6.3.

The influence of the spark energy on the kernel growth rate, which is expressed through the plasma velocity s_{pl} in Eq. 6.62, can be evaluated in a straightforward manner, if the simplifying assumption is made that the temperature inside the ignition kernel is spatially uniform and equal to the adiabatic flame temperature of the air-fuel mixture. This assumption is justifiable, because the breakdown phase, that is characterized by significantly higher temperatures within the plasma channel, lasts only for a few nanoseconds. Thus, the majority of the spark energy is transferred in the arc and glow discharge modes that span over a much longer time period (millisecond range) and are characterized by temperatures that do not differ too much from the adiabatic flame temperature. Since typical time increments in engine CFD simulations are on the order of one microsecond, it is impossible to temporally resolve the breakdown phase. It is thus reasonable to assume an isothermal ignition kernel of adiabatic flame temperature at the end of the first numerical time step.

The initial radius $r_{k,0}$ of the ignition kernel after the first time step can be derived from an energy balance [59]. Assuming that the energy necessary to heat the gas within the kernel from unburned to adiabatic flame temperature is supplied by the sum of the electrical energy W_{sp} of the spark discharge and the chemical energy released within the kernel,

$$\frac{4}{3}\pi r_{k,0}^3 \rho_k c_p \left(T_{ad} - T_u\right) = \dot{W}_{sp}\Delta t_0 + \frac{4}{3}\pi r_{k,0}^3 \rho_k \text{LHV}_{mix}, \tag{6.66}$$

the radius at time Δt_0 after spark onset becomes

$$r_{k,0} = \left[\frac{3\dot{W}_{sp}\Delta t_0}{4\pi\rho_k\left(c_p(T_{ad}-T_u)-\text{LHV}_{mix}\right)}\right]^{1/3}, \tag{6.67}$$

where LHV$_{mix}$ represents the lower heating value of the air-fuel mixture per gram mixture.

In the subsequent time steps the kernel growth is still primarily governed by the electrical energy supply to the spark gap and by heat conduction within the gas. As in Eq. 6.66, it is again assumed that the chemical energy stored within the ignition kernel is immediately released once it is heated to approximately adiabatic flame temperature. Hence, an energy balance in analogy to Eq. 6.66 can be established for the growing ignition kernel. However, in this case the volume change of the kernel is the product of the kernel surface area and the change in kernel radius dr_k. Additionally, an energy transfer efficiency from the spark plug to the gas η_{sp} has to be introduced, that is significantly less than unity in the arc and glow discharge modes. As noted above, it ranges between approx. 50% and 30%:

$$4\pi r_k^2 \frac{dr_k}{dt}\rho_u c_p\left(T_{ad}-T_u\right) = \eta_{sp}\dot{W}_{sp} + 4\pi r_k^2 \frac{dr_k}{dt}\rho_u \text{LHV}_{mix}. \tag{6.68}$$

The change in the kernel radius dr_k/dt is commonly referred to as the plasma velocity s_{pl}. Rearranging the above equation for this quantity yields

$$s_{pl} = \frac{dr_k}{dt} = \frac{\eta_{sp}\dot{W}_{sp}}{4\pi r_k^2 \rho_u\left(c_p(T_{ad}-T_u)-\text{LHV}_{mix}\right)}. \tag{6.69}$$

Table 6.3. Typical parameters in SI engine flames. $\phi = 1.0$, $n = 1500$ rpm [18]

Combustion Parameter	Quantity
Turbulence intensity, u'	2 m/s
Turbulent Reynolds No., Re_t	300
Damköhler No., Da	20
Karlovitz No., Ka	0.2
Integral length scale, l_I	2 mm
Kolmogorov Scale, l_K	0.03 mm
Laminar flame thickness, δ_l	0.02 mm
Laminar flame speed, s_l	0.5 m/s
Ratio u'/s_l	4
Ratio s_t/s_l	4
Mean flame radius of curvature, r_k	2 mm

The plasma velocity is inversely proportional to r_k^2 and thus, decreases rapidly as the ignition kernel is growing. As a consequence, the effective flame velocity, i.e. the sum of the plasma and turbulent flame velocities, has a characteristic history that is schematically displayed in Fig. 6.13. At the spark initiation it starts out from a very large value because of the plasma effect. Its influence decreases rapidly as the kernel becomes larger, and the effective flame velocity reaches a minimum, that is approximately equal to the laminar flame speed. For typical SI engine flow conditions and ignition systems, this minimum is observed about 0.2 ms after spark onset. Thereafter, the enhancement of the flame speed by turbulent wrinkling becomes more and more important and the effective burning velocity increases until it reaches the value of a fully developed turbulent flame.

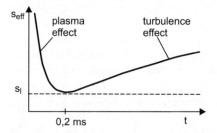

Fig. 6.13. Characteristic course of effective flame speed after ignition

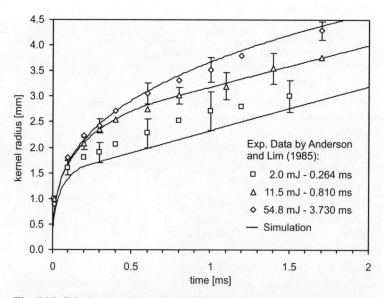

Fig. 6.14. Calculated and experimental flame kernel growth in a quiescent propane-air mixture, $\phi = 0.7$, $T_0 = 300$ K, $p_0 = 101.3$ kPa [59]

Stiesch et al. [59] implemented the above ignition model into the CFD-code KIVA-3V and validated it against experimental data presented by Anderson and Lim [2]. The results are displayed in Fig. 6.14, which shows the ignition kernel growth vs. time in a quiescent propane-air mixture of equivalence ratio 0.7 for three different spark energies and durations. The solid lines represent the simulation results that have been obtained with an energy transfer efficiency of $\eta_{sp} = 0.46$ in Eq. 6.69, while the symbols represent the experimental data. The error bars indicate the standard deviation of the measurements.

Spark Plug Protrusion

The electrodes and parts of the spark plug body represent a flow obstacle that may have a significant effect on the gas motion in the combustion chamber. This is especially the case in stratified charge DISI engines, where spark plugs have to reach further into the cylinder, because the mixture is often too lean close to the combustion chamber walls, and thus not suited for ignition. However, it is hardly possible to take account of these flow obstacles during the pre-processing (grid generation) of an engine simulation. This is because the computational grid is typically too coarse in order to accurately reproduce the shape of a spark plug.

The problem can be evaded by representing the spark plug by solid particles that are distributed according to the shape of the plug and independently of the computational grid arrangement [11, 58]. The particles are fixed in space and characterized by an extremely high drag coefficient of $C_D = 1000$. Thus, the displacement of the gas at the position of the solid spark plug can still not be reproduced correctly, but the effect on the gas flow is effectively accounted for. This is shown in Fig. 6.15, which shows the mean flow velocities and directions within a dome shaped combustion chamber of a two-stroke DISI engine shortly after the end of the scavenging process. The clockwise tumble motion is obstructed and redirected by the spark plug within the cylinder dome.

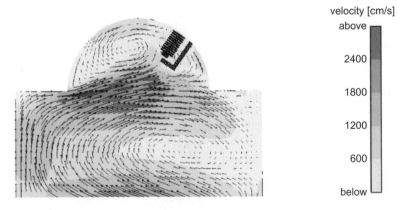

Fig. 6.15. Velocity field in a two-stroke DISI engine after scavenging

6.3 Premixed Combustion

6.3.1 The Flamelet Assumption

Combustion models for premixed combustion in homogeneously operated spark ignition engines are most often based on the so-called flamelet assumption [41]. That is, the flame front, that is wrinkled due to interaction with eddies in a turbulent flow, is nevertheless assumed to behave as a laminar flame on a local scale. There is a general agreement that the flamelet concept is applicable in the region of large Damköhler numbers with turbulent scales larger than the flame thickness. This region is represented by the lower right part of the Borghi diagram (Fig. 6.4), specifically by regimes 2 and 3. For turbulent Reynolds numbers less than unity, i.e. in regime 1 of the Borghi diagram, the flamelet approach correctly evolves to a laminar premixed flame.

The validity of the flamelet assumption can be explained by the fact, that for turbulent eddies larger than the flame thickness (Ka < 1) the flame front is not yet torn into distributed reaction zones, but that a coherent flame front can still be identified. In fact, Dinkelacker and co-workers showed, that the flamelet assumption is valid even for Karlovitz numbers somewhat greater than unity, because the thin reaction zone is led by a broader preheat zone [9, 56]. Within this preheat zone the smallest eddies tend to be dissipated due to an increase in the temperature dependent kinematic viscosity. Furthermore, for large Damköhler numbers the chemical time scales are short compared to the turbulent time scales. Thus, the combustion rate is most significantly governed by turbulence properties rather than by reaction chemistry, and the central problem in the flamelet regime is quantifying the effective flame speed. Table 6.3 suggests that the conditions for the flamelet assumption are typically satisfied in homogeneous charge SI engines.

6.3.2 Eddy-Breakup Models

Probably the simplest combustion model based on the flamelet assumption is the semi-empirical eddy-breakup model. The intrinsic idea behind this model type is that the combustion rate is determined by the rate, at which parcels of unburned mixture within the turbulent flame brush are broken down into smaller ones, such that there is sufficient interfacial area between the unburned mixture and hot gases to permit reaction.

Bilzard and Keck [4] described the combustion process as occurring in two subsequent stages. First, unburned mixture has to be entrained into the turbulent flame brush. This process is related to the turbulent flame speed by

$$\dot{m}_e = \rho_u A_t s_t, \qquad (6.70)$$

where A_t and s_t are the turbulent flame front area and speed as defined in Fig. 6.12. The turbulent flame speed may be calculated either by Eq. 6.64, or with one of the simpler relations presented in Chap. 3 (Eqs. 3.143 or 3.145).

In the second step, burn up within the flame brush occurs at a rate governed by the amount of unburned mass within the flame divided by a characteristic burning time:

$$\dot{m}_b = \frac{m_e - m_b}{\tau_b} + \rho_u A_t s_l. \tag{6.71}$$

The burning time is taken to be the ratio of the characteristic size of the entrained turbulent eddies to the laminar flame speed, where the constant C is of order unity:

$$\tau_b = C \cdot \frac{l_e}{s_l}. \tag{6.72}$$

Note, that the same model is also used in phenomenological combustion models where, because of a lack of knowledge of more detailed turbulence properties, the length scale is commonly related to the lift of the intake valve, Eq. 3.153. However, when used within the framework of CFD-codes, the length scale l_e is more often taken to be the local value of the Taylor microscale l_T at the position of the flame front [18]:

$$l_e = l_T \approx \sqrt{\frac{15}{\text{Re}_t}} \cdot l_I. \tag{6.73}$$

Both with phenomenological and with CFD simulations, the eddy-breakup model requires a geometric model for the flame front. Often, a simple spherical assumption for the flame envelope surface is made.

The Magnussen combustion model [27] is very similar to the above formulation as the combustion rate is primarily governed by a turbulence related time scale, too. Specifically, the fuel consumption rate is expressed on the base of a single-step reaction as

$$\dot{w}_f = \frac{A \cdot \min\left(Y_f, \frac{Y_{O_2}}{af_{st}}, B \frac{Y_p}{1 + af_{st}}\right)}{\tau_t}, \tag{6.74}$$

where the Y_i denote the local mean mass fractions of fuel, oxygen and products, and A and B are empirical constants. The stoichiometric air-fuel ratio enters the equation through af_{st}, and the turbulent time scale is modeled as the ratio of the turbulent kinetic energy k and its dissipation rate ε:

$$\tau_t = k / \varepsilon. \tag{6.75}$$

Eddy-breakup models have been and are still widely used in engine calculations, primarily due to their simplicity and because more comprehensive combustion models are not without problems themselves. It should be noted though, that one major deficiency exists in the above formulations, in that the turbulence related time scales, Eqs. 6.72 or 6.75, reduce to zero as a combustion chamber wall is approached. This leads to the unphysical prediction that the burning velocity tends to increase towards the wall [68].

In order to overcome this shortcoming, Abraham et al. [1] proposed several variations to the Magnussen model. The mass burning rate, expressed in terms of the change in species density ρ_i, is determined by dividing the deviation of the momentary species density from its equilibrium value by a characteristic time scale necessary to reach the equilibrium state:

$$\frac{d\rho_i}{dt} = -\frac{\rho_i - \rho_i^{eq}}{\tau_c}. \tag{6.76}$$

This allows to include multi-step reaction mechanisms in the analysis, which is of special importance in rich mixtures, where carbon monoxide und unburned hydrocarbons are present. Additionally, and most importantly, the characteristic time scale is now composed of both a turbulent and a laminar time scale,

$$\tau_c = f\tau_t + \tau_l, \tag{6.77}$$

where f is a delay coefficient that is initially zero and approaches unity later in the combustion process:

$$f = 1 - \exp\left[-(t - t_{sp})/\tau_d\right]. \tag{6.78}$$

Here t_{sp} is the spark timing, and $\tau_d = c_{m1} l_I / s_l$. The constant c_{m1} is of order unity.

The turbulent time scale is again related to the turbulence properties,

$$\tau_t = c_{m2} k / \varepsilon, \tag{6.79}$$

and thus reduces to zero as a wall is approached. The constant is typically chosen to be $c_{m2} \approx 0.05$. However, in the vicinity of walls the laminar time scale becomes important because of the reduced temperatures and prohibits an unphysical acceleration of the flame speed. For iso-octane, the laminar time scale is defined as

$$\tau_l = \frac{3.09 \cdot 10^{-12}}{\left[1.27(1 - 2.1 \cdot f_R)\right]^2} \cdot T \cdot \left(\frac{p}{p^\circ}\right)^{0.75} \cdot \exp\left[\frac{E(1 + 0.08 \cdot |\phi - 1.15|)}{T}\right], \tag{6.80}$$

where f_R is the residual mass fraction, p° the standard pressure of 101.3 kPa, and E the activation temperature of approx. 15,100 K.

While it has been shown that this model is capable of reasonably predicting global heat release rates and cylinder pressure traces [23], it cannot be used in order to make assessments of the local flame structure. Specifically, because of the diffusive term in the species conservation equations, the flame width (or flame thickness) tends to increase during the course of combustion [67]. Such a behavior is obviously unphysical and cannot be observed in experiments of premixed, homogeneous combustion.

6.3.3 Flame Area Evolution Models

The Weller / G-Equation Model

A more comprehensive combustion model than the eddy-break up model is the flame area evolution model presented by Weller et al. [67, 68], also referred to as the G-equation model [42]. It describes the combustion progress by a so-called regress variable b, which may be viewed as a non-dimensional species concentration of the combustion reactants. Thus, b has limits of unity and zero, in the unburned and fully burned states, respectively. Its temporal and spatial evolution can be described by the transport equation

$$\frac{\partial \overline{\rho}\tilde{b}}{\partial t} + \frac{\partial (\overline{\rho}\tilde{u}_i \tilde{b})}{\partial x_i} - \frac{\partial}{\partial x_i}\left(\overline{\rho}\tilde{D}_b \frac{\partial \tilde{b}}{\partial x_i}\right)$$
$$= -\left[\overline{\rho}\xi + (\overline{\rho}_u - \overline{\rho})\cdot \min(\xi, \xi_{eq})\right] \cdot s_l \left|\nabla \tilde{b}\right| , \quad (6.81)$$

where the overbar (–) denotes ensemble- or time-averaged quantities, and the tilde overbar (~) denotes Favre- or density-weighted averages. For example,

$$\tilde{u} = \frac{\overline{\rho u}}{\overline{\rho}} . \quad (6.82)$$

The differential operator ∇ in the last term of Eq. 6.81 describes the gradient of the regress variable. The local flame wrinkle factor, i.e. the wrinkled flame surface area per unit projected area, is denoted by ξ. It is equal to the turbulent to laminar flame speed ratio and related to the wrinkled flame area per unit volume Σ by

$$\xi = \frac{s_t}{s_l} = \frac{\Sigma}{\left|\nabla \tilde{b}\right|} . \quad (6.83)$$

There are two possibilities how to determine this flame wrinkle factor. It can either be derived from its own transport equation, including both a term for generation of wrinkles by turbulence and a term for destruction of wrinkles due to flame propagation [67]. Or, the simpler and more widely applied method is to assume that local equilibrium between wrinkle generation and removal prevails, i.e. $\xi = \xi_{eq}$. In this case the right hand side of Eq. 6.81, which represents the mean reaction rate reduces to

$$\frac{\partial \overline{\rho}\tilde{b}}{\partial t} + \frac{\partial (\overline{\rho}\tilde{u}_i \tilde{b})}{\partial x_i} - \frac{\partial}{\partial x_i}\left(\overline{\rho}\tilde{D}_b \frac{\partial \tilde{b}}{\partial x_i}\right) = -\overline{\rho}_u \xi_{eq} s_l \left|\nabla \tilde{b}\right| . \quad (6.84)$$

The quantity of the equilibrium flame wrinkle factor ξ_{eq}, i.e. the turbulent to laminar flame speed ratio, can be estimated based on the relations discussed in previous sections. Often the relation by Herweg and Maly [16], Eq. 6.64, is utilized for this purpose.

It is noteworthy about Eq. 6.84, that the reaction rate, i.e. the term on the right hand side, is not simply related to the turbulence scales such as in the eddy-breakup models, but that it is rather a function of both the turbulence properties and the combustion regress variable, for that a separate transport equation is solved. Thus, non-local effects can be accounted for that may become important in actual combustion engines. However, due to the diffusive term of the regress variable, there is still an unphysical increase in the flame front thickness predicted, albeit not as dramatically as in the eddy breakup models.

The Coherent Flame Model

The coherent flame model is in principle very similar to the above G-equation model, in that it characterizes the flame structure by a local effective wrinkling factor. Moreover, the evolution of a combustion progress variable is modeled by a transport equation that contains the local mean reaction rate as a source term. This is in analogy to Eq. 6.84. Consequently, both model types are referred to as flame area evolution models.

In the case of the coherent flame model, the flame surface area per unit volume, defined as

$$\Sigma = \frac{A_l}{V}, \qquad (6.85)$$

is utilized in order to determine the mean turbulent reaction rate,

$$\overline{\dot{w}_t} = \dot{w}_l \cdot \Sigma, \qquad (6.86)$$

where w_l is the consumption rate of fuel per unit of laminar flame area:

$$\dot{w}_l = \rho_u Y_f s_l. \qquad (6.87)$$

Note, that Eqs. 6.86 and 6.87 are equivalent to the term on the right hand side of Eq. 6.84. The only difference is, that the present analysis is based on the fuel mass, expressed through the fuel mass fraction Y_f in the unburned mixture, whereas the reaction rate in Eq. 6.84 is based on the total mass of unburned mixture.

The local flame surface density Σ is now estimated from its own evolution equation [7]:

$$\frac{\partial \Sigma}{\partial t} + \frac{\partial (u_i \Sigma)}{\partial x_i} - \frac{\partial}{\partial x_i}\left(\frac{v_t}{\sigma_\Sigma}\frac{\partial \Sigma}{\partial x_i}\right) = \alpha K_t \Sigma - \beta \frac{\dot{w}_l \left(1 + a\sqrt{k}/s_l\right)}{\rho \overline{Y}_f} \Sigma^2. \qquad (6.88)$$

In this equation the parameters a, α, β and σ_Σ are fixed at $a = 0.1$, $\alpha = 2.1$, $\beta = 1$, $\sigma_\Sigma = 1$. The turbulent diffusivity v_t may be modeled as discussed in Chap. 4, and ρ and $\overline{Y}_f$ are the local means of the density and the fuel mass fraction, respectively.

The first term on the right hand side of Eq. 6.88 denotes the production of flame surface density due to turbulence, and the second term represents the destruction due to mutual annihilation of adjacent flamelets. As has been pointed out by Boudier et al. [7], several methods of modeling the turbulent stretch rate K_t, which is defined as the relative change rate of the flame surface density ($K_t = 1/\Sigma \, d\Sigma/dt$), have been proposed in the literature. Typically, it is specified as a function of the most important turbulence and flame parameters. For example, the formulation

$$K_t = \sqrt{\varepsilon/v} \qquad (6.89)$$

may be applied, where ε represents the dissipation rate of turbulent kinetic energy, and v is the molecular viscosity. However, this equation represents only a coarse approximation of the stretch rate, and a more accurate correlation has been derived based on detailed results of both direct numerical simulations and experiments by Meneveau and Poinsot [32]. They suggested a rather complex equation in terms of the turbulence parameters ε, k, u', l_t and the laminar flame parameters s_l and δ_l. It will only be summarized briefly here. The ratio of the net stretch rate to the reciprocal turbulent time scale becomes

$$\frac{K_t}{\varepsilon/k} = \Gamma_K - \frac{3}{2}\left(\frac{l_t}{\delta_l}\right)\left(\frac{u'}{s_l}\right)^{-1} \ln\left(\frac{1}{1-P_q}\right), \tag{6.90}$$

where Γ_K and P_q are determined as follows:

$$\log_{10} \Gamma_K = -\frac{1}{(s+0.4)} e^{-(s+0.4)} + \left(1 - e^{-(s+0.4)}\right)\left(\sigma_1\left(u'/s_l\right)s - 0.11\right), \tag{6.91}$$

$$P_q = \frac{1}{2}\left[1 + \tanh\left(\text{sgn}(x)x^2\right)\right], \tag{6.92}$$

where

$$\sigma_1\left(u'/s_l\right) = \frac{2}{3}\left(1 - \frac{1}{2}e^{(u'/s_l)^{1/3}}\right), \tag{6.93}$$

$$s = \log_{10}\left(l_t/\delta_l\right), \tag{6.94}$$

$$x = \frac{\log_{10}\left(u'/s_l\right) - g\left(l_t/\delta_l\right)}{\sigma\left(l_t/\delta_l\right)}, \tag{6.95}$$

$$g\left(l_t/\delta_l\right) = \left(0.7 + 1/s\right)e^{-s} + \left(1 - e^{-s}\right)\left(1 + 0.36s\right), \tag{6.96}$$

$$\sigma\left(l_t/\delta_l\right) = 0.04\log_{10}\left(l_t/\delta_l\right). \tag{6.97}$$

Note, that Eqs. 6.91 to 6.97 are curvefits to numerical solutions of more fundamental representations of Γ_K and P_q [32].

Because of their resemblance, both the coherent flame and the G-equation model can be expected to show similar behavior under various possible boundary conditions. In general, they both allow satisfactory simulations of homogeneous charge combustion applications, and they provide more detailed insight into flame structures than the computationally more efficient eddy breakup models.

6.3.4 The Fractal Model

In the fractal model the degree of flame front wrinkling that accelerates the flame speed is not obtained by solving a transport equation but rather by assuming a self similarity between scales of different size [18]. This results in a power-law scaling between the wrinkled flame area and the inner and outer cutoff length scales, that are typically assumed to be the integral and Kolmogorov length scales, respectively:

$$\frac{s_t}{s_l} = \frac{A_l}{A_t} = \left(\frac{l_K}{l_I}\right)^{2-D}. \tag{6.98}$$

To use this model an equation for the fractal dimension D is required. Santavicca et al. [53] proposed the relation

$$D = 2.0 \frac{s_l}{u' + s_l} + 2.35 \frac{u'}{u' + s_l}, \tag{6.99}$$

based on experimental data that showed that D increases from 2.0 for a laminar flame (i.e. $s_t/s_l = 1$) to about $D = 2.35$ for highly turbulent flames.

6.4 Diffusion Combustion

The term diffusion combustion is commonly used to describe non-premixed flames as they occur, for example, in direct injection diesel engines. It stems from the fact that in these kind of flames the mixing time scales are dominant for the overall combustion process, and that both molecular and turbulent diffusion are of utmost importance for the mixing rate. Nevertheless, the phrase diffusion combustion can be somewhat misleading in that diffusion plays an important role in the propagation of premixed flame fronts as described in the foregoing section as well. The name non-premixed combustion would be more precise for the processes taking place in diesel engines, but as mentioned above, it is still common practice to refer to it as diffusion combustion.

In general, the mixing between fuel and oxidizer is the governing process for this combustion type, and reaction chemistry is much faster. Thus, the assumption *mixed = burned* is often made. However, there are some important exceptions to this rule, especially for extreme lean or rich mixing ratios close to the inflammability limits and for low temperatures late in the expansion stroke. These effects become particularly important in the formation of pollutants, see Chap. 7.

6.4.1 The Characteristic Time Scale Model

The laminar and turbulent characteristic time scale model discussed in Sect. 6.3.2 has been adopted to model non-premixed combustion in diesel engines by Patterson et al. [39]. In principle, the same set of equations that has been used for premixed combustion, Eqs. 6.76 to 6.80, is used for non-premixed combustion, albeit with some modifications. The change in species density is again expressed in terms of the actual and equilibrium mixture compositions as well as the characteristic time scale:

$$\frac{d\rho_i}{dt} = -\frac{\rho_i - \rho_i^{eq}}{\tau_c} = -\frac{\rho_i - \rho_i^{eq}}{\tau_l + f\tau_t}. \tag{6.100}$$

However, the delay factor f is now defined as a function of the local composition,

$$f = \left(1-e^{-r}\right)/0.632 ,\qquad (6.101)$$

where r is the ratio of the amount of products to that of total reactive species (except N_2), i.e.

$$r = \frac{Y_{CO_2} + Y_{H_2O} + Y_{CO} + Y_{H_2}}{1 - Y_{N_2}} .\qquad (6.102)$$

Thus, the parameter r indicates the local completeness of combustion. Its value varies from zero to unity for unburned and completely burned mixture, respectively. As a consequence, the delay factor f also changes from zero to unity depending on the local conditions. The interpretation of this behavior is, that the turbulence driven microscale mixing becomes more and more important compared to chemistry effects, the further combustion has proceeded.

For diesel fuel which is approximated as tetradecane, the laminar (chemistry) time scale is given as

$$\tau_l = A^{-1}[C_{14}H_{30}]^{0.75}[O_2]^{-1.5}\exp(E_A/RT) ,\qquad (6.103)$$

where the pre-exponential constant and the activation energy are $A = 7.68 \cdot 10^8$, and $E_A = 77.3$ kJ/mol, respectively. The turbulent (mixing) time scale has been specified as

$$\tau_t = 0.142 \cdot k/\varepsilon .\qquad (6.104)$$

Once the change rates of the various species densities are known, the amount of heat release can readily be estimated based on the respective enthalpies of formation of the involved species i:

$$\frac{dQ_{ch}}{dt} = -\sum_i \frac{d\rho_i}{dt}\frac{1}{MW_i}\tilde{h}^0_{f,i} .\qquad (6.105)$$

In most studies a two step reaction mechanism has been considered that describes the oxidation of fuel via CO to CO_2, and includes the following seven species: C_nH_{2n+2}, O_2, CO, CO_2, H_2, H_2O, N_2. Thus, the model is capable of predicting the reduced amount of energy that is released in fuel rich regions, where the equilibrium is shifted towards CO instead of CO_2. This phenomenon may be explained by the fact, that in the partial oxidation to CO 50% of the oxygen is consumed compared to complete oxidation to CO_2, but only about 30% of the total thermal energy is released. As a result, the heat release rate is delayed compared to the fuel consumption rate in the early stages after autoignition, when the level of mixture stratification is still very high [66].

In the original studies, the identical characteristic time scale was utilized for all species. However, in a later work this simplification was dropped in order to account for different kinetic reaction rates of the involved reactions and species [71]. Specifically, different characteristic time scales were introduced for the species fuel, CO and H_2. While the time scale for fuel remained unchanged, more realistic results were obtained by setting the CO and H_2 time scales to the fractional value:

$$\tau_{c,CO} = \tau_{c,H_2} = 0.2 \cdot \tau_{c,fuel} \,. \tag{6.106}$$

The characteristic time scale combustion model has been applied in numerous computational studies on diesel engine combustion, and considering the relative simplicity of the model, very reasonable results have been obtained in general. In fact, the time scale combustion model seems better suited to describe non-premixed combustion without a distinct flame front in diesel engines, than to describe premixed combustion in SI engines. This suggests, that the turbulent mixing of reactants that is important in diesel combustion is indeed primarily governed by the dissipation of the large scale eddies, whereas additional effects may become important in the propagation of a premixed, wrinkled flame sheet.

6.4.2 Flamelet Models

Model Concept

The intrinsic idea behind flamelet models is that even in non-premixed turbulent diffusion flames, a substantial fraction of the chemical reactions takes place in thin layers that may locally be treated as laminar reaction sheets [40, 41]. This concept is displayed in Fig. 6.16, and it can be justified by the fact that chemical time scales are typically short compared to diffusion and convection time scales. The overall turbulent flame brush is consequently viewed as an ensemble or superposition of numerous laminar flamelets, that is subject to a statistical probability distribution in analogy to the turbulent fluctuations of the flow field.

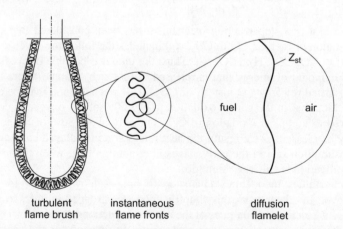

Fig. 6.16. Schematic illustration of the laminar flamelet concept

Within the thin laminar flamelets, combustion can be treated as a one-dimensional process that depends only on the locally one-dimensional mixing ratio between fuel and oxidizer. This mixing ratio is uniquely defined in terms of the mixture fraction Z, which represents the ratio of the fuel mass flow rate to the total mass flow rate:

$$Z = \frac{\dot{m}_{fuel}}{\dot{m}_{fuel} + \dot{m}_{oxidizer}}. \quad (6.107)$$

Thus, the mixture fraction has limiting values of unity in the pure fuel stream and zero in the pure oxidizer (air) stream. It should be noted, that Z is proportional to the mass fractions of the elements. Therefore, it is not affected by combustion but only by mixing.

Because of the one-dimensional nature of the flamelet, the combustion chemistry can be separated from the three-dimensional turbulent flow field in the mathematical description. This is desirable as it results in a significant reduction in the computational expenditure. Specifically, the following procedure is pursued: The changes in species densities due to chemical reactions are solved in a one-dimensional form as a function of the mixture fraction only. This calculation can be performed as a pre-processing for many different mixture fractions and other important flamelet parameters such as temperature, pressure and the scalar dissipation rate, that will be discussed in more detail below. The results of these calculations can then be stored in look-up tables (so-called flamelet libraries) and are readily available when the actual CFD calculation is performed. As the chemistry has to be solved only once for each set of flamelet boundary conditions, the consideration of more detailed reaction mechanisms becomes possible without the drawback of excessive CPU times.

The important effects of the turbulent fluctuations on combustion are accounted for only within the 3D-CFD calculations. This is done by weighting the one-dimensional flamelet results with the probability that the respective flamelet is present within a computational cell. An integration of these weighted flamelet solutions is then executed over all possible flamelets in order to obtain the overall solution for the new composition within each CFD grid cell:

$$\tilde{Y}_i = \int_0^1 P(Z) \cdot Y_i(Z) \cdot dZ. \quad (6.108)$$

In the above equation, $\tilde{Y}_i$ indicates the Favre-averaged mass fraction of species i within a CFD grid cell.

Hence, two major steps are necessary in employing a flamelet model for turbulent non-premixed combustion: (i) a description of the one-dimensional laminar flamelet structure in order to determine $Y_i(Z)$, and (ii) a modeling of the probability density function $P(Z)$, that describes the probability of the presence of a specific flamelet within the turbulent flow field. These steps will be discussed in more detail in the following sections.

3D Mixture Fraction Distribution

The turbulent mixing state at each position and point in time can be characterized by the mean and the variance of the mixture fraction Z. This is illustrated in Fig. 6.17, which compares three different mixing states and the according probability density functions (pdf's). All the three states have the same value for the mean mixture fraction ($\tilde{Z} = 0.5$), but a different variance. It becomes obvious that the variance (or more specifically, the second moment) is a measure of the turbulence level. A large variance represents a highly turbulent, inhomogeneous mixture, whereas a small variance indicates a more homogeneous state, where turbulent dissipation has already led to a more homogenous mixture.

In order to solve for the position- and time-dependent probability distribution within the combustion chamber, transport equations can be established for the mean and the second moment of the mixture fraction [46]:

$$\frac{\partial(\bar{\rho}\tilde{Z})}{\partial t} + \frac{\partial(\bar{\rho}\tilde{u}_i\tilde{Z})}{\partial x_i} = \frac{\partial}{\partial x_i}\left(\frac{\mu}{Sc_{\tilde{Z}}}\frac{\partial \tilde{Z}}{\partial x_i}\right) + \dot{\rho}^s, \qquad (6.109)$$

$$\frac{\partial(\bar{\rho}\widetilde{Z''^2})}{\partial t} + \frac{\partial(\bar{\rho}\tilde{u}_i\widetilde{Z''^2})}{\partial x_i} = \frac{\partial}{\partial x_i}\left(\frac{\mu}{Sc_{\widetilde{Z''^2}}}\frac{\partial \widetilde{Z''^2}}{\partial x_i}\right) + \frac{2\mu}{Sc_{\widetilde{Z''^2}}}(\nabla\bar{Z})^2 - \bar{\rho}\tilde{\chi}, \qquad (6.110)$$

where ∇ is the gradient operator, and $\tilde{\chi}$ is the mean scalar dissipation rate that can be modeled following Jones and Whitelaw [19]:

$$\tilde{\chi} = C_\chi \frac{\tilde{\varepsilon}}{\tilde{k}}\widetilde{Z''^2}. \qquad (6.111)$$

As will be discussed below, $\tilde{\chi}$ proves to be a parameter of paramount importance in flamelet modeling. The two Schmidt numbers for the mean and the second moment of the mixture fraction are assumed to be both equal to 0.9, and $c_{\tilde{\chi}}$ is about 2.0 [46].

However, in order to be able to integrate Eq. 6.108, it is not sufficient to know just the mean and the second moment of the mixture fraction distribution. In fact, the entire pdf needs to be known, and most often a beta-function is assumed for this purpose. It is defined as

$$P_Z(Z) = \frac{\Gamma(\alpha+\beta)}{\Gamma(\alpha)+\Gamma(\beta)} Z^{\alpha-1}(1-Z)^{\beta-1}, \qquad (6.112)$$

where the gamma-function of a generic term a is

$$\Gamma(a) = \int_0^\infty t^{a-1} e^{-t} dt, \qquad (6.113)$$

and the pdf-parameters α and β can be determined from the knowledge of $\tilde{Z}$ and $\widetilde{Z''^2}$:

$$\tilde{Z} = \frac{\alpha}{\alpha+\beta}; \qquad \widetilde{Z''^2} = \frac{\alpha\beta}{(\alpha+\beta)^2(\alpha+\beta+1)}. \qquad (6.114)$$

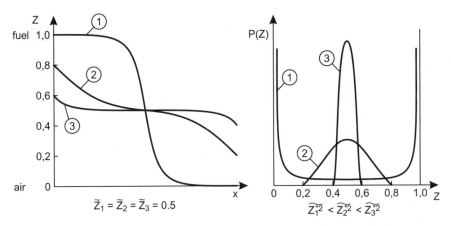

Fig. 6.17. Description of the spatial diffusion profile with a pdf in mixture fraction space

1D Flamelet Structure

Considering a locally defined coordinate system, where the coordinates x_2 and x_3 are within the surface of stoichiometric mixture and the coordinate x_1 is normal to it, x_1 can be replaced by the mixture fraction Z, if Z is considered a conserved scalar, i.e. if its transport equation does not contain any source terms. This assumption is obviously not exactly valid during the evaporation of a diesel spray, when there is definitely a source in the transport equation of the mixture fraction, arising from the phase change. Nevertheless, the concept is still utilized, and transport equations for mass and energy can be established in terms of the newly defined coordinates Z, x_2 and x_3. Because property gradients in Z direction, i.e. perpendicular to the stoichiometric mixture fraction iso-surface, are significantly greater than within the iso-surface, the derivatives with respect to x_2 and x_3 can be neglected. The transport equations reduce to a one-dimensional form in Z-direction, i.e. normal to the surface of stoichiometric mixture. The mass and energy equations become

$$\rho \frac{\partial Y_i}{\partial t} = \rho \frac{\chi}{2} \frac{\partial^2 Y_i}{dZ^2} + \dot{w}_i, \qquad (6.115)$$

$$\rho \frac{\partial T}{\partial t} = \rho \frac{\chi}{2} \frac{\partial^2 T}{dZ^2} + \frac{1}{c_p} \frac{\partial p}{\partial t} - \sum_{i=1}^{n} \frac{h_i}{c_p} \dot{w}_i . \qquad (6.116)$$

In the above equations the production rates of the species i are determined from

$$\dot{w}_i = MW_i \sum_k v_{ik} \tilde{w}_k , \qquad (6.117)$$

where index k specifies the various chemical reactions, and the molar reaction rate $\tilde{w}_k$ of a single reaction is obtained from an Arrhenius formulation as described by Eqs. 6.21 and 6.22. The instantaneous scalar dissipation rate is defined as:

$$\chi = 2D\left(\frac{\partial Z}{\partial x}\right)^2. \tag{6.118}$$

It can be interpreted as the inverse of a characteristic diffusion time. In general, it increases due to turbulent straining and decreases due to diffusion.

Two aspects should be noted about the above formulation. Firstly, there are no convective terms in the mass and energy transport equations, Eqs. 6.115 and 6.116. This can be explained by the fact that all scalar quantities are subject to the same convective transport and thus, there is no convection relative to Z. The second aspect is that the instantaneous scalar dissipation rate χ represents the only direct coupling between the turbulent flow field and the chemistry. It accounts for strain effects and is a crucial parameter in flamelet modeling.

Scalar Dissipation Rate

Since the calculation of the three-dimensional turbulent flow field only provides the mean of the scalar dissipation rate (Eq. 6.111) but not its instantaneous value, an additional modeling becomes necessary. In general, χ is a function of both the mixture fraction Z and the turbulent fluctuations. The Z-dependence has been analytically shown to be

$$\chi = \frac{a_s}{\pi} \exp\left(-2\left[erfc^{-1}(2Z)\right]^2\right), \tag{6.119}$$

where $erfc^{-1}$ indicates the inverse (not the reciprocal!) complimentary error function [40]. The parameter a_s is the stagnation point velocity gradient in a laminar diffusion flame, and may be viewed as a characteristic strain rate of the reacting flow. It is the only parameter in the above equation and may therefore be determined from the knowledge of only one χ-Z combination, e.g. the one at stoichiometric mixture fraction, i.e. $\chi_{st} = \chi(Z_{st})$. This is demonstrated below.

Equation. 6.119 applies to any Z within the mixture fraction space between zero and one, and can thus be rearranged to the form

$$\chi = \chi_{st}\frac{f(Z)}{f(Z_{st})}. \tag{6.120}$$

The expression $f(Z)$ indicates the exponential term in Eq. 6.119 and does not contain any unknowns variables. Thus, the definition of the statistical mean of the scalar dissipation rate can be written as

$$\tilde{\chi} = \int_0^1 \chi(Z)\cdot P_Z(Z)\cdot dZ = \chi_{st}\int_0^1 \frac{f(Z)}{f(Z_{st})}\cdot P_Z(Z)\cdot dZ. \tag{6.121}$$

Substitution of Eq. 6111 yields the relation

$$\chi_{st} = \frac{C_\chi \frac{\tilde{\varepsilon}}{\tilde{k}} \widetilde{Z''^2}}{\int_0^1 \frac{f(Z)}{f(Z_{st})} P_Z(Z) dZ}, \qquad (6.122)$$

which is utilized in order to quantify χ_{st}. Since the stoichiometric mixture fraction Z_{st} is a known quantity for any specific fuel, Eq. 6.119 can now be solved for a_s.

Venkatesh et al [63] systematically investigated the influence of different strain rates on the heat release rate in a diesel-type combustion within a constant volume chamber. Instead of applying Eq. 6.111 in order to obtain the mean scalar dissipation rate, they fixed χ_{st} to a constant value for the entire simulation. Three arbitrary values for χ_{st} were investigated, and it turned out that a reduction from $\chi_{st} = 1.0 \text{ s}^{-1}$ to $\chi_{st} = 0.1 \text{ s}^{-1}$ had only a minor effect. Conversely, the heat release rate was significantly reduced by increasing χ_{st} to a value of 10.0. This may be interpreted as a local quenching due to excessive strain rates, that have the effect that the heat conduction to both the lean and the rich sides of the reaction zone cannot be balance by heat production within the reaction zone.

Finally, the effect of the turbulent fluctuations on the instantaneous scalar dissipation rate is commonly accounted for by an additional pdf in analogy to the modeling of the mixture fraction distribution. Note, that in a strict mathematical sense a joint pdf for both the mixture fraction Z and the scalar dissipation rate χ would have to be established. For simplicity however, statistical independence is often assumed between Z and χ, and the joint pdf may be decomposed into two independent pdf's:

$$P_{Z\chi}(Z,\chi) \approx P_Z(Z) \cdot P_\chi(\chi) \qquad (6.123)$$

In contrast to the beta-function used for the mixture fraction pdf, Eq. 6.112, the χ-pdf is typically assumed to follow a log-normal distribution:

$$P_\chi(\chi) = \frac{1}{\chi \sigma \sqrt{2\pi}} \exp\left[-\frac{(\ln \chi - \mu)^2}{2\sigma^2}\right]. \qquad (6.124)$$

The variance parameter of this pdf has been found experimentally to be approximately equal to $\sigma^2 = 2.0$ for several different flow configurations. The parameter μ may be obtained from equating the mean of the log-normal distribution to the respective value determined from Eq. 6.111:

$$\tilde{\chi} = \exp\left(\mu + \frac{1}{2}\sigma^2\right) \equiv C_\chi \frac{\tilde{\varepsilon}}{\tilde{k}} \widetilde{Z''^2}. \qquad (6.125)$$

Re-Transformation

With the above formulations, laminar flamelet solutions can be estimated for various boundary conditions such as temperature, pressure and scalar dissipation rate. These solutions contain the information about the one-dimensional structure of the flamelet in mixture fraction space. The re-transformation into three-dimensional physical space is then executed by integrating the flamelet solutions for the species concentrations and the temperature over the entire ranges of both the mixture fraction and the scalar dissipation rate. Assuming statistical independence between the mixture fraction and the scalar dissipation rate, the complete form of Eq. 6.108 reads

$$\tilde{Y}_i = \int_0^1 \int_0^\infty P_Z(Z) \cdot P_\chi(\chi) \cdot Y_i(Z,\chi,p,T) \cdot d\chi \cdot dZ. \tag{6.126}$$

The new cell temperature can be calculated in an analogous way, based on the solution of Eq. 6.116 for the one-dimensional profile $T(Z)$.

It should be noted, that the above formulation, Eq. 6.122, assumes infinitely fast chemistry, i.e. the assumption *mixed = burned* is utilized. While this is a reasonable measure for the major species during diffusion combustion, it falls short of describing the chemically slow processes of autoignition and pollutant formation. A possible solution to this problem is to store not the steady-state species mass fractions Y_i but only their source terms, i.e. the reaction rates, within a flamelet library. The concentrations c_i of the species in question can then be estimated based on the transport equation:

$$\frac{\partial(\overline{\rho}\tilde{c}_i)}{\partial t} + \frac{\partial(\overline{\rho}\tilde{u}_i\tilde{c}_i)}{\partial x_i} - \frac{\partial}{\partial x_i}\left(\overline{\rho}D\frac{\partial \tilde{c}_i}{\partial x_i}\right) = \dot{\overline{Q}}_i(Z,\chi)$$

$$= \int_0^1 \int_0^\infty P_Z(Z) \cdot P_\chi(\chi) \cdot \dot{Q}_i(p,T,Z,\chi) \cdot d\chi \cdot dZ. \tag{6.127}$$

Discussion

Venkatesh et al. [63] carried out an extensive study that compared the performance of a flamelet model with the performance of a time scale combustion model such as discussed in Sect. 6.4.1. The results for a diesel-type combustion in a constant volume chamber showed, that the integral heat release rates predicted by the two models are very similar. This suggests, that combustion is indeed primarily limited by mixing processes and chemical kinetics play only a minor role after the autoignition period. This interpretation is further supported by the finding, that the heat release rates are not significantly different when different fuels (of comparable heating value) are used with the flamelet model. However, in the same study noticeable differences were found between the flamelet model and the time scale model with respect to the rate of nitrogen oxides formation. As this is known to be a process where the kinetics are slow compared to the mixing and turbulence time scales, see Chap. 7, the consideration of more detailed kinetics as well as the influence of strain rates are likely to be important for the formation and decomposition of pollutants as well as for autoignition. These information can only be provided by flamelet models.

However, it should be noted that the scalar dissipation rate, which is a parameter of paramount importance in flamelet models, needs to modeled by the semi-empirical approach of Eq. 6.111. Thus, the turbulent time scale k/ε, which is also the most crucial parameter in the time scale combustion models, is of great importance in flamelet models as well. Consequently, uncertainties that may arise in the turbulence modeling do affect the flamelet combustion model, albeit not as directly as they affect the time scale combustion models.

Representative Interactive Flamelets (RIF)

Mauss et al.[31] have found that the inner flame structure is not able to follow fast changes in the scalar dissipation rate, as they are observed in the highly unsteady development of the mixing field in DI diesel engines. Although the chemistry can be considered to follow the flow field changes instantaneously, diffusive fluxes and the formation of a new diffusion-production-balance are not infinitely fast. Therefore, the history of the scalar dissipation rate is important for the transient flamelet evolution, and the use of steady flamelet libraries as they have been discussed above can be expected to be a source of inaccuracies.

The RIF concept attempts to eliminate these problems by interactively solving the one-dimensional flamelet chemistry with the three-dimensional CFD code [46]. The parameters and boundary conditions that govern the unsteady evolution of these flamelets are extracted from the CFD code by statistical averaging over a representative domain of interest. Figure 6.18 shows a schematic diagram of the interaction between the CFD-code and the flamelet-code.

There are two aspects that are noteworthy about the RIF concept shown in Fig. 6.18. Firstly, the energy equation in the CFD-code is formulated in terms of the total enthalpy H_t, that inherently combines the information about the temperature and composition by means of the enthalpies of formation. This has the advantage, that there are no chemical source terms in the CFD-equations. The new cell temperature is rather determined in an iterative manner by comparing the total enthalpy obtained from the CFD code with the composition obtained from the flamelet code. The second aspect is that the numerical separation of the two codes allows for a resolution of the chemical time scales in the flamelet code by subcycling a time step of the CFD code. This is important since the respective timescales may deviate by about three orders of magnitude.

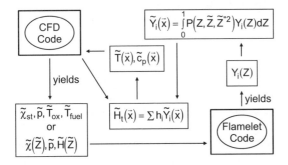

Fig. 6.18. Code structure of the representative interactive flamelet concept [46]

The RIF concept represents a promising way to describe the possible interactions between chemistry and turbulence in the non-premixed combustion of DI diesel engines. However, there appear to be some unresolved issues that need to be further investigated. Because the computational expenditure increases significantly when interactive flamelets are solved instead of using steady flamelet libraries, the number of representative flamelets that can be considered is limited to a small number. Therefore, one needs to determine whether the simulation results converge for an increasing number of representative flamelets, and how many flamelets are needed in order to yield such convergence. Furthermore, there is some uncertainty left in the question about which criterion should be applied to distinguish the various representative flamelets, even though it seems reasonable to relate this criterion to the scalar dissipation rate.

6.4.3 pdf-Models

Motivation

In Chap. 4 the closure problem has been discussed, that arises with the use of the Reynolds-Averaged Navier-Stokes equations (RANS). It has the effect that additional terms, including the velocity fluctuations, appear in the momentum and energy transport equations. As noted above, these terms have to be modeled because they cannot be derived from first principles. Similar closure problems can be observed for almost all field-variables in turbulent combustion as well. They are especially important for chemical reaction rates, as the rate coefficients typically have a nonlinear temperature dependency, e.g. [22]. Estimating the reaction rates based on a time-averaged temperature instead of taking into account the entire spectrum of fluctuations would therefore result in a significant underprediction of the actual mean reaction rate.

Generally speaking, the closure problem states that new unknown terms including higher moments appear in the transport equations of lower moments of a variable. A typical solution to such closure problems is the semi-empirical modeling of higher moments in terms of the transported lower moments, e.g. the description of the velocity fluctuations (second moments) in the RANS equations as a function of time averaged quantities (first moments). With respect to the estimation of reaction rates in turbulent flows, such semi-empirical modeling assumptions are made as well. For example, in the eddy-breakup models, the problem of describing the turbulence effects on the kinetic reaction rates is bypassed entirely by directly modeling the mass burning rate in terms of the turbulent dissipation, which itself depends on time-averaged quantities as well as on several empirical constants, see Chap. 4. And even the more comprehensive flamelet model described in the foregoing section relies on a semi-empirical modeling of the mean scalar dissipation rate and, in most applications, assumes infinitely fast chemistry. While those semi-empirical closure assumptions are useful for many engineering combustion applications, they may fall short under some boundary conditions, and more accurate ways of describing turbulent reactive flows become desirable.

A method for avoiding the closure problem is the statistical treatment of the fluctuating quantities by probability density functions, e.g. [25, 48]. This method is commonly referred to as the direct pdf-closure. For example, at a spatial location $\vec{x}$, the mean production rate of a species i can be written as

$$\bar{\dot{w}}_i = \int_0^1 ... \int_0^1 \int_0^1 \int_0^\infty \int_0^\infty \dot{w}_i \, P(\rho, T, Y_1, Y_2, ..., Y_n; \vec{x}) \, d\rho \, dT \, dY_1 \, dY_2 ... dY_n, \quad (6.128)$$

where P is a multivariable joint pdf, that contains the entire information about the nature of the turbulent fluctuations. In particular, it describes the statistical probability that a specific fluid state, characterized by density ρ, temperature T and mass fractions Y_1 to Y_n, is found at location $\vec{x}$. So far there are no assumptions or empirical correlations involved in the formulation of Eq. 6.128.

However, a major problem exists in that the exact form of the probability density function needs to be known in order to be integrated. This matter is further complicated by the fact that an a-priori parameterized pdf, similar to the beta-pdf for the mixture fraction in the flamelet model, cannot be formulated for other variables. This is related to the fact that most variables are, in contrast to the mixture fraction, not conserved scalars. They are rather subject to source terms originating from combustion. Nevertheless, primarily two possibilities have been followed in order to determine the multivariable pdf's. These are, namely, the transported and the presumed pdf methods. They will be described in the following.

Transported pdf

In principle, the temporal evolution of a joint scalar pdf is governed by fluid mixing and reaction chemistry. Thus, a transport equation for the pdf can be derived based on the Navier-Stokes equations along with conservation principles for chemical species [65]. Let the most general joint probability density function

$$P(u_x, u_y, u_z, \psi_1, ..., \psi_N; x, y, z, t) \, du_x \, du_y \, du_z \, d\psi_1 ... d\psi_N \quad (6.129)$$

indicate the probability, that at time t and at the singular spatial position x, y and z, the fluid has velocity components in the range between u_i and $u_i + du_i$ and values of the scalar quantities in the range between ψ_α and $\psi_\alpha + d\psi_\alpha$, where the ψ_α denote the density, enthalpy and mass fractions. Then the transport equation for the evolution of the pdf reads [48, 49]:

$$\rho(\vec{\psi})\frac{\partial P}{\partial t} + \rho(\vec{\psi})\sum_{j=1}^{3}\left(u_j\frac{\partial P}{\partial t}\right) + \sum_{j=1}^{3}\left[\left(\rho(\vec{\psi})g_j - \frac{\partial \bar{p}}{\partial x_j}\right)\frac{\partial P}{\partial u_j}\right]$$

$$+ \sum_{\alpha=1}^{N}\left(\frac{\partial}{\partial \psi_\alpha}\left[\rho(\vec{\psi})S_\alpha(\vec{\psi})P\right]\right)$$

$$= \sum_{j=1}^{3}\left(\frac{\partial}{\partial u_j}\left[\left\langle\frac{\partial p'}{\partial x_j} - \sum_{i=1}^{3}\frac{\partial \tau_{ij}}{\partial x_i}\middle|\vec{u},\vec{\psi}\right\rangle P\right]\right) + \sum_{\alpha=1}^{N}\left(\frac{\partial}{\partial \psi_\alpha}\left[\sum_{i=1}^{3}\left\langle\frac{\partial J_i^\alpha}{\partial x_i}\middle|\vec{u},\vec{\psi}\right\rangle P\right]\right). \quad (6.130)$$

In the foregoing equation, the x_i denote the x, y and z coordinates, and the u_i and g_i denote the velocity components and gravitational forces in these directions, respectively. The N-dimensional vector $\vec{\psi}$ denotes the N different scalars ψ_α, and the S_α denote the source terms for the respective scalars (e.g. the chemical source terms). The τ_{ij} are the components of the stress tensor, and the J_i^α are the components of the molecular flux (i.e. diffusion or heat conduction) of scalar α in direction i. Finally, the term $\langle a|\vec{u},\vec{\psi}\rangle$ indicates the conditional average of a function a for fixed values of $\vec{u}$ and $\vec{\psi}$. In physical terms, the conditional averages describe the mean molar fluxes for given values of the velocity and the scalars.

However, it is not possible to integrate the pdf-transport equation, Eq. 6.130, by conventional numerical integrating techniques. This stems from the high dimensionality of the joint pdf, which depends not only on position and time, but also on the velocity components as well as on the scalars ψ_α. It is obvious, that the problem becomes worse, when detailed chemical reaction mechanisms are employed, and the number of different species mass fractions that enter the pdf is increased.

The problem of integrating Eq. 6.130 is very similar to the problem encountered when solving the so-called spray equation, Eq. 5.4, in two-phase flows that has been discussed above. Consequently, the same solution technique, i.e. the Monte-Carlo method, is applied in pdf-combustion models as well as in spray calculations. In this computational approach, the pdf is represented by a large number of stochastic particles. These particles evolve in time according to convection, chemical reaction, molecular transport, and body forces, and thus mimic the evolution of the pdf. However, in the context of combustion models, the same kind of problems arise with the use of the Monte-Carlo method, that have already been discussed in Chap. 5. In order to assure statistical convergence, an extremely large number of particles is necessary, even for two-dimensional systems with strongly reduced reaction mechanisms. Thus, computational expenditure is prohibitive for 3D engine applications, and the transported pdf approach is limited to fundamental studies with simpler geometries and boundary conditions than typically encountered in diesel engine combustion chambers. Nevertheless, applications of the transported pdf method to simpler boundary conditions have yielded very good results, and thus contribute to a better understanding of fundamental combustion phenomena, that are of importance for internal combustion engines as well.

Presumed pdf

An alternative and computationally much more efficient way to determine the multivariable joint pdf is to assume statistical independence between the involved scalars. Under this premise the joint pdf can be decomposed into several single-variable pdf's of presumed structure [13]. This is in analogy to the treatment of the Z-χ-pdf in the flamelet model, Eq. 6.123:

$$P(\psi_1,\psi_2,...,\psi_N) \approx P_1(\psi_1)\cdot P_2(\psi_2)\cdot ...\cdot P_N(\psi_N). \tag{6.131}$$

Note, that each P in Eq. 6.131 identifies a different probability function.

Each of the new single-variable pdf's $P_\alpha(\psi_\alpha)$ is now empirically constructed as a parameterized analytical function of presumed shape. An example is the beta-function that has been used for the mixture fraction distribution in the flamelet model. Any presumed pdf can be uniquely identified if its moments are known. In fact, the number of moments that need to be determined in order to identify a presumed pdf is equal to the number of shape parameters contained in that function. For example, the two parameter beta-function described above (Eq. 6.112) requires the knowledge of the first and the second moment. These moments can be estimated by solving their own transport equations that can be derived from the Navier-Stokes equations, compare Eqs. 6.109 and 6.110.

Much progress has been made with the use of presumed analytical pdf's in combustion modeling, e.g. [25]. Nevertheless, some significant problems remain that are inherently coupled with this approach. First of all, statistical independence between the scalars involved in the original joint pdf is not always a reasonable assumption in engine combustion processes. Moreover, it is hardly practical to solve transport equations for more than the first two moments. However, often the actual pdf's have features that are poorly represented by two-parameter functions. Last but not least, the shape and thus the general function describing the pdf may change due to source terms arising from combustion processes. In this respect the mixture fraction Z, that has been utilized in the flamelet model, is the only variable that can be treated as a conserved scalar. Its transport equation has no source terms, since the mixture fraction is not affected by combustion.

6.5 Partially Premixed Combustion in DISI Engines

6.5.1 Flame Structure

When direct injection spark ignition (DISI) engines are operated in the stratified charge mode, combustion is likely to proceed in partially premixed flames. As schematically shown in Fig. 6.19, combustion is initiated at the periphery of the fuel cloud. Ideally, the equivalence ratio at the spark plug position is close to stoichiometric at the thermodynamically best spark timing in order to assure a quick and stable ignition.

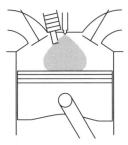

Fig. 6.19. Schematic illustration of a DISI engine in stratified charge mode

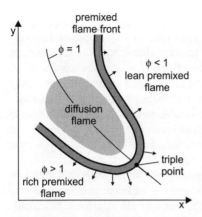

Fig. 6.20. Schematic illustration of a triple flame structure [20]

Once the spark-induced ignition kernel has reached a critical size, the heat production within the flame exceeds the heat losses, and a premixed self-sustained flame front starts to propagate from the spark plug position into the combustion chamber. This is very similar to the process observed in homogeneously operated SI engines with external mixture formation. However, there is one major difference in that the flame front may reach both lean and rich mixture zones in the stratified charge case. The speed and the amount of energy released by this flame front is influenced by turbulence and temperature, and most significantly by the local equivalence ratio of the mixture. Thus, the flame front will propagate fastest along a line of approximately stoichiometric air-fuel ratio and slower into both richer and leaner mixture. This behavior results in the nose-shaped structure shown in Fig. 6.20. The premixed flame front will eventually quench if the mixture becomes too lean at the periphery of the fuel cloud.

Behind the premixed flame front a secondary diffusion flame, similar to the flame type observed in non-premixed diesel combustion, may develop because of turbulent mixing of relatively hot residuals from the rich and lean flame branches. As a result, three different flame types are present at the leading edge of the flame: a rich and a lean premixed flame front as well as a diffusion flame. Therefore, the leading edge is commonly referred to as the triple point, and the entire flame structure is often denoted as a triple flame.

6.5.2 A Formulation based on Lagrangian Flame Front Tracking

A number of numerical studies on stratified charge DISI engines have utilized combustion models of the eddy-breakup (or characteristic time scale) kind, partly with remarkable success, e.g. [10, 47, 58, 60]. However, as has been pointed out before, these models are generally better suited to describe non-premixed diffusion combustion phenomena than to reflect the propagation of a premixed flame front. This assessment applies, even though the eddy breakup model was initially invented in order to describe the latter combustion type. Consequently, it appears that eddy breakup models represent a practical approximation for describing partially premixed flames, but that it would nevertheless be desirable to utilize more detailed models in order to more closely reproduce the triple flame structure discussed in the foregoing section.

Based on a work of Fan et al. [11], that describes the initial growth of a spark enhanced flame kernel by mass-less Lagrangian marker particles, Stiesch et al. [57, 59] proposed a mathematical formulation for partially premixed triple flames. It utilizes two separate submodels for the premixed flame fronts and for the secondary diffusion combustion taking place in the partly burned gas regions, respectively. The spark ignition process as well as the propagation of the premixed flame fronts is modeled with the help of discrete marker particles tracking the current position of the flame front. On the other hand, the secondary diffusion combustion is modeled by the characteristic time scale combustion model as described in Sect. 6.4.1.

The concept of the flame front tracking by marker particles is illustrated in Fig. 6.21. Directly after spark-onset, a large number of discrete particles is uniformly distributed about the two main steradians such that a spherical ignition kernel is described. About 5,000 to 10,000 particles are necessary for typical combustion chambers and grid resolutions. The initial radius of the ignition kernel is estimated following the energy balance derived in Sect. 6.2.3:

$$r_{k,0} = \left[\frac{3\dot{W}_{sp} \Delta t_0}{4\pi \rho_k \left(c_p (T_{ad} - T_u) - \text{LHV}_{mix} \right)} \right]^{1/3}. \quad (6.132)$$

Thereafter, the ignition kernel expands and its surface is continuously transformed into a self-sustained premixed flame front which is tracked by the marker particles. The movement of the particles is due to spark-enhanced ignition, self-sustained flame front propagation and convection by the turbulent gas motion. Because the latter two mechanisms depend on the local properties of the flow field, the flame kernel looses its spherical shape and follows the propagation of the premixed flame front in a three-dimensional manner.

The effective burning velocity of the flame front, which proceeds relative to the convective motion of the gas flow, is calculated as the maximum of the spark-induced plasma velocity and the turbulent flame velocity:

$$s_{eff} = \max(s_{pl}, s_t). \quad (6.133)$$

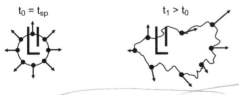

Fig. 6.21. Flame front tracking by Lagrangian marker particles [59]

The plasma velocity is obtained from the energy balance of the flame kernel, compare Eq. 6.69:

$$s_{pl} = \frac{dr_k}{dt} = \frac{\eta_{sp} \dot{W}_{sp}}{4\pi r_k^2 \rho_u \left(c_p(T_{ad} - T_u) - \text{LHV}_{mix}\right)}. \tag{6.134}$$

The turbulent flame velocity may be estimated based on the formulation of Herweg and Maly [16], Eq. 6.64, that implicitly describes the transition from a laminar flame to a fully developed turbulent flame by including strain rates and other effects. Note, that in Eq. 6.133 the greater value of either s_{pl} or s_t is chosen instead of adding the two quantities. This is necessary because chemistry effects are already accounted for in the energy balance that leads to the plasma velocity in Eq. 6.134.

The visible propagation speed of the flame front is accelerated because of thermal expansion of the burned gases. Thus, Eq. 6.133 needs to be multiplied by an expansion coefficient for the purpose of flame front tracking:

$$s_{FF} = f_{exp} \cdot s_{eff}. \tag{6.135}$$

Following Heywood [17], the expansion coefficient can be specified as

$$f_{exp} = \frac{\rho_u}{\rho_b}(1 - y_b) + y_b = \frac{\rho_u / \rho_b}{\left[(\rho_u / \rho_b) - 1\right]x_b + 1}, \tag{6.136}$$

where y_b denotes the burned volume fraction ($y_b = V_b/V_{cyl}$), and x_b denotes the burned mass fraction ($x_b = m_b/m_{cyl}$).

For simplicity, a single-step reaction mechanism from fuel and O_2 towards CO_2 and H_2O is assumed within the premixed flame front. The mass burning rate per computational cell becomes the product of fuel density $\rho_{f,u}$ within the unburned gas mixture, the flame speed s_{eff}, and the flame surface area within that cell $A_{FF,cell}$. The latter is estimated by assuming that all marker particles within the cell are positioned on a sphere with its origin at the spark gap, see Fig. 6.22:

$$\begin{aligned}\frac{dm_b}{dt} &= \min\left[\rho_{f,u}, x_{f,O2}\rho_{O_2,u}\right] s_{eff} A_{FF,cell} \\ &= \min\left[\rho_{f,u}, x_{f,O2}\rho_{O_2,u}\right] s_{eff} 4\pi r_{p,cell}^2 \frac{N_{p,cell}}{N_{p,tot}}.\end{aligned} \tag{6.137}$$

In case of rich equivalence ratios, the oxygen density becomes the limiting factor in the mass burning rate as opposed to the fuel vapor density. Thus, the minimum of these two species densities needs to be considered in Eq. 6.137, where $x_{f,O2}$ is the stoichiometric fuel-oxygen mass ratio. Once the change in composition has been determined, the heat release and the new cell temperatures can be estimated by balancing the total enthalpy including the specific enthalpies of formation.

The secondary diffusion flame taking place in the partly burned region behind the premixed flame front is modeled by the characteristic time scale combustion model. As noted above, this model is best suited for describing mixing controlled flames, and therefore seems an appropriate choice for the present purpose. A two-step chemical mechanism, including carbon monoxide, is considered in the diffusion combustion, and the change rates of the various species densities are obtained by dividing the deviation from the equilibrium composition by a characteristic time scale composed of a laminar (chemical) and a turbulent (mixing) time:

$$\frac{d\rho_i}{dt} = -\frac{\rho_i - \rho_i^{eq}}{\tau_c} = -\frac{\rho_i - \rho_i^{eq}}{\tau_l + \tau_t}. \qquad (6.138)$$

The magnitude of the two time scales are given in Eqs. 6.79 and 6.80. However, it should be noted that the delay coefficient f in Eq. 6.78 is assumed to be always equal to unity in the present context. This accounts for the fact that the non-premixed part of combustion has already been treated by the separate flame front model. Furthermore, the time scale combustion model is only activated in computational grid cells that have either been swept by the premixed flame front, or in cells that where an onset-temperature has been reached that is greater than the one typically chosen for diesel combustion. A value of 1300 K has been chosen in ref. [57].

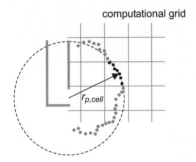

Fig. 6.22. Modeling of the turbulent flame surface within a computational grid cell [59]

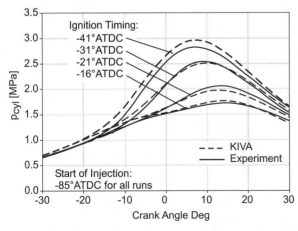

Fig. 6.23. Pressure traces for an ignition timing variation of a DISI engine [57]

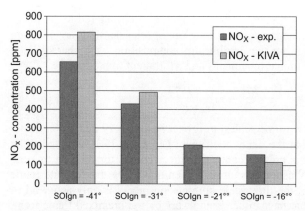

Fig. 6.24. NO_x-emissions for an ignition timing variation of a DISI engine [57]

In the same reference, the model has been implemented into the CFD-code KIVA-3V and applied to simulate an ignition timing variation in a small DISI two-stroke engine operated at part load conditions. The simulation results for pressure traces and nitrogen oxides emissions are compared against experimental data in Figs. 6.23 and 6.24, respectively. While there remain some deviations between calculations and measurements, especially with respect to the NO_x-emissions, the model seems generally capable of reflecting the governing combustion mechanisms in partially premixed flames well.

6.5.3 A Formulation based on the G-Equation

Kech et al. [20, 51] presented a combustion model for partially premixed flames that is based on the principal concept of the G-equation model for fully premixed flames, compare Sect. 6.3.3. Consequently, the reaction state is described by the favre-averaged regress variable $\tilde{b}$. It is transported according to

$$\frac{\partial \overline{\rho}\tilde{b}}{\partial t} + \frac{\partial\left(\overline{\rho}\tilde{u}_i\tilde{b}\right)}{\partial x_i} - \frac{\partial}{\partial x_i}\left(\overline{\rho}\tilde{D}_b \frac{\partial \tilde{b}}{\partial x_i}\right) = -\overline{\rho}_u s_t \left|\nabla \tilde{b}\right|, \qquad (6.139)$$

where the source term on the right hand side represents the mean reaction rate $\tilde{\dot{w}}$.

However, compared to the original G-equation model [42, 68], several variations need to be introduced in order to account for the fact that the mixture in DISI engines is only partially premixed (or still partially non-premixed, for that matter). Hence, it is not sufficient to specify a gas state just by the instantaneous local value of the combustion regress variable, but additional information must be provided that characterizes the mixing state between fuel and oxidizer. For this purpose transport equations are solved for the species mass fractions of fuel, O_2, N_2, CO_2, H_2O and $CO_{2,res}$, as one component of the residual gas:

$$\frac{\partial \bar{\rho}\tilde{Y}_i}{\partial t} + \frac{\partial \left(\bar{\rho}\tilde{u}_i\tilde{Y}_i\right)}{\partial x_i} - \frac{\partial}{\partial x_i}\left(\bar{\rho}\tilde{D}_{Yi}\frac{\partial \tilde{Y}_i}{\partial x_i}\right) = \tilde{Y}_i^c . \tag{6.140}$$

The source term on the right hand side is due to combustion and is related to the reaction rate of an assumed single-step mechanism from fuel to CO_2 and H_2O. For example, for fuel vapor it becomes

$$\tilde{Y}_{fuel}^c = -\bar{\rho}_u s_t \left|\nabla \tilde{b}\right| \tilde{Y}_{fuel,u} , \tag{6.141}$$

and the other species are treated accordingly. With the knowledge of both the total CO_2 mass fraction Y_{CO2} and the one resulting from the residuals $Y_{CO2,res}$, the mixing state as well as the combustion state are uniquely identified. In fact, the regress variable can now be reconstructed from the species mass fractions as

$$\tilde{b} = \frac{\tilde{Y}_{fuel}^*}{\tilde{Y}_{fuel}^* + x_{f,CO2}\left(\tilde{Y}_{CO2} - \tilde{Y}_{CO2,res}\right)} , \tag{6.142}$$

where

$$\tilde{Y}_{fuel}^* = \min\left[\tilde{Y}_{fuel}, x_{f,O2}\tilde{Y}_{O2}\right], \tag{6.143}$$

and $x_{f,CO2}$ and $x_{f,O2}$ denote the stoichiometric fuel-CO_2 and fuel-O_2 mass ratios, respectively.

The turbulent flame speed s_t, that is required in order to solve for the mean reaction rate in Eq. 6.141, is again based on the turbulent to laminar flame speed ratio proposed by Herweg and Maly [16], Eq. 6.64. The laminar flame speed is modeled following Keck and co-workers [33, 34, 52],

$$s_l = s_{l,0}\left(\frac{T_u}{T_0}\right)^{\alpha}\left(\frac{p}{p_0}\right)^{\beta} , \tag{6.144}$$

where the parameters $s_{l,0}$, α and β have been specified in Eqs. 3.140 to 3.142 and Table 3.4. However, a new difficulty arises in that the above relation for the laminar flame velocity applies only for homogeneously premixed fuel-air mixtures which may not be present in stratified charge DISI engines. Therefore, an approach is followed that is very similar to the flamelet modeling of non-premixed diffusion flames that has been discussed in Sect. 6.4.2 [44].

In order to account for the effect of the turbulent fluctuations of the partially premixed gases on the mean burning velocity, a probability density function is utilized for the mixture fraction Z and its variance:

$$\bar{s}_l\left(\tilde{Z}\right) = \int_0^1 s_l\left(Z\right) \cdot P_Z\left(\tilde{Z}, \widetilde{Z''^2}\right) \cdot dZ . \tag{6.145}$$

The mixture fraction is defined as in Eq. 6.107, and the joint pdf $P_Z\left(\tilde{Z}, \widetilde{Z''^2}\right)$ is assumed to be the same beta-function that has been utilized in the flamelet model, Eqs. 6.112 to 6.114. Again, the transport equations for the mean and the second moment of the mixture fraction, Eqs. 6.109 and 6.110, have to be solved in order to determine the shape parameters of the beta-function that identify the pdf.

The above set of equations represents a closed formulation that can be solved in order to describe both the mixing and the combustion progress in partially premixed triple flames. However, the numerical integration of Eq. 6.145 can be tedious. To overcome this problem and shorten CPU times, Kech et al. [20] suggested the following curve-fit:

$$\overline{s}_l(\tilde{Z}) = s_l(Z_{st}) \cdot \left[P_{Z_{st}}(\tilde{Z}, \widetilde{Z''^2}) \right]^{2/3} \cdot C_{Z_{st}}. \qquad (6.146)$$

Here, Z_{st} denotes the stoichiometric mixture fraction, and P_{Zst} is the discrete probability of locating this stoichiometric mixture fraction in an ensemble of fluctuations, if the mean and the second moment of the mixture fraction are given by $\tilde{Z}$ and $\widetilde{Z''^2}$, respectively. A comparison between the curve-fit in Eq. 6.146 and the respective solution that results from numerical integration of Eq. 6.145 is shown in Fig. 6.25 for a model constant of $C_{Zst} = 0.9$.

Figure 6.26 shows an application of the combustion model to a conceptual study where the flame propagation has been calculated in a stratified mixture field within a closed vessel. The mixture fraction distribution varies from the lean to the rich flammability limits in vertical direction as shown in the right part of the figure. The ignition location is at $Z = Z_{st}$. Iso-lines of constant reaction rates are included in Fig. 6.26 in order to illustrate the flame structure. Both lean and rich flame fronts can be observed, that propagate into the unburned zones at the top and the bottom of the vessel. The high iso-line densities indicate that the thickness of the flame fronts is fairly thin. The curvature of the premixed flame fronts is an artifact that is due to thermal expansion of the burned gases within the closed vessel. The secondary diffusion flame is observed in the middle of the combustion vessel at approximately stoichiometric mixture, where the iso-lines indicate an increased reaction rate in the burned region behind the actual premixed flame fronts. These results suggest that the model is capable of predicting the triple flame structure in partially premixed flames that has been discussed above.

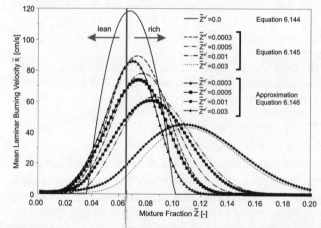

Fig. 6.25. Mean laminar flame speed, $p = 8$ bar, $T = 700$ K, iso-octane [20]

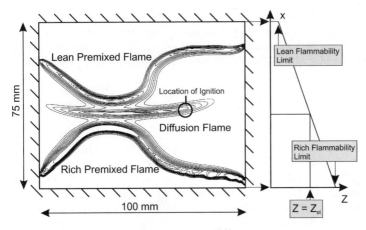

Fig. 6.26. Computed triple flame structure [20]

References

[1] Abraham J, Bracco FV, Reitz RD (1985) Comparisons of Computed and Measured Premixed Charge Engine Combustion. Combustion and Flame, vol 60, pp 09–322
[2] Anderson RW, Lim MT (1985) Experimental Study of a Developing Spark Kernel. 8th Int Conf on Gas Discharges and their Applications, pp 511–514 Leeds University Press, UK
[3] Baulch DL, Cobos CJ, Cox RA, Frank P, Hayman G, Just T, Kerr JA, Murrells T, Pilling MJ, Troe J, Walker RW, Warnatz J (1992) Evaluated Kinetic Data for Combustion Modeling. J Phys Chem Ref Data, vol 21, pp 411–429
[4] Blizard NC, Keck JC (1974) Experimental and Theoretical Investigation of Turbulent Burning Model for Internal Combustion Engines. SAE Paper 740191
[5] Bollig M, Pitsch H, Hewson JC, Seshadri K (1996) Reduced n-Heptane Mechanism for Non-Premixed Combustion with Emphasis on Pollutant-Relevant Intermediate Species. 26th Symp (Int) Combust, pp 729–737, The Combustion Institute, Pittsburgh, PA
[6] Borghi R (1984) On the Structure of Turbulent Premixed Flames. in: Bruno C, Casci C (eds) Recent Advances in Aeronautical Science. Pergamon
[7] Boudier P, Henriot S, Poinsot T, Baritaud T (1992) A Model for Turbulent Flame Ignition and Propagation in Spark Ignition Engines. 24th Symp (Int) Combust, pp 503–510, The Combustion Institute, Pittsburgh, PA
[8] Chevalier C, Loussard P, Müller UC, Warnatz J (1990) A Detailed Low-Temperature Reaction Mechanism of n-Heptane Auto-Ignition. 2nd Int Symp COMODIA 90, Kyoto, p 93
[9] Dinkelacker F (2001) Struktur turbulenter Vormischflammen. Habilitationsschrift, University Erlangen-Nuremberg, Germany
[10] Fan L, Reitz RD (2001) Multi-Dimensional Modeling of Mixing and Combustion of a Two-Stroke Direct-Injection Spark-Ignition Engine. SAE Paper 2001-01-1228

[11] Fan L, Li F, Han Z, Reitz RD (1999) Modeling Fuel Preparation and Stratified Combustion in a Gasoline Direct Injection Engine. SAE Paper 1999-01-0175
[12] Frenklach M, Wang H, Rabinowitz MJ (1992) Optimization and Analyses of Large Chemical Kinetic Mechanisms Using the Solution Mapping Method – Combustion of Methane. Prog Energy Combust Sci, vol 18, pp 47–73
[13] Gutheil E, Bockhorn H (1987) The Effect of Multi-Dimensional PDF's in Turbulent Reactive Flows at Moderate Damköhler Number. Physicochemical Hydrodynamics, vol 9, p 525
[14] Halstead M, Kirsch L, Quinn C (1977) The Autoignition of Hydrocarbon Fuels at High Temperatures and Pressures – Fitting of a Mathematical Model. Combust Flame, vol 30, pp 45–60
[15] Haworth CD, Poinsot TJ (1992) Numerical Simulations of Lewis Number Effects in Turbulent Premixed Flames. J Fluid Mech, vol 244, pp 405–436
[16] Herweg R, Maly RR (1992) A Fundamental Model for Flame Kernel Formation in SI Engines. SAE Paper 922243
[17] Heywood JB (1988) Internal Combustion Engine Fundamentals. McGraw-Hill, New York, NY
[18] Heywood JB (1994) Combustion and its Modeling in Spark-Ignition Engines. 3rd Int Symp COMODIA 94, pp 1–15
[19] Jones WP, Whitelaw JH (1982) Calculation Methods for Reacting Turbulent Flows: A Review. Combust Flame, vol 48, p 1
[20] Kech JM, Reissing J, Gindel J, Spicher U (1998) Analyses of the Combustion Process in a Direct Injection Gasoline Engine. 4th Int Symp COMODIA 98, pp 287–292
[21] Kong SC, Han Z, Reitz RD (1995) The Development and Application of a Diesel Ignition and Combustion Model for Multidimensional Engine Simulations. SAE Paper 950278
[22] Kuo KK (1986) Principles of Combustion. Wiley, New York, NY
[23] Kuo TW, Reitz RD (1992) Three-Dimensional Computations of Combustion in Premixed-Charge and Direct-Injected Two-Stroke Engines. SAE Paper 920425
[24] Kusaka J, Yamamoto T, Daisho Y (2000) Simulating the Homogeneous Charge Compression Ignition Process using a Detailed Kinetic Model for N-Heptane Mixtures. Int J Engine Research, vol 1, no 3, pp 281–289
[25] Libby PA, Williams FA (1994) Turbulent reacting flows. Academic Press, New York, NY
[26] Lovas T, Mauss F, Hasse C, Peters N (2002) Modeling of HCCI Combustion Using Adaptive Chemical Kinetics. SAE Paper 2002-01-0426
[27] Magnussen BF, Hjertager BH (1976) On Mathematical Modeling of Turbulent Combustion with Special Emphasis on Soot Formation and Combustion. 16th Symp (Int) Combust, pp 719–729, The Combustion Institute, Pittsburgh, PA
[28] Maly R, Vogel M (1978) Initiation and Propagation of Flame Fronts in Lean CH_4-Air Mixtures by the Three Modes of the Ignition Spark. 17th Symp (Int) Combust, pp 821–831, The Combustion Institute, Pittsburgh, PA
[29] Mantel T (1992) Three Dimensional Study of Flame Kernel Formation Around a Spark Plug. SAE Paper 920587
[30] Mauss F, Johansson B (2001) Homogeneous Charge Compression Ignition Engines: A Review on Experiments and Numerical Investigations. 5th Congress on Engine Combustion Processes, pp 5–19, Haus der Technik, Essen, Germany

[31] Mauss F, Keller D, Peters N (1990) A Lagrangian Simulation of the Flamelet Extinction and Re-Ignition in Turbulent Jet Diffusion Flames. 23rd Symp (Int) Combust, pp 693–698, The Combustion Institute, Pittsburgh, PA
[32] Meneveau C, Poinsot T (1991) Stretching and Quenching of Flamelets in Premixed Turbulent Combustion. Combust Flame, vol 86, pp 311–332
[33] Metghalchi M, Keck JC (1980) Laminar Burning Velocity of Propane-Air Mixtures at Hight Temperature and Pressure. Combust Flame, vol 38, pp 143–154
[34] Metghalchi M, Keck JC (1982) Burning Velocities of Mixtures of Air with Methanol, Iso-octane, and Indolene at High Pressure and Temperature. Combust Flame, vol 48, pp 191–210
[35] Miller RJ, Rupley FR (1989) CHEMKIN-II: A Fortran Chemical Kinetics Package for the Analysis of Gas-Phase Chemical Kinetics. Report SAND 89-8009
[36] Moran MJ, Shapiro HN (1992) Fundamentals of Engineering Thermodynamics. 2nd edn, Wiley, New York, NY
[37] Otto F (2001) Fluid Mechanical Simulation of Combustion Engine Processes. Class Notes, University of Hanover, Germany
[38] Otto F, Dittrich P, Wirbeleit F (1998) Status of 3D-Simulation of Diesel Combustion. 3rd Int Indicating Symp, pp 289–308, Mainz, Germany
[39] Patterson MA, Kong SC, Hampson GJ, Reitz RD (1994) Modeling the Effects of Fuel Injection Characteristics on Diesel Engine Soot and NOx Emissions. SAE Paper 940523
[40] Peters N (1984) Laminar Diffusion Flamelet Models in Non-Premixed Turbulent Combustion. Prog Energy Combust Sci, vol 10, pp 319–339
[41] Peters N (1986) Laminar Flamelet Concepts in Turbulent Combustion. 21st Symp (Int) Combust, pp 1231–1250, The Combustion Institute, Pittsburgh, PA
[42] Peters N (1992) Fifteen Lectures on Laminar and Turbulent Combustion. Ercoftac Summer School, RWTH Aachen, Germany
[43] Peters N (1993) Flame Calculation with Reduced Mechanisms – An Outline. in Peters N, Rogg B (eds): Reduced Kinetic Mechanisms for Applications in Combustion Systems. Lecture Notes in Physics, Springer, Berlin, Germany
[44] Peters N (1997) Four Lectures on Turbulent Combustion. Ercoftac Summer School, RWTH Aachen, Germany
[45] Pischinger S, Heywood JB (1990) A Model for Flame Kernel Development in a Spark-Ignition Engine. 23rd Symp (Int) Combust, pp 1033–1040, The Combustion Institute, Pittsburgh, PA
[46] Pitsch H, Barths H, Peters N (1996) Three-Dimensional Modeling of NOx and Soot Formation in DI-Diesel Engines Using Detailed Chemistry Based on the Interactive Flamelet Approach. SAE Paper 962057
[47] Pontoppidan M, Gaviani G, Bella G, de Maio A, Rocco V (1999) Experimental and Numerical Approach to Injection and Ignition Optimization of Lean GDI-Combustion Behavior. SAE Paper 1999-01-0173
[48] Pope SB (1985) PDF Methods for Turbulent Reactive Flows. Prog Energy Combust sci, vol 11, pp 119–192
[49] Pope SB (1990) Computations of Turbulent Combustion: Progress and Challenges. 23rd Symp (Int) Combust, pp 591–612, The Combustion Institute, Pittsburgh, PA
[50] Ramos JI (1989) Internal Combustion Engine Modeling. Hemisphere, New York, NY

[51] Reissing J, Peters H, Kech JM, Spicher U (2000) Experimental and Numerical Analyses of the combustion Process in a Direct Injection Gasoline Engine. Int J Engine Research, vol 1, no 2, pp 147–161
[52] Rhodes DB, Keck JC (1985) Laminar Burning Speed Measurements of Indolene-Air-Diluent Mixtures at High Pressures and Temperature. SAE Paper 850047
[53] Santavicca DA, Liou D, North GL (1990) A Fractal Model of Turbulent Flame Kernel Growth. SAE Paper 900024
[54] Sazhina EM, Sazhin SS, Heikal MR, Babushok VI, Johns R(2000) A Detailed Modelling of the Spray Ignition Process in Diesel Engines. Combust Sci and Tech, vol 160, pp 317–344
[55] Semenov NN (1935) Chemical kinetics and Chain Reactions. Oxford University Press, London, UK
[56] Soika A, Dinkelacker F, Leipertz A (1998) Measurement of the Resolved Flame Structure of Turbulent Premixed Flames with Constant Reynolds Number and Varied Stoichiometry. Proc Combust Inst, vol 27, pp 785–792
[57] Stiesch G, Merker GP (2002) A Simplified Model for Description of Triple Flames in Stratified Charge Gasoline Engines. Proc 12th Int Multidim Engine Modeling Users Group Meeting, Detroit, MI
[58] Stiesch G, Tan Z, Merker GP, Reitz RD (2001) Modeling the Effect of Split Injections on DISI Engine Performance. SAE Paper 2001-01-0965
[59] Stiesch G, Pagel S, Merker GP (2002) Modeling Spark Ignition and Partially Premixed Flame Propagation in DISI Engines. 12th Int Heat Transfer Conf, pp 881–886, Grenoble, France
[60] Subramaniam MN, Stiesch G, Reitz RD (2001) Simulation of a 2-Stroke GDI Engine in KIVA-3V. Proc 14th ILASS Americas Conf, Dearborn, MI
[61] Tekawa K, Aoki O, Nomura H, Ujiie Y, Tsue M, Kono M (1998) Numerical Analysis of the Effect of Spark Components on Ignition Process in a Quiescent Methane-Air Mixture. 4th Int Symp COMODIA 98, pp 191–196
[62] Thiele M, Warnatz J, Maas U (1999) 2D-Simulation of Ignition Induced by Electrical Discharges. SAE Paper 1999-01-1178
[63] Venkatesh G, Abraham J, Magi V (2002) A Comparison of Mixing-Controlled and Flamelet Models for Diesel Combustion. SAE Paper 2002-01-1116
[64] Warnatz J (1984) Critical Survey of Elementary Reaction Rate Coefficients in the C/H/O System. in: Gardiner WC (ed) Combustion Chemistry, Springer, New York, NY
[65] Warnatz J, Maas U, Dibble RW (2001) Combustion: Physical and Chemical Fundamentals, Modeling and Simulation, Experiments, Pollutant Formation. 3rd edn, Springer, Berlin, Germany
[66] Weisser GA (2001) Modelling of Combustion and Nitric Oxide Formation for Medium-Speed DI Diesel Engines. Ph.D. Thesis, ETH Zurich, Switzerland
[67] Weller HG (1993) The Development of a New Flame Area Combustion Model Using Conditional Averaging. Thermo-Fluids Section Report TF/9307, Imperial College, London, UK
[68] Weller HG, Uslu S, Gosman AD, Maly RR, Herweg R, Heel B (1994) Prediction of Combustion in Homogeneous-Charge Spark-Ignition Engines. 3rd Int Symp COMODIA 94, pp 163–169

[69] Westbrook CK, Dryer FL (1981) Simplified Reaction Mechanisms for the Oxidation of Hydrocarbon Fuels in Flames. Combust Sci Tech, vol 27, pp 31–43
[70] Williams FA (1985) Combustion Theory, 2nd edn, Benjamin/Cummings, Menlo Park, CA
[71] Xin J, Montgomery D, Han Z, Reitz RD (1997) Multidimensional Modeling of Combustion for a Six-Mode Emissions Test Cycle on a DI Diesel Engine. J of Engineering for Gas Turbines and Power, vol 117, pp 683–691
[72] Zheng J, Yang W, Miller DL, Cernansky NP (2002) A Skeletal Chemical Kinetic Model for the HCCI Combustion Process. SAE Paper 2002-01-0423

7 Pollutant Formation

7.1 Exhaust Gas Composition

In the ideal case of complete combustion of a hydrocarbon fuel with stoichiometric air, the exhaust gas would be composed of the chemical species carbon dioxide (CO_2), water (H_2O) and molecular nitrogen (N_2) only. For lean equivalence ratios, molecular oxygen (O_2) could be observed among the products as well. However, in real combustion systems there are two reasons that inhibit complete combustion: (i) Elementary chemical reactions never proceed completely into one direction, but they always approach an equilibrium state between products and reactants. Thus, at least a small amount of reactants will remain. (ii) Local boundary conditions such as mixture distribution, temperature and turbulence level are often non-ideal. Therefore, flame extinction, accompanied with unburned or partially burned species, or the formation of entirely new products, e.g. soot or nitrogen oxides, may occur. Consequently, additional components are present in the exhaust gases of combustion engines. These components are carbon monoxide (CO), unburned hydrocarbons (HC), nitrogen oxides (NO_x) and particulate matter which is often approximated as soot. Depending on the quality of the fuel there may also be traces of sulfur oxides (SO_x) within the exhaust gas.

In Fig. 7.1 typical engine-out exhaust compositions (without exhaust gas aftertreatment) of both gasoline and diesel engines are illustrated. It becomes obvious, that the pollutants that are caused by imperfect combustion, amount for only a very small fraction of the entire exhaust stream. In SI engines their volume fraction ranges around 1%, and in diesel engines it is even less. Consequently, the formation of pollutants is irrelevant from a thermodynamic point of view, i.e. it does not have a noticeable effect on the total amount of heat released from combustion. The strong focus on the reduction of pollutant emissions in today's development of internal combustion engines is thus not caused by an attempt to further improve fuel economy. The driving force is rather given by the fact that both humans and nature react extremely sensitive to some of the hazardous pollutants. Therefore, even small concentrations in the ppm-range can be of great importance.

Having said the above, there is nevertheless often a strong dependency between the indicated engine efficiency and engine parameters, that also affect the quantity of emissions. One example for such a parameter is the injection timing in diesel engines, that affects both fuel economy and exhaust emissions. However, the dependency of the efficiency on such parameters is not caused by the total amount of heat released from the chemical reactions, but rather by the timing and the rate at which the energy is released in relation to the engine cycle.

Figure 7.1 also breaks down the pollutants into the various chemical species. For SI engines, CO accounts for the greatest fraction, and HC and NO_x are also observed in noticeable quantities. All these quantities can however be effectively removed from the exhaust gas by a three-way catalytic converter. In the diesel exhaust, NO_x and soot represent the major challenges, especially because an effective exhaust gas aftertreatment represents a much greater difficulty for a combustion concept subject to excess air [7].

With respect to the mathematical modeling of pollutant formation in engine simulations, nitrogen oxides and soot are basically the solitary points of focus. This is certainly due to the fact these components represent the toughest challenges in meeting today's and future emission legislations, especially for diesel and direct injection SI engines. An additional, more trivial reason however, is that the governing mechanisms leading to the formation of NO_x and soot are relatively well understood. The remaining two exhaust gas components of importance, namely CO and HC, are more difficult to quantify in numerical simulations as they both result from similar mechanisms, i.e. from incomplete combustion. With the simulation tools available today, it is hardly possible to predict at which stage of the multi-step oxidation process a reaction will abort, when boundary conditions become unfavorable for complete combustion. Thus, it is not feasible to quantify the fractions of carbon monoxide and unburned hydrocarbons in numerical simulations, or to even break down the HC into the different partially oxidized components, e.g. into aldehydes and ketones.

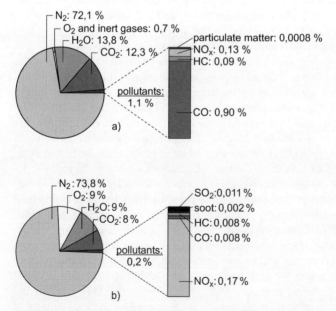

Fig. 7.1. Engine-out exhaust gas composition (volume fractions). a) Spark ignition engine b) Diesel engine [24]

7.2 Nitrogen Oxides

7.2.1 Reaction Paths

Nitrogen Oxides are abbreviated by the symbol NO_x and comprise both nitric oxide (NO) and nitrogen dioxide (NO_2). Whereas the NO_2/NO ratio is typically negligibly small in SI engines, it can range between 10 and 30% in diesel engines [15]. However, even if the NO_2-fraction within the total oxides of nitrogen is close to the upper limit of this range, it is almost exclusively formed via the NO molecule, e.g. by the reaction:

$$NO + HO_2 \rightarrow NO_2 + OH \,. \tag{7.1}$$

For this reason, the modeling of the total nitrogen oxides formation in combustion engines is most often reduced to the formation of NO.

Four different mechanisms lead to the formation of nitric oxide in combustion systems. The *thermal NO* is formed at high temperatures under slightly lean conditions within the burned products. The involved nitrogen and oxygen stem from the combustion air. The *prompt NO* path describes the reaction of N_2 from combustion air with hydrocarbon radicals in fuel-rich regions. Because of the need of hydrocarbon radicals, this path is followed directly within the reaction zone. The third mechanisms is the *fuel NO*, which refers to the formation of NO from fuel-bound nitrogen as it may be contained in coal and heavy distillates of petroleum. Finally, there is a reaction path characterized by the *N_2O-intermediate*. This route is activated at lower temperatures than the thermal NO in a fuel-lean and high pressure environment. It can become important in gas turbine combustion, but typically, it has only a minor effect in diesel engine combustion [30].

It is widely accepted that both within diesel and SI engines, the major fraction of NO is formed via the thermal path. The precise percentage of its contribution to the total NO formation is still subject to disputes in the literature, but in all studies known to the author it is attributed to account for a fraction of at least 80 to 95%. For this reason many numerical studies are solely based on the thermal NO formation and neglect the remaining mechanisms.

The second NO formation path that may have a noticeable effect is the prompt mechanism within the flame region. It is though to contribute between 5 and 20% to the total NO_x concentration in internal combustion engines. This mechanisms has also been utilized in a number of numerical studies. However, its consideration is only feasible when the entire combustion process is described by a fairly detailed chemical mechanism as well. A strongly reduced reaction mechanism, that incorporates only one or two global reaction steps for the oxidation of the fuel molecules to CO_2 and H_2O, cannot provide enough information about intermediate species, that are necessary for the prompt NO mechanism to yield useful results.

Both the fuel NO and the N_2O mechanisms are hardly accounted for in combustion engine modeling, because they only contribute to negligible amounts of nitrogen oxides anyway. The following discussion will therefore concentrate on the thermal and prompt mechanisms.

7.2.2 Thermal NO

The thermal formation of nitric oxide is described by the so-called extended Zeldovich mechanism. It describes the breakup of a nitrogen molecule by an oxygen atom, and the subsequent oxidation of the N atoms:

(i) $\quad N_2 + O = NO + N, \quad k_{i,f} = 7.6 \cdot 10^{13} \exp[-38{,}000/T]\ cm^3/(mol\ s) \quad$ (7.2)

(ii) $\quad N + O_2 = NO + O, \quad k_{ii,f} = 6.4 \cdot 10^9 \exp[-3{,}150/T]\ cm^3/(mol\ s) \quad$ (7.3)

(iii) $\quad N + OH = NO + H, \quad k_{iii,f} = 4.1 \cdot 10^{13} \qquad\qquad cm^3/(mol\ s) \quad$ (7.4)

The name of the mechanism stems from Y.B. Zeldovich [32], who was the first to postulate the importance of reactions (i) and (ii) in 1946. Due to underestimations of NO levels, the third NO-creating elementary reaction involving OH radicals was later added to the mechanism by Lavoie et al. [20]. Thus, the name extended Zeldovich mechanism. The rate coefficients of the forward reactions specified in Eqs. 7.2 to 7.4 have been taken from Bowman [6]. It should be noted however, that the pre-exponential factors, especially that of reaction (i), are subject to uncertainties by as much as a factor of two.

The reaction path is referred to as thermal NO formation, because the first reaction, which initiates the overall mechanism by production of nitrogen atoms, proceeds only at high temperatures. This becomes obvious from the high activation energy (38,000 K), that is caused by the stable triple-bond of the N_2-molecule. Consequently, considerable amounts of thermal NO are produced only in the hot products regions, where gas temperatures are well above 2000 K. But even in these regions, chemistry is still relatively slow, such that the chemical equilibrium is not reached. This effect is illustrated in Fig. 7.2, where an ideal NO history based on the equilibrium assumption is compared to a realistic course of NO concentrations vs. crank angle in an engine combustion chamber. After the start of combustion, the increase in the actual NO concentration falls behind the ideal equilibrium case ($\Delta 1$), because the slow chemistry cannot follow the fast increase in gas temperature due to combustion. Later in the engine cycle the equilibrium NO concentration decreases almost to zero because gas temperatures are reduced again during the expansion stroke. However, the realistic NO concentration does not follow its equilibrium value. It rather "freezes" at a concentration much greater than the equilibrium value corresponding to exhaust conditions ($\Delta 2$). This freezing of the reverse reactions of the mechanism specified in Eqs. 7.2 to 7.4 is partly caused by the temperature dependence of the respective rate coefficients, and partly, because the concentrations of the N-, O-, and H-atoms, that are necessary for the reverse reactions to proceed, are drastically reduced when the temperature decreases during expansion.

As a consequence, the thermal NO formation in combustion engines needs to be estimated as a rate controlled process in order to enable the prediction of realistic emission levels. According to Sect. 6.1.2, the formation rate of NO can be written as

$$\frac{d[NO]}{dt} = k_{i,f}[N_2][O] + k_{ii,f}[N][O_2] + k_{iii,f}[N][OH]$$
$$- k_{i,r}[NO][N] - k_{ii,r}[NO][O] - k_{iii,r}[NO][H], \quad (7.5)$$

and $d[N]/dt$ becomes

$$\frac{d[N]}{dt} = k_{i,f}[N_2][O] - k_{ii,f}[N][O_2] - k_{iii,f}[N][OH]$$
$$- k_{i,r}[NO][N] + k_{ii,r}[NO][O] + k_{iii,r}[NO][H]. \quad (7.6)$$

Because reactions (ii) and (iii) proceed very rapidly compared to reaction (i), quasi-steadiness of the nitrogen atom can be assumed:

$$\frac{d[N]}{dt} \approx 0. \quad (7.7)$$

Thus, summation of Eqs. 7.5 and 7.6 yields

$$\frac{d[NO]}{dt} = 2k_{i,f}[N_2][O] - 2k_{i,r}[NO][N], \quad (7.8)$$

and the unknown concentration of the nitrogen atoms is obtained from rearranging Eq. 7.6 under the constraint of Eq. 7.7:

$$[N] = \frac{k_{i,f}[N_2][O] + k_{ii,r}[NO][O] + k_{iii,r}[NO][H]}{k_{i,r}[NO] + k_{ii,f}[O_2] + k_{iii,f}[OH]}. \quad (7.9)$$

With this simplification, the differential equation 7.8 contains only known variables and can be integrated numerically: [NO] is the integrated quantity itself, the nitrogen concentration [N$_2$] is easily obtained from the equivalence ratio, and the remaining species concentrations may be obtained from the assumption of partial equilibrium of the OHC-system, compare Chap. 6. The rate coefficients of the reverse reactions are determined from the respective equilibrium constants, Eq. 6.25.

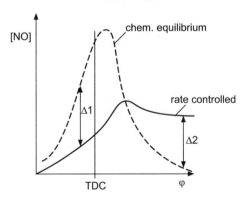

Fig. 7.2. In-cylinder NO concentrations assuming chemical equilibrium and under realistic slow chemistry conditions

As the influence of the reverse reaction in Eq. 7.8 is typically small, it is entirely neglected in many studies. In this case the NO change rate reduces to

$$\frac{d[\text{NO}]}{dt} = 2k_{i,f}[\text{N}_2][\text{O}], \qquad (7.10)$$

where the concentration of the oxygen atom is again obtained from partial equilibrium assumptions. It should be noted however, that with the above relation a decomposition of NO during the late phase of the engine cycle can generally not be reproduced. The right hand side of Eq. 7.10 remains always greater than zero.

The thermal NO formation represents the mechanism that can be predicted with the best accuracy of all pollutant formation mechanisms in combustion engines. However, since the kinetic reaction rate shows the extreme temperature dependence discussed above, the quality of the NO prediction is closely coupled to the prediction quality of the heat release profile, that governs the pressure trace as well as the temperature distribution. Generally speaking, an agreement to about ±20% can be expected between NO predictions and measurements, provided that the actual pressure trace is reproduced reasonably well, too. A typical example for estimated and measured NO emissions is displayed in Fig. 6.24 for a spark timing variation of a DISI engine. The corresponding pressure traces are shown in Fig. 6.23.

7.2.3 Prompt NO

The mechanism leading to prompt NO formation was postulated by C.P. Fenimore [8], and is hence commonly referred to as the Fenimore mechanism. It was later refined by Miller and Bowman [25]. As noted above, prompt NO is formed directly within the reaction zone. The reaction mechanism is more complicated than the one for thermal NO, because the prompt NO results from the radical CH, which was previously thought to be an unimportant intermediate species. The rate-limiting step in the mechanism is the reaction of the CH radical with molecular nitrogen to hydrocyanic acid and a nitrogen atom:

$$\text{CH} + \text{N}_2 \rightarrow \text{HCN} + \text{N}, \quad k_f = 4.4 \cdot 10^{12} \exp[-11{,}060/T] \text{ cm}^3/(\text{mol s}) \qquad (7.11)$$

The nitrogen can thereafter be converted to NO by reactions (ii) and (iii) of the extended Zeldovich mechanism, Eqs. 7.3 and 7.4. The HCN undergoes further reactions that lead to the formation of NO via the intermediate NCO, e.g.:

$$\text{HCN} + \text{OH} = \text{CN} + \text{H}_2\text{O}, \qquad (7.12)$$

$$\text{CN} + \text{O}_2 \rightarrow \text{NCO} + \text{O}, \qquad (7.13)$$

$$\text{HCN} + \text{O} \rightarrow \text{NCO} + \text{H}, \qquad (7.14)$$

$$\text{NCO} + \text{O} \rightarrow \text{NO} + \text{CO}, \qquad (7.15)$$

$$\text{NCO} + \text{OH} \rightarrow \text{NO} + \text{CHO}. \qquad (7.16)$$

The reactions summarized in Eqs. 7.12 to 7.16 represent only an example of the reaction path. The actual mechanism is more complex in that more reactions and additional intermediate species are involved.

Several aspects are noteworthy about the prompt NO mechanism. Since the rate-limiting reaction step, Eq. 7.11, involves the CH radical, the prompt NO path becomes important only under fuel-rich conditions where noticeable CH concentrations can be detected, compare Figs. 6.2 and 6.3. Moreover, the activation energy in Eq. 7.11 is significantly less than the one in the rate-limiting step of the thermal NO mechanism (11,060 K compared to 38,000 K). Therefore, the temperature dependency of the prompt NO is not as pronounced as the one of the thermal NO, and the relative contribution of the prompt path to the overall NO formation becomes more important for reduced combustion temperatures. However, the uncertainty in the rate coefficient of Eq. 7.11 is greater than the respective uncertainties in the thermal NO formation. Thus, the predictions of prompt NO utilizing the Fenimore mechanism are usually less accurate [31]. Finally, it should be noted that the estimation of fairly accurate CH concentrations within the flame zone represents a major challenge in applying the Fenimore mechanism. This task requires the use of detailed reaction mechanisms for the entire combustion process, which are very expensive in CPU time. This and the fact, that prompt NO has only a limited contribution to the total NO formation in combustion engines, are the main reasons, why the Fenimore mechanism has so far been accounted for in relatively few studies. Nevertheless, closed formulations of such complete mechanisms are available in the literature, e.g. [4, 5, 13, 14, 18].

7.3 Soot

7.3.1 Phenomenology

It is widely accepted that the formation of soot particles is preceded by the presence of polycyclic aromatic hydrocarbons (PAH) [3]. This mechanism is schematically shown in Fig. 7.3. The first step is the growth of the PAH by *conglomeration* of molecules. Typically, they are referred to as PAH as long as the molecules are arranged in a two-dimensional structure. Once they extend into three-dimensional space, they are referred to as particles (*particle inception* or *nucleation*). The particles are then subject to *surface growth* by addition of mainly acetylene (C_2H_2), and they further increase in size by *coagulation*. During the entire process hydrogen is continuously abstracted, such that the resulting soot particles are finally characterized by a very high carbon fraction. The particles vary in size, and typically distributions between 10 and 1000 nm for the diameter are observed, with a maximum number density at about 100 nm. The average density of soot particles is thought to be about 2000 kg/m^3. Due to their porous nature, the soot particles are characterized by a high surface area to mass ratio. Thus, they are susceptible for adsorption and condensation of additional hydrocarbon molecules, even after the end of combustion.

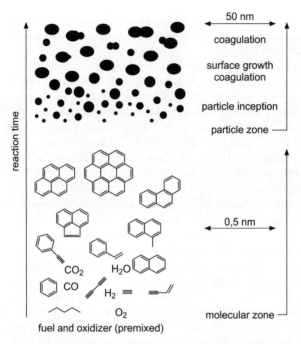

Fig. 7.3. Schematic reaction path leading soot formation in premixed flames [3]

Soot formation rates in combustion systems depend strongly on temperature and on the carbon to oxygen ratio, i.e. the fuel-air ratio. This aspect is illustrated in Fig. 7.4, which shows that the relative soot yield strongly increases for air-fuel equivalence ratios richer than $\lambda \approx 0.5$ ($\phi \approx 2.0$). For air-fuel ratios that are less rich, the hydrocarbon molecules are primarily converted to carbon monoxide instead of soot. Soot formation is further limited to a temperature range between about 1200 and 2000 K. This can be explained by the fact, that soot formation is needing radical precursors such as C_3H_3, that are not present at lower temperatures. Conversely, these precursors are pyrolized and oxidized at elevated temperatures [31]. There is also a pressure dependence of the soot formation rate, which is however not as distinct as the T- and ϕ-dependencies. Generally, a moderate increase in the soot yield is observed for an increasing pressure.

With respect to internal combustion engines, soot emissions are most significant in diesel engines and may also become important in stratified charge operation of DISI engines. In gasoline engines with external mixture formation soot is hardly an issue because the equivalence ratio is homogeneously distributed and close to stoichiometric under most operating conditions, compare Fig. 7.1. However, in the stratified mixture of diesel and DISI engines large amounts of soot are formed under locally rich conditions early after ignition. Only later in the combustion process sufficient oxygen is mixed into to previously rich zones such that a substantial fraction of the formed soot is oxidized again. This is possible because the global equivalence ratio in both diesel and DISI engines is lean.

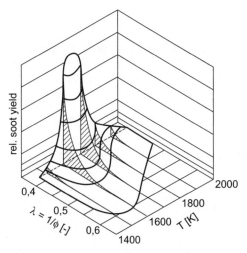

Fig. 7.4. Relative soot yield vs. air-fuel equivalence ratio and temperature [27]

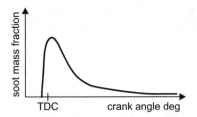

Fig. 7.5. Schematic history of the soot mass fraction within a diesel engine

It should be noted though, that not the complete amount of soot formed can be oxidized again and a certain fraction is emitted into the exhaust. This is indicated in Fig. 7.5, which shows a schematic history of the soot mass fraction within a diesel engine combustion chamber. At some point during the expansion stroke, the oxidation of previously formed soot comes to a stop because the concentration of O-atoms and OH-radicals that are important in the oxidation process becomes too small to attack the soot particles. This "freezing" of the oxidation occurs approximately between 1300 and 1400 K.

Typically, at least 90% and often up to 99% of the formed soot is oxidized again such that the soot concentration in the exhaust gas is only a small fraction of the maximum in-cylinder concentration during combustion. This represents a principal mathematical problem in the modeling of engine-out soot emissions. Since the soot mass in the exhaust is only a very small difference between two large quantities, i.e. between the formation and oxidation, a significant relative error will result if there is only a slight deviation in either the production or oxidation rate. Consequently, soot models can today only be utilized in order make qualitative assessments, but relative errors can easily sum up to 100% and more.

7.3.2 Semi-Global Mechanisms

Semi-global soot models describe the formation and oxidation of soot particles by two or several global reaction steps. Probably the earliest and simplest approach is the two-step model proposed by Hiroyasu et al. [16]. It describes the change in the soot mass by the Equation

$$\frac{dm_s}{dt} = \frac{dm_{s,f}}{dt} - \frac{dm_{s,ox}}{dt}, \qquad (7.17)$$

where the first term and the second term on the right hand side denote the rates of soot formation and oxidation, respectively:

$$\frac{dm_{s,f}}{dt} = A_f m_{f,v} p^{0.5} \exp\left[-\frac{E_{s,f}}{\tilde{R}T}\right], \qquad (7.18)$$

$$\frac{dm_{s,ox}}{dt} = A_{ox} m_s \frac{p_{O2}}{p} p^{1.8} \exp\left[-\frac{E_{s,ox}}{\tilde{R}T}\right]. \qquad (7.19)$$

Thus, the rate of soot production is proportional to the mass of vaporized fuel $m_{f,v}$, and the oxidation rate depends on the soot mass m_s itself as well as on the partial pressure of molecular oxygen p_{O2}. The activation energies have been set to $E_{s,f}$ = 52,335 kJ/kmol and $E_{s,ox}$ = 58,615 kJ/kmol, respectively, and the pre-exponential constants A_f and A_{ox} need to be adjusted in order to reproduce experimentally determined soot emissions of a particular engine. A review of the parameter settings used in various studies has been presented by Kleemann [19].

In many recent engine studies that include the estimation of soot emissions, the soot balance equation, Eq. 7.17, as well as the formation rate of soot, Eq. 7.18, are recovered, but the oxidation rate of Eq. 7.19 is replaced by an approach suggested by Nagle and Strickland-Constable [26]. This model relates carbon oxidation to the surface chemistry involving more reactive sites A and less reactive sites B. Three types of reactions are assumed: the oxidation of both A and B sites as well as their thermal rearrangement/annealing:

(i) $A + O_2 \rightarrow A + 2\,CO$, $\dot{w}_i = \dfrac{k_A p_{O2}}{1 + k_Z p_{O2}} x$ (7.20)

(ii) $B + O_2 \rightarrow A + 2\,CO$, $\dot{w}_{ii} = k_B p_{O2}(1-x)$ (7.21)

(iii) $A \rightarrow B$ $\dot{w}_{iii} = k_T x$ (7.22)

The fraction of A sites on the carbon surface is denoted by x. Assuming quasi-steadiness for x, i.e. $\dot{w}_{ii} = \dot{w}_{iii}$, it becomes:

$$x = \left(1 + \frac{k_T}{k_B p_{O2}}\right)^{-1}. \qquad (7.23)$$

The various reaction rates are in units (g-atom/cm^2-s), and the respective rate constants are specified as follows:

$$k_A = 20.0\exp[-15{,}100\,\text{K}/T], \tag{7.23}$$

$$k_B = 4.46\cdot 10^{-3}\exp[-7{,}650\,\text{K}/T], \tag{7.24}$$

$$k_T = 1.51\cdot 10^{5}\exp[-48{,}800\,\text{K}/T], \tag{7.25}$$

$$k_Z = 21.3\exp[+2{,}060\,\text{K}/T]. \tag{7.26}$$

The soot mass oxidation rate replacing Eq. 7.19 now becomes

$$\frac{dm_{s,ox}}{dt} = \frac{6 MW_C}{\rho_s D_s} m_s R_{tot}, \tag{7.27}$$

where MW_C is the carbon molecular weight, ρ_s is the soot density, D_s is a characteristic soot particle diameter, and R_{tot} denotes the total soot oxidation rate, i.e. the sum of Eqs. 7.20 and 7.21:

$$R_{tot} = \left(\frac{k_A p_{O2}}{1+k_Z p_{O2}}\right)x + k_B p_{O2}(1-x). \tag{7.28}$$

Several different values for the soot particle properties have been suggested in the literature. Typically, the soot density and the characteristic particle diameter are assumed to be approximately equal to 2000 kg/m³ and $2.5\cdot 10^{-8}$ m, respectively.

Schubiger et al. [28] also utilized a two-step semi-global chemical mechanism to model the soot formation and oxidation in analogy to Eq. 7.17. While the soot formation rate in this model is similar to Eq. 7.18 in that an empirical Arrhenius equation is applied, the soot oxidation rate is now described in terms of a characteristic turbulent mixing time in order to account for the turbulence effects on the soot oxidation:

$$\frac{dm_{s,ox}}{dt} = A_{ox}\frac{1}{\tau_{trb}}m_s\left(\frac{p_{O2}}{p_{O2,ref}}\right)^{1.3}\exp\left[-\frac{15{,}000\,\text{K}}{T}\right]. \tag{7.29}$$

Within the framework of phenomenological combustion models, the turbulent time scale can be established in analogy to the procedure discussed in Sect. 3.4.2, that was proposed by the same group of authors. If the soot model is to be used within CFD codes, the turbulent time scale may be directly related to the turbulence properties, e.g. by Eq. 6.104. The pre-exponential factor A_{ox} needs to be adjusted to experimental soot emissions again.

Fusco et al. [11] presented a more detailed but still quasi-global mechanism that is composed of eight reaction steps. As illustrated in Fig. 7.6, vaporized fuel can be converted by pyrolysis into either a generic precursor radical species (1) or into a surface growth species (2), which is assumed to by C_2H_2. Both intermediate species can either be oxidized directly (3, 4), or they can contribute to the formation of soot. Inception of soot particles occurs from the precursor radicals (5), and the growth species leads to an increase in the size of individual particles (6). Finally, the existing soot particle are either oxidized into inert products again (7), or they are subject to coagulation such that the particle size is increased whereas the particle number is reduced.

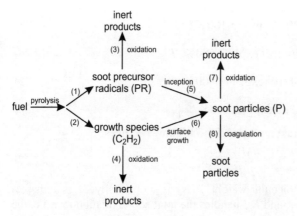

Fig. 7.6. Schematic illustration of the 8-step soot mechanism by Fusco et al. [11]

The kinetic rates of the eight reaction steps are summarized in Table 7.1. The quantity S_P denotes the accumulated surface area of all particles within the control volume of interest, and D_P is the particle diameter, which is determined from the soot volume fraction and the particle number under the assumption that the particles are ideally spherical. Note, that the particle oxidation (7) is identical to the soot oxidation rate by Nagle and Strickland-Constable, Eqs. 7.19 to 7.28.

Balancing the four quantities of interest,

- $[N_P]$ particle number density in $1/cm^3$,
- $[PR]$ precursor radical concentration in mol/cm^3,
- $[C_2H_2]$ growth species concentration in mol/cm^3,
- f_v soot volume fraction in cm^3/cm^3,

leads to the following differential equations, that can be integrated numerically:

$$\frac{d[N_P]}{dt} = N_A r_5 - r_8, \qquad (7.30)$$

$$\frac{d[PR]}{dt} = r_1 - r_3 - r_5, \qquad (7.31)$$

$$\frac{d[C_2H_2]}{dt} = r_2 - r_4 - r_6, \qquad (7.32)$$

$$\frac{df_v}{dt} = \frac{1}{\rho_s}\left(r_5 MW_{PR} + r_6 MW_C - r_7 MW_C\right). \qquad (7.33)$$

In the above equations MW_{PR} and MW_C denote the molecular weights of the precursor radicals and the carbon atom, respectively. The soot density is assumed to be $\rho_s = 1800$ kg/m^3, and N_A represents the Avogadro-constant ($6.023 \cdot 10^{23}$ mol^{-1}).

Table 7.1. Reaction rates of the 8-step soot mechanism by Fusco et al. [11]

Process (i)	Chem. Reaction	Reaction Rate r_i	A_i [mol, cm, s]	$E_{A,i}$ [J/mol]
(1) radical formation	$C_mH_n \rightarrow m/2 \, PR$	$r_1 = \dfrac{m}{2} A_1 \exp\left(\dfrac{-E_{A,1}}{\tilde{R}T}\right) \cdot [\text{fuel}_g]$	$0.7 \cdot 10^{12}$	502,400
(2) C_2H_2 formation	$C_mH_n \rightarrow m/2 \, C_2H_2$	$r_2 = \dfrac{m}{2} A_2 \exp\left(\dfrac{-E_{A,2}}{\tilde{R}T}\right) \cdot [\text{fuel}_g]$	$2.0 \cdot 10^8$	205,200
(3) radical oxidation	$PR + O_2 \rightarrow \text{products}$	$r_3 = A_3 \exp\left(\dfrac{-E_{A,3}}{\tilde{R}T}\right) \cdot [PR] \cdot [O_2]$	$1.0 \cdot 10^{12}$	167,500
(4) C_2H_2 oxidation	$C_2H_2 + O_2 \rightarrow \text{products}$	$r_4 = A_4 \exp\left(\dfrac{-E_{A,4}}{\tilde{R}T}\right) \cdot [C_2H_2] \cdot [O_2]$	$6.0 \cdot 10^{13}$	209,000
(5) particle inception	$PR \rightarrow P$	$r_5 = A_5 \exp\left(\dfrac{-E_{A,5}}{\tilde{R}T}\right) \cdot [PR]$	$1.0 \cdot 10^{10}$	209,000
(6) surface growth	$P + C_2H_2 \rightarrow P$	$r_6 = A_6 \exp\left(\dfrac{-E_{A,6}}{\tilde{R}T}\right) \cdot [C_2H_2] \cdot S_P^{0.5}$ where $S_P = \pi D_P^2 N_P$	$4.2 \cdot 10^4$	50,200
(7) particle oxidation	$P + O_2 \rightarrow \text{products}$	$r_7 = S_P \left[x \dfrac{k_A p_{O2}}{1 + k_Z p_{O2}} + (1-x) k_B p_{O2} \right]$ where $x = \left(1 + \dfrac{k_T}{k_B p_{O2}}\right)^{-1}$ $k_A = A_A \exp\left(\dfrac{-E_{A,A}}{\tilde{R}T}\right)$ (g atom cm^{-2} atm^{-1}) $k_B = A_B \exp\left(\dfrac{-E_{A,B}}{\tilde{R}T}\right)$ (g atom cm^{-2} atm^{-1}) $k_T = A_T \exp\left(\dfrac{-E_{A,T}}{\tilde{R}T}\right)$ (g atom cm^{-2}) $k_Z = A_Z \exp\left(\dfrac{-E_{A,Z}}{\tilde{R}T}\right)$, (atm^{-1})	$2.0 \cdot 10^1$ $4.46 \cdot 10^{-3}$ $1.51 \cdot 10^5$ $2.13 \cdot 10^1$	125,600 63,640 406,100 -17,200
(8) coagulation	$x \cdot P \rightarrow P$	$r_8 = k_k \cdot T^{0.5} \cdot f_v^{1/6} \cdot N_P^{11/6}$, where $k_k = 1.05 \cdot 10^{-7}$, (cm mix s^{-1} K$^{-1/2}$)		

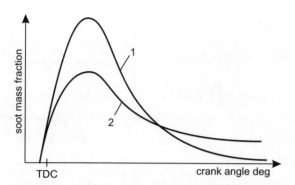

Fig. 7.7. Qualitative behavior of eight-step (1) two-step (2) soot mechanisms

The above semi-global soot mechanisms have been and are still widely used in both phenomenological and CFD combustion studies. Provided that the set of empirical parameters contained in the mathematical formulations is properly adjusted to a specific investigated engine, their performance is typically such, that the qualitative effects of various engine parameters on the soot emission can be estimated. That is, it is possible to predict whether a specific change in an operating parameter will increase or decrease the soot concentration in the exhaust. However, because of a lack in chemical reaction details and, more importantly, because of the principal mathematical problem that the final engine-out soot emission is only a small difference between two large quantities, it is not possible to make any quantitative predictions about absolute soot concentrations in the exhaust gas. In fact, relative errors in the predicted engine-out soot emissions of up to 100% are not unusual.

It should further be noted, that there seems to be an inherent difference between the performances of the two-step mechanisms and the eight-step mechanism, respectively. This behavior is schematically illustrated in Fig. 7.7. While the eight-step mechanism tends to predict high rates of oxidation that result in a soot emission of only about 1% of the maximum in-cylinder concentration during combustion, the various two-step mechanisms tend to predict significantly reduced oxidation rates that result in final soot emissions still averaging as much as 10 to 20% of the maximum in-cylinder concentration [29]. This principal behavior can hardly be altered even if the empirical parameters of the two-step mechanism are changed in order to adjust the absolute value of the final concentration in the exhaust. In this respect, recent crank angle resolved experimental data of in-cylinder soot concentrations suggest, that the soot concentration profiles predicted by the eight-step mechanism are more realistic for modern DI diesel engines equipped with high pressure injection systems, e.g. [12, 23, 28].

A reasonable prediction of the entire temporal soot concentration profile becomes especially important, when advanced heat transfer models are utilized, that consider the influence of radiative heat transfer in addition to convective heat transfer. In this case both the maximum as well as the exhausted soot concentrations are of interest and the more detailed eight-step mechanism seems to be more appropriate. On the other hand, the adjustment of the eight-step mechanism is much more difficult than it is for the two-step mechanism, since the former contains as many as 21 empirical parameters. Proper adjustment is thus only possible with a very large number of experimental data points. As a consequence, the two-step model is still most often used in engine calculations. It is commonly adjusted to produce reasonable exhaust emissions, and the possibility of unrealistically low soot concentrations during combustion is tolerated.

7.3.3 Detailed Chemistry Mechanisms

A number of comprehensive reaction mechanisms have been proposed that describe the principal procedure of soot formation and oxidation in much greater detail than the semi-global mechanisms discussed in the forgoing section, e.g. [10, 22]. These comprehensive models, which are in general computationally expensive, include both gas-phase kinetics in order to describe the chemical reactions on a molecular scale, as well as particle dynamics in order to describe processes such as particle inception, surface growth, coagulation and oxidation on the particle scale, compare Fig. 7.3.

The formation of the first aromatic ring (benzene) from linear hydrocarbon molecules is subject to controversy in the literature. Basically two possible mechanisms have been suggested. The first one is started by C_3H_4 decomposition or by reaction of CH or CH_2 with C_2H_2 to C_3H_3 [31]. Two C_3H_3-radicals may then perform a self-reaction and, after rearrangement of hydrogen atoms, lead to the first aromatic ring (C_6H_6), see Fig. 7.8. The alternative mechanism has been described by Franklach and Wang [10]. As illustrated in Fig. 7.9, it is driven by the combination of acetylene as well as by addition of H and abstraction of H_2. Typically, a high and a low temperature route are distinguished.

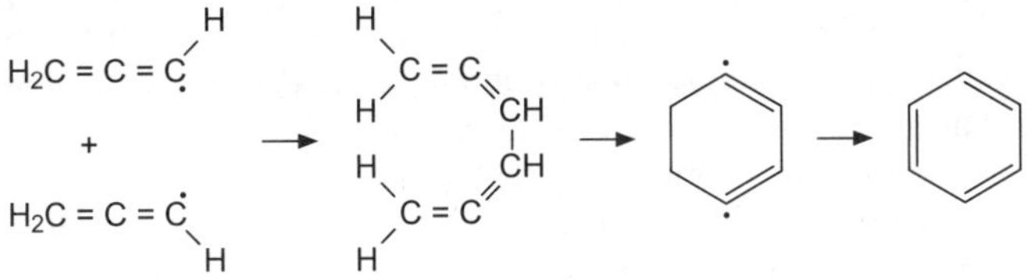

Fig. 7.8. Formation of the first aromatic ring by C_3H_3 self-reaction [31]

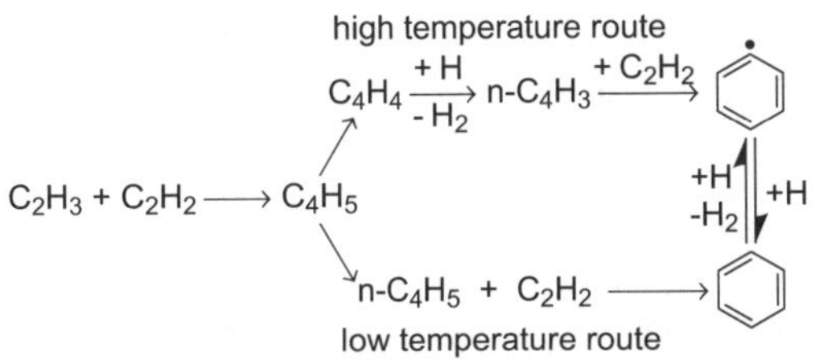

Fig. 7.9. Pathways to the first aromatic ring by addition of C_2H_2 [10]

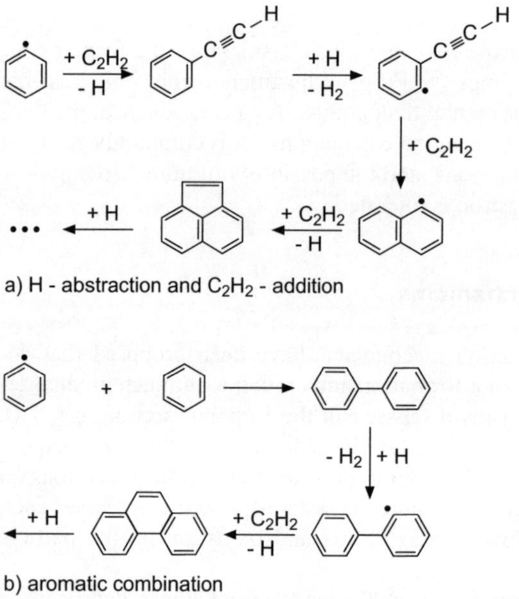

a) H - abstraction and C₂H₂ - addition

b) aromatic combination

Fig. 7.10. PAH growth by the HACA sequence and by aromatic combination [10]

After the initial formation of single aromatic rings, larger polycyclic aromatic hydrocarbons (PAH) are created by the so-called HACA sequence (H-abstraction, C_2H_2-addition), see Fig. 7.10 a). Additionally, single aromatic rings can directly combine to form more complex aromatic structures as shown in part b) of Fig. 7.10. PAH growth beyond four aromatic rings is subject to a literally infinite number of polymerization steps, resulting in soot particles of different structure and size. Instead of explicitly modeling all these steps, Frenklach's linear lumping technique is utilized in order to reduce the computational effort [9]. With this mathematically rigorous method, a statistical description of the PAH size distribution is chosen, such that only a small number of transport equations needs to be solved in order to determine the moments of the distribution function. This method is in general similar to the characterization of the mixture fraction pdf in the flamelet combustion model that has been discussed in detail in Chap. 6. The same approach is also used to identify the size distribution of the actual soot particles. In this case the source terms that are included in the transport equations of the statistical moments of the distribution function account for particle dynamics like nucleation, condensation, coagulation, surface growth and oxidation. The latter process accounts for the attack of soot particles by both O_2 molecules and OH radicals.

A very similar detailed soot mechanism was adapted for use within a steady flamelet framework by Balthasar, Mauss and co-workers [1, 2, 17]. They successfully applied the model to laminar and turbulent diffusion flames as well as to diesel engine combustion. The reported agreement to experiments is of remarkable quality, while the computational cost could be restricted due to the numerical decoupling between chemistry and turbulent flow field enabled by the flamelet approach. As discussed in Chap. 6, this procedure allows to execute the chemistry calculations in one-dimensional space as a pre-processing and to store the results in so-called flamelet libraries for later interpolation by the 3D-CFD code.

To summarize the above, the various comprehensive soot models available today allow for a significantly better description of the chemical details in soot formation and oxidation compared to semi-global mechanisms. However, their computational expenditure is still enormous, and they cannot resolve the principal mathematical problem encountered in diesel engines, that soot formation and oxidation are of comparable size, whereas the remaining engine-out soot emission is only a small difference between those quantities. Consequently, a comprehensive soot model is still likely to result in considerable relative uncertainties with respect to engine-out soot emissions, even if it has proven to yield very good results in steady flame applications. For the above two reasons, detailed chemistry soot models are most useful in fundamental studies with simpler boundary conditions, but they are not yet a standard in three-dimensional diesel engine modeling [21].

References

[1] Bai XS, Balthasar M, Mauss F, Fuchs L (1998) Detailed Soot Modeling in Turbulent Jet Diffusion Flames. 27th Symp (Int) Combust, pp 1623–1630, The Combustion Institute, Pittsburgh, PA
[2] Balthasar M, Heyl A, Mauss F, Schmitt F, Bockhorn H (1996) Flamelet Modeling of Soot Formation in Laminar Ethyne/Air-Diffusion Flames. 26th Symp (Int) Combust, pp 2369–2377, The Combustion Institute, Pittsburgh, PA
[3] Bockhorn H (ed) (1994) Soot Formation in Combustion: Mechanisms and Models. Springer, Berlin, Germany
[4] Bockhorn H, Chevalier C, Warnatz J, Weyrauch V (1991) Experimental Investigation and Modeling of Prompt NO Formation in Hydrocarbon Flames. In: Santoro RJ, Felske JD (eds), HTD-Vol 166, Heat Transfer in Fire and Combustion Systems, Book No G00629-1991
[5] Bollig M, Pitsch H, Hewson JC, Seshadri K (1996) Reduced n-Heptane Mechanism for Non-Premixed Combustion with Emphasis on Pollutant-Relevant Intermediate Species. 26th Symp (Int) Combust, pp 729–737, The Combustion Institute, Pittsburgh, PA
[6] Bowman CT (1975) Kinetics of Pollutant Formation and Destruction in Combustion. Prog Energy Combust Sci, vol 1, pp 33–45
[7] Eastwood PG (2000) Critical Topics in Exhaust Gas Aftertreatment. Research Studies Press, Baldock, UK
[8] Fenimore CP (1971) Formation of Nitric Oxide in Premixed Hydrocarbon Flames. 13th Symp (Int) Combust, pp 373–380, The Combustion Institute, Pittsburgh, PA
[9] Frenklach M (1985) Computer Modeling of Infinite Reaction Sequences – a Chemical Lumping. Chem Engineering Sci, vol 40, no 10, pp 1843–1849

[10] Franklach M, Wang H (1994) Detailed Mechanism and Modeling of Soot Particle Formation. In: Bockhorn H (ed) Soot Formation in Combustion. pp 165–192, Springer, Berlin, Germany
[11] Fusco A, Knox-Kelecy AL, Foster DE (1994) Application of a Phenomenological Soot Model to Diesel Engine Combustion. 3rd Int Symp COMODIA 94, pp 315–324
[12] Greis AE, Grünefeld G, Becker M, Pischinger S (2002) Quantitative Measurements of the Soot Distribution in a Realistic Common-Rail DI Diesel Engine. 11th Int Symp Application of Laser Techniques to Fluid Mechanics
[13] Hewson JC (1995) Reduced Mechanisms for Hydrocarbon and Nitrogen Chemistry in Diffusion Flames. CECR Report 95-01, Center for Energy and Combustion Research, University of California, San Diego
[14] Hewson JC, Bollig M (1996) Reduced Mechanisms for NOx Emissions form Hydrocarbon Diffusion Flames. 26th Symp (Int) Combust, pp 2171–2179, The Combustion Institute, Pittsburgh, PA
[15] Heywood JB (1988) Internal Combustion Engine Fundamentals. McGraw-Hill, New York, NY
[16] Hiroyasu H, Kadota T, Arai M (1983) Development and Use of a Spray Combustion Model to Predict Diesel Engine Efficiency and Pollutant Emissions. Part 1: Combustion Modeling. Bull JSME, vol 26, no 214, pp 569–575
[17] Karlsson A, Magnusson I, Balthasar M, Mauss F (1998) Simulation of Soot Formation Under Diesel Engine Conditions Using a Detailed Kinetic Soot Model. SAE Paper 981022
[18] Klaus P, Warnatz J (1995) A Contribution towards a Complete Mechanism for the Formation of NO in Flames. Joint Meeting of the French and German Sections of the Combustion Institute, Mulhouse, France
[19] Kleemann AP (2001) CFD Simulation of Advanced Diesel Engines. Ph.D. Thesis, Imperial College, University of London, UK
[20] Lavoie GA, Heywood JB, Keck JC (1970) Experimental and Theoretical Investigation of Nitric Oxide Formation in Internal Combustion Engines. Combust Sci Tech, vol 1, pp313–326
[21] Maly RR, Stapf P, König G (1998) Progress in Soot modeling for Engines. 4th Int Symp COMODIA 98, pp 25–34
[22] Mauss F, Trilken B, Breitbach H, Peters N (1994) Soot Formation in Partially Premixed Diffusion Flames. In: Bockhorn H (ed) Soot Formation in Combustion. pp 325–349, Springer, Berlin, Germany
[23] Mayer K, Spicher U (2000) Optical Investigations on Combustion in a DI Diesel Engine with an Endoscopic System and the Two-Color-Method. ASME ICE Spring Congress, San Antonio, TX
[24] Merker GP, Stiesch G (1999) Technische Verbrennung – Motorische Verbrennung. Teubner, Stuttgart, Germany
[25] Miller JA, Bowman CT (1989) Mechanism and Modeling of Nitrogen Chemistry in Combustion. Prog Energy Combust Sci, vol 15, pp 287–338
[26] Nagle J, Strickland-Constable RF (1962) Oxidation of Carbon between 1000–2000°C. Proc 5th Conf on Carbon, vol 1 pp154–164, Pergamon Press, London, UK
[27] Pischinger F, Schulte H, Jansen J (1988) Grundlagen und Entwicklungslinien der dieselmotorischen Brennverfahren. VDI Berichte no 14, VDI Verlag, Düsseldorf, Germany

[28] Schubiger RA, Boulouchos K, Eberle MK (2002) Rußbildung und Oxidation bei der dieselmotorischen Verbrennung. Motortechnische Zeitschrift MTZ, vol 63, no 5, pp 342–353
[29] Stiesch G (1999) Phänomenologisches Multizonen-Modell der Verbrennung und Schadstoffbildung im Dieselmotor. Ph.D. Thesis, University of Hanover, Germany
[30] Stiesch G, Merker GP (1999) A Phenomenological Model for Accurate and Time Efficient Prediction of Heat Release and Exhaust Emissions in Direct-Injection Diesel Engines. SAE Paper 1999-01-1535
[31] Warnatz J, Maas U, Dibble RW (2001) Combustion: Physical and Chemical Fundamentals, Modeling and Simulation, Experiments, Pollutant Formation. Springer, Berlin, Germany
[32] Zeldovich YB (1946) The Oxidation of Nitrogen in Combustion and Explosions. Acta Physicochimica, USSR, vol 21, pp 577–628

8 Conclusions

In the above text the state of the art in modeling in-cylinder processes of internal combustion engines has been presented and discussed. The development and application of such mathematical formulations is of great importance in today's research and development of combustion engines for several reasons. Firstly, simulation models, that have been properly adjusted to a specific range of boundary conditions, can be utilized to execute extensive parametric studies. In this context, simulation models are much more time and cost efficient than the alternative execution of experiments. Secondly, and maybe most importantly, numerical simulation tools can provide detailed information about any process variable at any point in time and space, that would be impossible to obtain with the sole execution of experiments. Consequently, a much better basis for the interpretation of complex results will be available, if both numerical and experimental studies are conducted in parallel. This aspect is of special importance, as combustion engines become more and more sophisticated and the task of further improving their performance becomes more and more complex. Last but not least, numerical simulations allow to perform conceptual studies with extreme boundary conditions, that could not be realized in experiments because of either too large or too small length and time scales, or because a dangerous outcome prohibits the execution of the respective experiment.

Three categories of combustion models can be distinguished. The thermodynamic (or zero-dimensional) models treat the combustion chamber as a homogeneously mixed control volume at any point in time. The phenomenological (or quasi-dimensional) models divide the combustion chamber into several zones of different composition and temperature, and they comprise more detailed submodels for several important spray and combustion subprocesses. Thus, they allow a prediction of characteristic quantities such as heat release and pressure histories, as well as an estimation of the pollutant formation, that cannot be obtained with the simpler thermodynamic models. Finally, the CFD-codes contain the most detailed submodels and solve for the multi-dimensional, turbulent flow field on the basis of the conservation equations of mass, energy and momentum.

Each of the model categories has its own field of application, and all of them are necessary to provide assistance in today's engineering tasks. Obviously, the thermodynamic models are utilized when computational expenditure is crucial, e.g. in extensive parametric studies. And as computer power is likely to further increase, they may become attractive even for real-time applications as part of control systems. Multi-dimensional CFD-codes on the other hand are necessary in order to obtain more detailed information about the physical and chemical subprocesses taking place during the combustion process. Moreover, since they represent the only model category that resolves the turbulent flow field, they are most useful in investigating the influence of geometric changes in the intake manifold or combustion chamber designs. However, whenever the major interest is put on the prediction of global quantities such as the integral heat release rate or the engine-out NO_x emission, phenomenological models may be capable of producing results of comparable accuracy with only a fraction of the computational expenditure. It should be noted however, that as computer power increases, the focus shifts more and more towards the more comprehensive models.

Much research activity is presently conducted in an attempt to further improve the quality and capabilities of spray and combustion models, with a strong emphasis on multi-dimensional CFD-codes. This concentration seems obvious, since this most comprehensive model category offers the greatest potential for further improvements. Based on the critical analysis of such models given in this text, the following aspects represent the most crucial limitations in the present state of combustion modeling, and consequently, further research should be directed into these directions.

- A major challenge is still given by the fact, that spray calculations utilizing the Lagrangian formulation to describe the evolution of the liquid phase are subject to strong dependencies on the numerical grid resolution. The results do not converge for continuously refined grids because of statistical convergence issues, and often the model parameters have to be set to unphysical values in order to account for these difficulties.

- A similar problem of strong grid dependencies is observed in modeling the convective heat transfer between gas and cylinder walls. Most often the logarithmic wall functions are utilized for this purpose, which are however limited to a certain range within the thermal and velocity boundary layers. Since the thickness of each of the boundary layers varies dramatically due to spray and combustion effects, an adaptive grid would be necessary in order to stay within the range of applicability. A more desirable possibility would be a modeling approach, that is entirely unaffected by the grid resolution.

- A further improvement in turbulence modeling seems necessary, because the turbulence properties greatly affect both the spray and mixture field evolution as well as the combustion process. The available turbulence models are mostly semi-empirical in nature, and consequently rely on substantial parameter tuning for each specific flow configuration.

- The primary breakup of sprays, taking place in the direct vicinity of the nozzle orifice, is still subject to significant uncertainties. This is partly due to the fact, that the spray is optically very dense at this location, and thus hardly accessible by optical measuring techniques. As a result, a comprehensive understanding of the absolute contributions of cavitation, liquid-phase turbulence and aerodynamic effects on the spray disintegration has not been achieved to date.

8 Conclusions   277

- Particle-particle interactions such as droplet collisions are often subject to problems with respect to statistical convergence within the Eulerian-Lagrangian framework of spray simulations. A solution to this problem would require a dramatic increase in both the grid resolution and in the number of stochastic particles representing the spray, that seems hardly feasible in the near future.

- As mentioned above, combustion predictions are greatly affected by the turbulence model. Relatively simple approaches like the eddy-breakup models can provide useful results, but they tend to predict unphysical effects in some details of the flame, e.g. an excessive flame thickness or a flame acceleration towards a wall. The more detailed models based on the laminar flamelet assumption represent an attractive approach to combine detailed chemistry with complex, turbulent flows. However, they are again strongly influenced by the (uncertain) turbulence predictions via the scalar dissipation rate. Finally, the most detailed pdf-combustion-models have shown to yield remarkable results in relatively simple, two-dimensional flow configurations, but they do not seem to be applicable to complex, three-dimensional, turbulent engine combustion studies in the near future.

- The quantitative prediction of soot emissions seems impossible at the present time. This is partly due to a lack in detailed chemistry models. But maybe of even greater importance is the fact, that the engine-out soot concentration is only a small difference between two large quantities, namely the formation and subsequent oxidation of soot. Thus, a significant error relative to the small concentration in the exhaust gas has to be expected, even if a detailed soot chemistry model has proven to yield good results in steady flame configurations.

- Having in mind that a long-term trend in future engine development is directed towards the "design" and utilization of synthetic fuels and towards new combustion systems such as homogeneous charge compression ignition (HCCI), multi-component fuel models become more and more desirable, that are capable of realistically describing the evaporation process of such fuels. Moreover, since the HCCI concept relies on the control of relatively long ignition delays, detailed ignition chemistry models that can predict the multi-stage ignition of realistic fuels with good accuracy become necessary as well.

To summarize all the above, the predictive quality of the present spray and combustion models has already reached a level, that makes their use extremely helpful and a necessity in the various engineering tasks related to combustion engine development. Nevertheless, further improvements are still desirable and also necessary in the fields mentioned above.

Index

air entrainment (see entrainment)
air-fuel mixing 119
Arrhenius equation 197
atom balance 195
atomization 60, **130**
   sheet~ **146**
autoignition 94, **209**

Biot number 173
blob-injection model 137
blowby 12
Borghi diagram 203
boundary layer
   thermal~ 115
   thickness 84
   velocity~ 113
Boussinesq approximation 108
burning rate (see heat release rate, mass burning rate)

cavitation 132, 139
charge purity 29
chemical equilibrium **193**, 197
chemical explosion 206
chemical potential 194
closure problem 108, 238
coherent flame model 226
compression work 12
computational expenditure 41
conservation equations 101
continuity equation (see mass conservation)
continuous thermodynamics 174
continuum droplet model 124
control volume 5, 102
cool flames 208, 210
crank mechanism 13
Cummins engine model 47
cutoff length scale 106, 227
cycle-by-cycle variations 107, 216

Damköhler number 203
delivery ratio 29
diffusion combustion 63, 67, **228**
direct injection engines 3, 119
direct numerical simulation 104
discharge coefficient 9, 148
discrete droplet model 124, 185, 186
DISI engines 241
drag
   coefficient 127
   force 121, 126, 159
droplet
   breakup **153**
   breakup regimes 153
   breakup time 138, 149, 156, 158
   coalescence 121, **161**
   collision 121, **161**, 184
   diffusion-limit model 173, 180
   distortion 127, 155
   evaporation **60**, **172**
   infinite-diffusion model 173
   kinematics 126
   size 119, 137, 138, 150, 157, 170
   vortex model 173

eddy-breakup models 222
   (see also time scale models)
eddy-diffusion model 108
eddy-viscosity model 108
Einstein convention 102
emission formation (see pollutant formation)
end gas 92, 94
energy conservation 6, 7, **103**, 117
ensemble-averaging 225
enthalpy of formation 194
entrainment
   model 91, 222
   rate 51, 59
equation of state 6

equilibrium constant 195, 197
exhaust gas aftertreatment 256
exhaust gas composition 255
expansion work 12
explosion diagram 207

Favre-averaging 225
flame
  laminar surface area 217
  regimes 202, 204
  surface area 88, 92
  surface density 226
  turbulent stretch 204, 226
  turbulent surface area 217
  turbulent wrinkling 204, 217, 225
  types (see flame regimes)
flame area evolution models 224
flame front 72, 88, 202, 242
  fractal model 227
flame speed 216
  laminar 89, 247
  thermal expansion effects 244
  turbulent 90, 217
flamelet models **230**
  flamelet library 231, 237
  mixture fraction pdf 232
  representative interactive flamelets 237
  χ-pdf 235
flamelet regime 222
flammability limits 207
flash boiling 180
fuel
  composition **174**
  concentration 48, 55
  evaporation 61, 119, **171**, 180
    (see also droplet evaporation)
  injection 11
  multi-component~ 174, 181, 277

gas exchange 8
gas jet theory 47
G-equation model **224**, 246
Gibbs free energy 193
grid dependencies 116, **181**, 276
  reduction of ~ 183

HACA sequence 270
heat flux profiles 87
heat release 200
  analysis 35
  rate 16, 35, **43**, 44, 56, 62, 65, **88**
    (see also mass burning rate)
heat transfer 6, **14**, 52, 63, **77**, **112**
  coefficient 14, 83
  conduction 104
  convective 14, 81, 112
  radiative 77, 84
  soot layer insulation 80
hollow-cone spray 146

ICAS-model 184
ideal gas 6, 23, 25
ignition **205**
  autoignition 94, **209**
  delay 21, 62, 70, 72, 208, 209
  induction time 206
  integral 23, 62
  spark ignition **213**
ignition kernel **214**, 243
incompressible fluid 102
injection
  rate profile 11
  velocity 58, 148
integral length scale (see turbulent scales)
internal energy 23

Karlovitz number 203
Kelvin-Helmholtz model 135, 158, 160
knock 88, **92**, 212
  induction time 95
  integral 94
Kolmogorov scales
  (see turbulent scales) 203
Kronecker delta 52
k-ε model 109

Laplace number 167
large eddy simulation 105
length scale limiter 183

marker particles 243
mass burning rate 67, 70, 88, 91, 215, 223, 244 (see also heat release rate)
mass conservation 5, 7, **101**, 117
mass diffusion 61
mixing stoichiometry 73
mixing-length model 109
mixture fraction 231
model categories 2, 41

momentum conservation 53, 59, **103**, 117
Monte-Carlo technique 124, 240
multidimensional models 2, 119, 193, 275

Navier-Stokes equations 103
negative temperature coefficient 208
Newtonian fluid 103
non-equilibrium turbulence model 111
non-premixed combustion (see diffusion combustion)
NO_x formation 32, 76, 236, **257**
   mechanisms 257
   prompt NO 260
   thermal NO 34, **258**
Nusselt number 14, 61
N-zone models 75

Ohnesorge number 132

packet model 57, 76
partial equilibrium 196, 200
partially premixed combustion 241
particulate matter (see soot)
pdf-models 238
   presumed pdf 240
   transported pdf 239
phenomenological models 2, **41**, 276
pilot-injection (see pre-injection)
plasma velocity 219, 244
pollutant formation 41, 72, 76, 199, 255
polycyclic aromatic hydrocarbons 261, 270
polygon-hyperbola combustion profile 19
pre-injection 65, 69
   timing 72
premixed combustion 57, 63, 65, 67, **222**
pressure swirl atomizer 146
probability density function 122, 231, 239
   joint pdf 239, 240

quasi-dimensional models (see phenomenological models)
quasi-steadiness 200, 259

Rayleigh-Taylor model 159

reaction
   kinetics 196
   mechanisms 198
   rate 197, 198
   rate coefficient 197
real gas effects 26
regress variable 225
restitution coefficient 170
Reynolds averaging 107
Reynolds number 14, 48, 127, 132, 154, 202
Reynolds stress models 111
Reynolds stress tensor 108
RNG k-ε model 111

Sauter mean diameter 60
scalar dissipation rate 232, 234, 236, 237
scavenging efficiency 11, 29
Shell-model 209
Sherwood number 61
single-zone model 6
soot
   comprehensive models 269
   eight-step model 266
   formation 77, **261**, 277
   oxidation 264
   particle properties 261
   T-/composition dependency 262
   temporal concentration profile 263
   two-step model 264
soot layer insulation 80
soot radiation 77, 84
source terms 116, 123
spark discharge 88, 213
spark ignition **213**, 243
spark plug protrusion 221
spatial resolution 182
spray **119**
   angle 48, 54, 60, 137, 141, 143, 156
   packets 58
   penetration 47, 54, 65, 67
   regimes 120
   velocity 54, 58
spray breakup
   breakup length 134, 137, 149, 161
   breakup time 58
   primary~ 131, 139, 276
      (see also atomization)
   regimes 131
   secondary~ (see droplet breakup)

spray equation 122
squish flow 82
statistical convergence 182, 184, 185, 240, 277
stochastic particle technique 124, 182
stoichiometric coefficient 193
Stokes' postulation 103
stratified charge 241
stress tensor 103
surface perturbations 136, 148
swirl 48, 53, 82
systems analysis 35

Taylor series 101
Taylor-analogy breakup model 127, **155**
thermal efficiency 88, 93, 255
thermal explosion 205
thermodynamic models 2, 5, 275
time scale models 67, 224, 228, 236, 243
transient system simulation 36
trapping efficiency 29
triple flame 242, 248
turbulence
    combustion induced 15
    dissipation 83, 110
    intensity 70, 90, 128, 143
    models 109, 184, 276
    production 68, 119
    scales (see turbulent scales)
turbulent
    conductivity 108
    dispersion 128
    fluctuations 106, 110, 235, 238, 247
    kinetic energy 45, 83, 91, 109, 128, 142
    mixing 74, 76
    strain rate 218, 235
    viscosity 108, 109, 110
turbulent scales
    dissipation length scale (see integral length scale)
    integral length scale 105, 128, 203
    Kolmogorov length scale 105, 203
    Kolmogorov time scale 203
    Taylor microscale 223
    turbulent time scale 67, 223, 229
two-phase flow 116
two-zone model 32, 35, 72

void fraction 121, 182

wall
    friction 82
    heat transfer (see heat transfer)
    impingement 60, 165
    impingement regimes 167
wall functions 112, 114, 276
water bell 152
wave-breakup model 135
Weber number 132, 153, 162, 167
well stirred reactor 204
Wiebe function 17

zero-dimensional models (see thermodynamic models)

Printing and Binding: Strauss GmbH, Mörlenbach